I0037958

Technical Peer Review Report

Report of the Review Panel

Spent Nuclear Fuel Canister Qualification Support

ASME International

Institute for Regulatory Science

DISCLAIMER

This report was prepared through the collaborative efforts of the American Society of Mechanical Engineers (ASME) Center for Research and Technology Development and the Institute for Regulatory Science (referred to thereafter with the collaborators as the Society) for the Office of Science and Technology Development of the U.S. Department of Energy (referred to hereafter as the Sponsor).

Neither the Society nor the Sponsor, or others involved in the preparation or review of this report nor any of their respective employees, members, or persons acting on their behalf, make any warranty, expressed or implied, or assume any legal liability or responsibility for the accuracy, completeness, or usefulness of any information, apparatus, product, or process disclosed, or represent that its uses would not infringe privately owned rights.

Information contained in this work has been obtained by the American Society of Mechanical Engineers from sources believed to be reliable. However, neither ASME nor its authors or editors guarantee the accuracy or completeness of any information published in this work. Neither ASME nor its authors and editors shall be responsible for any errors, omissions, or damages arising out of the use of this information. The work is published with the understanding that ASME and its authors and editors are supplying information but are not attempting to render engineering or other professional services. If such engineering or professional services are required, the assistance of an appropriate professional should be sought.

Statement from By-Laws: The Society shall not be responsible for statements or opinions advanced in papers . . . or printed in its publications. (7.1.3)

For authorization to photocopy material for internal or personal use under circumstances not falling within the fair use provisions of the Copyright Act, contact the Copyright Clearance Center (CCC), 222 Rosewood Drive, Danvers, MA 01923, Tel: 978-750-8400, www.copyright.com. Requests for special permission or bulk reproduction should be addressed to the ASME Technical Publishing Department.

ISBN No. 0-7918-0221-3

TABLE OF CONTENTS PAGE NO.

Preface

This report contains the results of a peer review performed jointly by the American Society of Mechanical Engineers (ASME) and the Institute for Regulatory Science (RSI). Based on a request from the Idaho National Engineering and Environmental Laboratory of the U.S. Department of Energy (DOE), a Review Panel (RP) was established to peer review "Spent Nuclear Fuel Canister Qualification Support". Consistent with the ASME/RSI process, the following RP was appointed by the Peer Review Committee for Energy and the Environment (PRCEE) of ASME:

Jacob Fish
Tom Hendrickson, Chair
Peter Turula
Norman Gerstein
Ratib Karam

During the period covered by this report, the ASME PRCEE overseeing the peer review consisted of the following individuals:

Charles O. Velzy, Member of Executive Panel (EP) and Chair
Ernest L. Daman, Member of EP
Nathan H. Hurt, Member of EP
A. Alan Moghissi, Member of EP; Principal Investigator of the Peer Review Program
Gary A. Benda
Erich W. Bretthauer
Irwin Feller
Robert A. Fjeld
William T. Gregory, III
Peter B. Lederman
Jeffrey A. Marqusee
Lawrence C. Mohr, Jr.
Goetz K. Oertel
Glen W. Suter, II
Cheryl A. Trottier

The supporting staff included the following individuals:

Michael Tinkelman: Director of Research at the Center for Research and Technology Development of ASME in Washington, DC; and Administrative Manager of the ASME PRCEE.

Betty R. Love: Executive Vice President, RSI, Columbia, MD; and Administrative Manager of the Peer Review Program.

Sorin R. Straja: Vice President for Science and Technology, RSI

M.C. Kirkland: Vice President for Southeastern Region, RSI; and Technical Secretary.

Sharon D. Jones: Director of Training Programs, RSI; and Manager of Review Panel Operations.

The biographical summaries of the members of the RP, the PRCEE, and the technical staff are located at the end of this report.

The Review Criteria were provided by the Project Team—consisting of principal investigators, project managers, and others involved in the project. These criteria were slightly revised by the Technical Secretary and approved by the Project Team. The RP received documents describing various aspects of the project. The summary of the project included in this report was prepared by the Technical Secretary using the same documents that had been provided to the Review Panel. The Project Team received the project summary for review and approval. In addition, the staff of RSI undertook the task of preparing a listing of acronyms. This list, as reviewed and approved by the Project Team, is also included in this report.

The RP met from December 1, 2003 through December 5, 2003, in Columbia, MD. At the beginning of the meeting, the RP and the Project Team were introduced to the ASME/RSI peer review process. The introduction was followed by presentations by the Project Team and a discussion period. The agenda of this meeting is shown in the Appendix. At the end of the meeting, the RP met in an executive session to develop its strategy for writing the report. Subsequently, the RP wrote its findings and recommendations. The *Report of the Review Panel* was then copyedited.

Consistent with the procedures established by the ASME/RSI process, the copyedited *Report of the Review Panel* was provided to DOE for identification of potential errors, misunderstandings, and areas of ambiguity. The Technical Secretary contacted the members of the RP reporting the comments received from DOE which were considered by the RP.

Charles O. Velzy
A. Alan Moghissi

Appendix

AGENDA

Monday, December 1, 2003

Hilton Columbia

8:00 a.m.   ASME/RSI Peer Review Process   A. Alan Moghissi, ASME/RSI

8:50 a.m.   Depart for Institute for Regulatory Science

Institute for Regulatory Science

9:00 a.m.   Background on NSNFP's Canister Program   Thomas J. Hill, INEEL
            • Hanford's MCO
            • Standardized DOE SNF Canister
            • ISFP Canister

10:00 a.m. Break

10:15 a.m. Repository Criteria for DOE Canisters   Thomas J. Hill

10:45 a.m. Canister Qualification Support Efforts   D. Keith Morton, INEEL

12:00 noon Lunch

1:00 p.m.   Completed Canister Qualification Efforts to Date   D. Keith Morton
            • Aging Studies
            • 18-inch Standardized DOE SNF Canister Drop Tests
            • Analysis Methodology
            • Preliminary Friction Parameters
            • Flaws Static Loads

2:30 p.m.   Break

2:45 p.m.   Completed Canister Qualification Efforts *(continued)*

4:00 p.m.   Break

4:15 p.m.   Future Qualification Efforts   D. Keith Morton
            • 24-inch and MCO Canister Drop Tests
            • Friction Parameters
            • Flaw Dynamic Loading
            • Analysis Methodology for Combined Synergistic Effects

5:30 p.m.   Adjournment

ASME/RSI Peer Review
Department of Energy Spent Nuclear Fuel Canister Qualification Support

Columbia, MD – December 1-5, 2003

AGENDA

Tuesday, December 2, 2003

Institute for Regulatory Science

   8:00 a.m.   Executive Session

   9:30 a.m.   Discussion

 10:30 a.m.   Break

 10:45 a.m.   Discussion *(continued)*

12:00 noon   Lunch

   1:00 p.m.   Executive Session

   2:30 p.m.   Break

   2:45 p.m.   Executive Session *(continued)*

   5:00 p.m.   Adjournment

Wednesday, December 3, 2003

Institute for Regulatory Science

   8:00 a.m.   Executive Session

   5:00 p.m.   Adjournment

Thursday, December 4, 2003

Institute for Regulatory Science

   8:00 a.m.   Executive Session

   5:00 p.m.   Adjournment

Friday, December 5, 2003

Institute for Regulatory Science

   TBD   Closeout Session

Executive
Summary

The National Spent Nuclear Fuel Program (NSNFP) developed standardized U.S. Department of Energy (DOE) spent nuclear fuel (SNF) canisters for the handling and interim storage of SNF at various DOE sites, as well as the transport, handling, and disposal of SNF at the repository. According to the DOE Preliminary Design Specification for Standardized Spent Nuclear Fuel Canisters, DOE/SNF/REP011, the types and amounts (proper fissile material limits) of SNF to be loaded into canisters of proper configuration using properly-designed internals (baskets, spacers, sleeves, dividers, cans, etc.) will preclude nuclear criticality concerns.

The Project Team (PT), as part of a licensing support effort, has developed a comprehensive plan to assist DOE to meet USNRC licensing requirements for DOE's SNF canisters. The key components of the plan were presented to the Review Panel (RP) as the DOE SNF Canister Qualification Support Chart. The chart is periodically updated to reflect program status.

The PT explained to the RP that:

1. The need to demonstrate that the canisters would not breach should any canister be dropped during transfer events, arises from the lack of adequate records on some types of DOE-owned SNF.
2. This lack of adequate records prevented the typical analysis of the radiation dose to onsite and offsite personnel should there be an accidental release of radioactive material from any of the canisters.
3. Thus, the PT considers it is necessary to show that the canisters will remain leak-tight should any canister be inadvertently dropped.

The strategy of the PT is to develop a canister designed as a "no breach" container under postulated accidental canister drop scenarios. To date, the PT has designed and constructed a number of canisters, and drop tested nine canisters at Sandia National Laboratory. The drop tests were performed with the canisters in various orientations ranging from vertical to horizontal, and from various heights from two feet up to 30 ft (9 m). The 30-ft (9 m) drop test is in excess of the current facility drop requirement of 23 ft (6.9 m). The drop tests were performed at Sandia National Laboratory under carefully-controlled conditions. Instrumentation devices were attached to the canisters during the tests to record strain and acceleration data, and the drop tests were visually recorded. Post drop pressure tests at 50 psig confirmed the structural integrity of the canisters. Also, a helium leak-test was performed on four of the most deformed canisters, and the test demonstrated that the canisters did not leak. The calculated canister stress-strain results exceeded ASME B&PV Code, Section III stress criteria. Therefore, a combined computational analysis and experimental proof-testing approach has been initiated.

In order to enhance accidental drop performance the standardized canisters include a skirt on both ends. The skirts act as energy-absorbing devices that protect the containment vessel from direct impact for most drop orientations.

In assessing the information provided by the PT, the RP found that the approach to the analysis and proof testing by the PT is technically sound, and the underlying bases for the analysis and proof testing are technically sound and defensible. Similarly, the underlying bases for the individual testing steps, as well as the integrated demonstration, are technically sound and defensible. The basis for the computational modeling was the standard canister loaded with rebar simulating the maximum content weight. The integrated basis for canister analysis and testing is documented in the DOE SNF Canister Qualification Support Chart presented by the PT. This chart, which is updated as the program proceeds, identifies components of the qualification support program including areas of technical concern regarding canister performance under postulated drop conditions.

The PT has, by and large, followed established scientific and engineering principles and standards. The PT has indeed thought in-depth about the steps required in the canister qualification support program. The overall plan for the canister qualification support includes four project tasks supporting the project goal, as follows:

1. Aging Studies
2. Accidental Drop Events
3. Flaw Evaluations and Testing
4. Material Considerations

The RP did find that some of the sub-tasks which were designated as "complete" lacked "completeness" in the sense that experiments which were performed raised additional questions about the issues involved. For example, the friction factor dependence on the angle of drop test and the lack of strain measurement associated with the drop test.

The PT has developed a DOE SNF Canister Qualification Support Program to demonstrate that the canisters will meet the licensing requirements of the U.S. Nuclear Regulatory Commission (USNRC). This ongoing program has already demonstrated that new canisters will withstand accidental drops without rupture, and is being extended to demonstrate the effects of aging, flaws, and material considerations. When complete, the program will have collected sufficient data to meet the licensing requirements of the USNRC.

Peer Review
Process

INTRODUCTION

There is consensus within the technical community on the definition, process, and key criteria for the acceptability of peer review. Peer review consists of a critical evaluation of a topic by individuals who—by virtue of their education, experience, and acquired knowledge—are qualified to be peers of an investigator engaged in a study. A peer is an individual who is able to perform the project, or the segment of the project that is being reviewed, with little or no additional training or learning.

Recognizing that peer review constitutes the core of acceptability of scientific and engineering information, virtually all professional societies of scientists and engineers have instituted formal procedures for peer review for their activities. The American Society of Mechanical Engineers (ASME), also known as ASME International, has over a century of experience in peer review. Consistent with its mission and tradition, ASME, in cooperation with the Institute for Regulatory Science (RSI), has established a peer review program devoted to the review of activities of various government agencies (ASME 2003, RSI 2003). The reports of the peer reviews resulting from this program have been published (ASME/RSI 1997, 1998, 1999, 2000, 2001a, 2001b, 2001c, 2002a, 2002b, 2002c, 2002d, 2002e, 2003a, 2003b, 2003c, 2004).

PEER REVIEW PROCESS

The structure of the peer review process established by the ASME/RSI team consists of a tiered system. For each specific area, the entire process is overseen by a committee. The review of specific topics is performed by Review Panels (RPs).

Peer Review Committee for Energy and the Environment

The Peer Review Committee for Energy and the Environment (PRCEE) is a standing committee of ASME formed to oversee peer review for one particular program in an agency. Its members are chosen on the basis of their education, experience, peer recognition, and contribution to their respective areas of competency. An attempt is made to ensure that all needed technical competencies and diversity of technical views are represented in the PRCEE. The members of the PRCEE must be approved by the Board on Research and Technology Development of the Council on Engineering of ASME. The PRCEE includes an Executive Panel (EP) that is responsible for the day-to-day operations of the PRCEE. Except for the EP, membership in ASME is not required for appointment to the PRCEE. As the overseer of the entire peer review process, the PRCEE enforces all relevant ASME policies, including compliance with professional and ethical requirements. A key function of the PRCEE is the approval of the appointment of members of RPs for a specific project.

Review Panels

The review of a project, a document, a technology, or a program is performed by an RP consisting of a small group of highly-knowledgeable individuals. Upon the completion of their task, the RPs are disbanded. The selection of reviewers is based on the competencies required for the specific review assignment. The number of individuals in an RP depends upon the complexity of the subject to be reviewed. The selection of a reviewer is based on the totality of that individual's qualifications. However, there are several generally-recognized and fundamental criteria for assessing qualifications of a reviewer. These are as follows:

1. Education: A minimum of a B.S. degree, preferably an advanced degree in an engineering or scientific field, is required for any peer reviewer.

2. Experience: In addition to education, the reviewer must have significant experience in the area that is being reviewed.

3. Peer recognition: Election to an office of a professional society, serving on technical committees of scholarly organizations, and similar activities are considered to be a demonstration of peer recognition.

4. Contributions to the profession: Contributions to the profession may be demonstrated by publications in peer-reviewed journals. In addition, patents, presentations at meetings where the papers were peer-reviewed, and similar activities are considered to be contributions to the profession.

5. Conflict of Interest: One of the most complex and contested issues in peer review is a set of subjects collectively called conflict of interest. The ideal reviewer is an individual who is intimately familiar with the subject and yet has no monetary interest in it. Despite this apparent difficulty, the ASME and similar organizations have successfully performed peer review without having a real or apparent conflict of interest. The guiding principle for conflict of interest is as follows: *Those who have a stake in the outcome of the review may not act as a reviewer or participate in the selection of reviewers.*

Due to the multidisciplinary nature of many projects reviewed by the ASME/RSI team, rapid identification of qualified peer reviewers and their availability to participate in the review process are key ingredients for a successful program. The process used for the identification of reviewers is multifaceted. The Administrative Manager of the Peer Review Program receives recommendations from sources within ASME; previous members of the RP; sister societies; other organizations and individuals; the U.S. Department of Energy (DOE); DOE contractors; and others. However, the selection of peer reviewers is based entirely on criteria identified by ASME. The details of various aspects of peer review, including conflict of interest, can be found in the ASME *Manual for Peer Review* (ASME 2003) and the associated procedures (RSI 2003).

COOPERATION WITH OTHER PROFESSIONAL SOCIETIES

The ASME is a large professional engineering society having in excess of 125,000 members. Although the predominant discipline of the members is mechanical engineering, there are members who—by virtue of their education, training, or experience—are competent in other disciplines. The Council on Engineering includes divisions ranging from classical mechanical engineering (design, heat transfer, and power) to solar engineering; environmental engineering; and safety and risk analysis. Despite the diverse competency within ASME, it is recognized that on occasion it will become necessary to peer review activities which include disciplines that are outside the areas of competency of ASME and its members. These disciplines may include geology, hydrology, toxicology, and ecology. Consequently, ASME has reached formal and informal agreements with its sister societies to identify qualified reviewers in areas outside of those covered by the membership of ASME.

PERFORMING ORGANIZATIONS

The Center for Research and Technology Development of ASME manages a number of scientific and engineering activities, including peer reviews. Because of ASME's conscious effort to maintain a small in-house staff, it relies upon other organizations to provide detailed project management services in its research, development, and similar activities. Accordingly, ASME and RSI joined forces in a collaborative effort to perform the peer review for the U.S. Department of Energy. While the ASME staff in Washington, DC provides the staff support for the PRCEE, the detailed management and staff support for the RPs is provided by RSI.

American Society of Mechanical Engineers

As one of the largest professional engineering societies, ASME has a long and distinguished history. Its activities are carried out primarily by members who volunteer their time in support of engineering and scientific advancement. For obvious reasons, ASME also has a paid staff to manage the day-to-day operations of such a large professional society. ASME has a detailed structure for its operation, consisting of councils, boards, divisions, and committees. The Council on Engineering has 38 divisions, including: Environmental Engineering; Solid Waste Processing; Nuclear Engineering; Safety Engineering; and Risk Analysis. The Council on Codes and Standards develops ASME codes and standards that are the backbone of many industries—including power production—worldwide. The Council on Codes and Standards is also responsible for the development of standards for activities such as certification of incinerator operators. The ASME was a founding member of the American Association of Engineering Societies and a founding member of the American National Standards Institute.

Institute for Regulatory Science

RSI is a not-for-profit organization chartered under section 501(c)3 of the Internal Revenue Service. It is dedicated to the idea that societal decisions must be based on the best available scientific and engineering information. According to the RSI mission statement, peer review is the foundation of the best available scientific and engineering information. Consequently, RSI has promoted peer review within government and industry as the single most important measure of reliability of scientific and engineering information. In its activities, RSI seeks the cooperation of scholarly organizations. Historically, a large number of RSI activities have been performed in cooperation with ASME. RSI is located in the Washington, DC metropolitan area.

Project
Summary

PRELIMINARY DESIGN SPECIFICATION FOR STANDARDIZED SPENT NUCLEAR FUEL CANISTERS OF U.S. DEPARTMENT OF ENERGY

INTRODUCTION

The Nuclear Waste Policy Act (NWPA 1982) and its subsequent amendments assigned U.S. Department of Energy (DOE) the responsibility for managing the disposal of spent nuclear fuel (SNF) and high-level waste (HLW) of domestic origin. It also designated Yucca Mountain, NV, as the candidate site for the geologic repository. This law assigned the responsibility for the development of generally-applicable environmental standards and the licensing of the repository to the U.S. Environmental Protection Agency and the U.S. Nuclear Regulatory Commission (USNRC) respectively. Within DOE, the responsibility for SNF and HLW disposal, including the development and licensing of a geologic repository, was assigned to DOE's Office of Civilian Radioactive Waste Management (OCRWM). Based on a Presidential (1985) decision, HLW and SNF resulting from defense activities will also be disposed in the civilian geologic repository.

The NWPA defines SNF as the fuel that has been withdrawn from a nuclear reactor following irradiation and not reprocessed. It is also defined to include non-fuel components as identified in 10 CFR Part 961, Appendix E (DOE 2003d). The NWPA defines HLW as the highly-radioactive material resulting from reprocessing SNF, including liquid waste produced directly in reprocessing; any solid material derived from such liquid waste that contains fission products in sufficient concentrations; and other highly-radioactive material that has been determined by the USNRC, consistent with law, to require permanent isolation. Consequently, the repository will accept SNF and solidified HLW.

The NWPA authorized the Secretary of Energy to site, construct, and operate one monitored retrievable storage (MRS) facility. The MRS, if built, will provide temporary storage for SNF until it is shipped to the geologic repository for permanent disposal. The MRS must be licensed by the USNRC (2003c), and may have an initial 40-year license term with the option for license renewal.

The National Spent Nuclear Fuel Program (NSNFP) of DOE is implementing certain requirements of NWPA by developing a set of standard canisters for SNF including its handling, interim storage, transportation, and disposal in the national repository. The mission of this program also includes to safely, reliably, and efficiently manage the DOE-owned SNF and SNF returned to the U.S. from foreign research reactors (FRR). This activity is cooperatively performed with the OCRWM, the Idaho National Engineering and Environmental Laboratory (INEEL), DOE Hanford Site, Oak Ridge National Laboratory, Argonne National Laboratory, and the Savannah River Site (SRS).

Within DOE, the Office of Environmental Management (EM) is responsible for the interim management and preparation for disposal of the DOE SNF (DOE 1999) which includes SNF generated from activities related to nuclear weapons production, research programs, and others. The EM establishes the methods to be employed in the treatment, handling, storage, and preparation for disposal of the DOE SNF, including both current inventory and expected receipts. The NSNFP, operating from the INEEL, assists EM in implementing these methods.

Consistent with its mission, the NSNFP has developed specification to provide the requirements and necessary information for designing the standardized canisters to be used for handling, interim storage, transportation, and disposal in the national repository of DOE SNF (DOE 1999). This design specification addresses two different outer diameter (OD) sizes of DOE SNF standardized canisters, including two different lengths for each canister OD, resulting in a total of four unique canister geometries. This design specification does not consider using these DOE SNF canisters for either U.S. Navy or commercial SNF or HLW (either commercial or defense) materials. Similarly, it does not provide any procurement-specific information.

Although using the same terminology, the standardized DOE SNF canister should not be confused with the containers used in current interim SNF storage systems for commercial spent nuclear fuels. Many independent spent fuel storage installation (ISFSI) or dry storage system vendors also call their SNF container a canister. However, the commercial nuclear industry storage canister (hereafter referred to as the storage industry canister) is typically 5 to 6 ft (1.5 to 1.8 m) in diameter. In contrast, the DOE SNF canisters are approximately 1.5 to 2 ft (0.45 to 0.60 m) in diameter. It should be noted that the design features of the Multi Canister Overpack (MCO) and the Idaho Spent Fuel Project (ISFP) are discussed later in this document and are also considered DOE SNF canisters.

The storage industry canister is intended to provide the secondary barrier for the commercial SNF. In contrast to industry canisters, the DOE SNF canisters are intended to constitute the primary barrier or cladding replacement for interim storage and transportation to the repository. Consequently, not only damaged or failed, but also intact DOE SNF are expected to be placed directly into the standardized DOE SNF canisters. The design requirements for DOE SNF canisters at the time of project initiation are listed in Table 1. Over time, these requirements have continued to evolve as the OCRWM nears its license application submittal to the USNRC. The Disposability Interface Specification has evolved into the Waste Acceptance System Requirements Document (WASRD).

FUNCTIONAL REQUIREMENTS

Project requester and end user

The development of the standardized DOE SNF canisters is guided under the direction of the NSNFP, as an implementor of the DOE SNF Program. The DOE SNF Program is the project requestor. The NSNFP personnel is located at the INEEL administered by the Bechtel BWXT Idaho (BBWI). The end user of the finished canisters will ultimately be the geologic repository where SNF will be permanently disposed. However, where applicable, the fuel custodians at the DOE sites (Hanford, INEEL, and SRS) will be responsible for loading, handling, and storing DOE SNF in these standardized canisters.

Purpose of the Standardized Canisters

The purpose of a standardized canister for DOE-owned SNF is as follows:

1. To provide an easy and standard handleable unit to confine DOE SNF materials
2. To be durable for storing SNF
3. To provide easily-transportable units
4. To be a unit for final disposal at the national repository, without the necessity of the DOE SNF being removed from the canister or reopening a sealed canister

Table 1. Design requirements matrix for SNF canisters (DOE 1999). Note that this table does not specifically address quality assurance requirements.

Requirements	Subject
Federal Regulations	
10 CFR Part 60.131 (h)	Criticality Control
10 CFR Part 60.135 (a) (1)	In situ chemical, physical, and nuclear properties do not compromise function of waste package
10 CFR Part 60.135 (a) (2)	Design considerations for waste package and content interactions
10 CFR Part 60.135 (b) (1)	Waste packages and its components shall not contain explosive, pyrophoric, or chemically reactive material
10 CFR Part 60.135 (c) (1)	DOE SNF shall be in solid form and placed in sealed containers
10 CFR Part 71.63 (b)	Double containment of certain SNF
10 CFR Part 72.122 (h) (1) & (5)	Confinement of SNF
10 CFR Part 72.122 (1)	Retrievability from storage
10 CFR Part 72.124	Criticality control
10 CFR Part 72.128 (a) (3)	Confinement during storage
DOE OCRWM Documents	
Disposability Interface Specification - July 1998 (Draft)	
Disposability Standard 2.1.20	Canister shell and lid materials shall be low-carbon austenitic stainless steel or stabilized austenitic stainless steel or other equally corrosion-resistant alloys
Disposability Standard 2.1.21	Canisters shall be vacuum dried, backfilled with an inert gas (e.g., helium) and sealed
Disposability Standard 2.1.26	Canisters shall have a unique alphanumeric identifier
Disposability Standard 2.1.27	Canisters not seal-welded shall have a tamper-indicating device
Disposability Standard 2.1.28	Damage or deformation to canisters shall be limited such that the canisters can (1) still be lifted and moved, (2) continue to meet the dimensional envelope required for disposal (loading into the disposal container), and (3) maintain a seal
Disposability Standard 2.2.20.3	Canisters shall have the capability to stand upright without support on a flat surface and be capable of being placed, without forcing, into a right-circular cylindrical cavity of the proper dimension
Disposability Standard 2.2.21.3	Canisters shall not exceed the total weight limits as specified (DIS conflicts with ICD values but ICD values are used)
Disposability Standard 2.2.22.2	Canisters shall be capable of being lifted vertically with remote handling fixtures
Disposability Standard 2.3.22	Canistered SNF shall be shown to have a calculated keff of 0.95 or less
Disposability Standard 2.4.21	Canisters shall comply with thermal output limits [currently vague on DOE SNF canisters]
Disposability Standard 2.4.22	Canisters shall comply with surface contamination limits
Disposability Standard 2.4.23	Canisters shall not exceed pressure limits
Disposability Standard 2.4.24	Canisters shall have no detectable leak rate at the time of receipt at the repository. At a minimum, the canister shall be leak tested using OCRWM-approved method or shown via an OCRWM-approved method of fabrication controls and volumetric inspections to be properly sealed. The canister shall be reevaluated prior to shipment, as required, if suspected of leaking.
Interface Control Document - March 1999	
ICD Section 10.1.1	Canisters are right-circular cylinders and after being filled with SNF can stand vertically on a flat surface
ICD Section 10.1.2	Canisters shall not exceed weight limits
NUREG-1617 (Draft) - March 1998	
Section 4.5.1.3	Packages designed for the transport of damaged SNF include packaging of the damaged fuel in a separate inner container (second containment system) that meets the requirements of 10 CFR 71.63 (b)

The standardized DOE SNF canisters will provide long-term advantages in four basic areas:

1. Handling and safety:

1.1. Ensure that the DOE SNF is directly handled (bare) only once during its interim storage, transportation, and final disposal handling stages
1.2. Promote standardized handling of materials at all facilities, interim-storage sites, and the repository
1.3. Improve human factors and performance efficiency of handling procedures
1.4. Provide contamination control to reduce safety risks
1.5. Reduce radiation exposure and risk to the workers and to the public, by not having to open the canisters for inspection and provide better automated, remote-handling possibilities

2. Storage and disposal:

2.1. Provide standardized storage configurations at the interim storage sites and the repository
2.2. Minimize the number of canisters which must be qualified for interim storage and receipt by the repository

3. Transportation:

3.1. Promote standard sizing configurations to fit into transportation casks or packagings for efficient packaging design and, consequently, more efficient loading and unloading procedures
3.2. Promote standardized sizes and compatibility for maximum utilization of transportation facilities and equipment

4. Economics:

4.1. Reduce overall handling, transportation, and repository emplacement costs in a simpler, more integrated operation since fewer modifications will be required for OCRWM to accept DOE SNF
4.2. Minimize system costs by reducing the design costs of multiple canister designs

General functional criteria

Figure 1 shows schematically the various paths that the current DOE SNF may follow from the point it begins its initial disposal handling until placement into the repository. This basic structure shows the route SNF can take from initial handling; through interim storage; to repository disposal. It can be used as a guide in the demonstration of the technical adequacy of the canister design as well as demonstrating compliance with regulatory requirements. The indicated process guides the general functional criteria of the standardized canisters. These functional criteria are:

1. The canisters shall be right circular cylinders.
2. Two different diameter-sized canisters shall be designed, with nominal outer diameters of 24.00 in (610 mm) and 18.00 in (457 mm), to accept a significant portion of the various DOE SNFs currently in existence.
3. Two overall lengths for each canister size, 3,000 mm (118.11 in) and 4,570 mm (179.92 in), will be considered maximum lengths from end to end inclusive of the cap ends, labeling, and any handling fixtures.

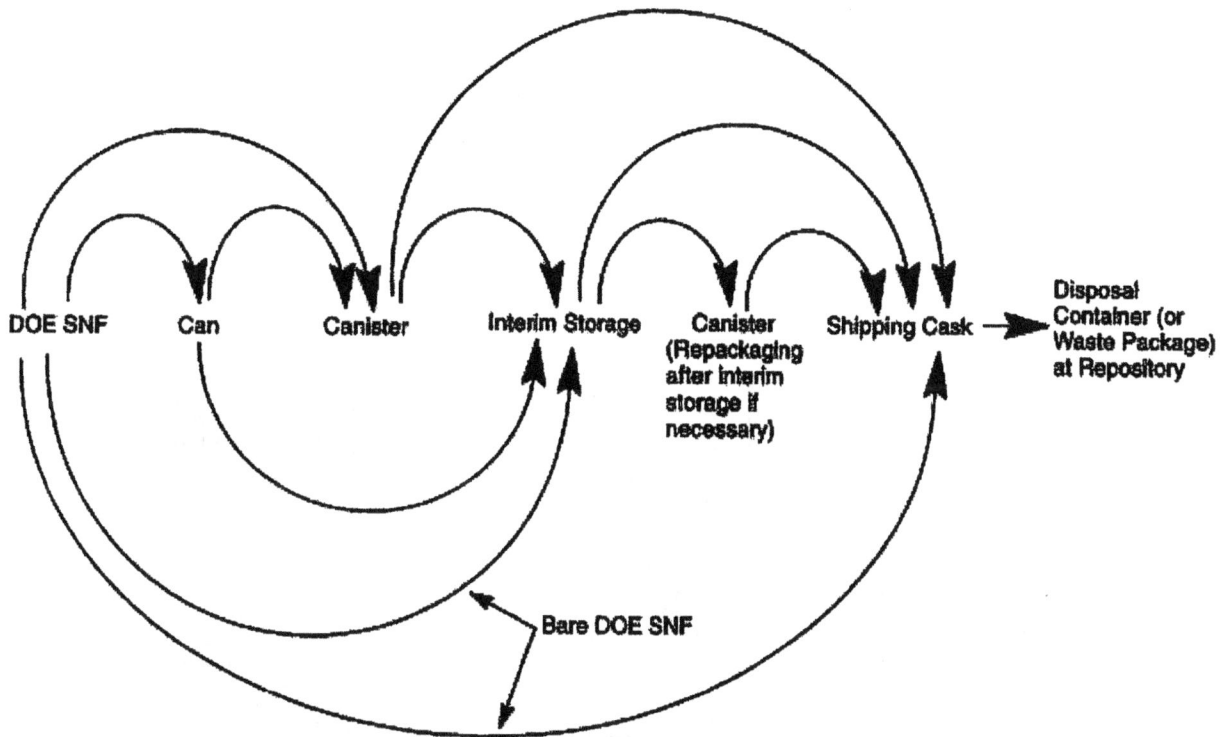

Fig. 1. Possible DOE SNF paths to repository disposal (DOE 1999).

4. The 610 mm (24.00 in) OD canister can be accommodated into a repository waste package as a replacement for one of the HLW canisters, providing a canister for large-sized DOE SNF where needed, and also to provide a potential overpack option for the 457 mm (18.00 in) OD canister.

5. The 457 mm (18.00 in) OD canister can be accommodated in the center hole of a five-pack waste package at the repository as well as being able to be stored in various facilities at the INEEL, Hanford, or SRS.

6. The canisters must perform as required while subjected to the most severely-anticipated environmental conditions and natural phenomenon postulated to occur for the entire service life of the canister.

7. The design of the standardized canisters shall be robust enough to accommodate the grappling and handling equipment configurations at the interfacing facilities when loaded to the weight limits of the canister.

8. Sealing of the canisters shall be accomplished by welding.

9. The canisters shall be designed to provide safe storage (in coordination with the storage facility design) of SNF at any location in the continental United States for a minimum of 40 years. The canisters, in coordination with the handling systems, interim storage systems, and transportation systems of many different facilities—as well as the repository disposal system—must provide confinement for the SNF under all anticipated normal, off-normal, and accident conditions.

10. A safety analysis will need to be performed commensurate with the potential consequences of any activity being performed in conjunction with these canisters.

Figure 2 illustrates the various anticipated uses of the DOE SNF canister.

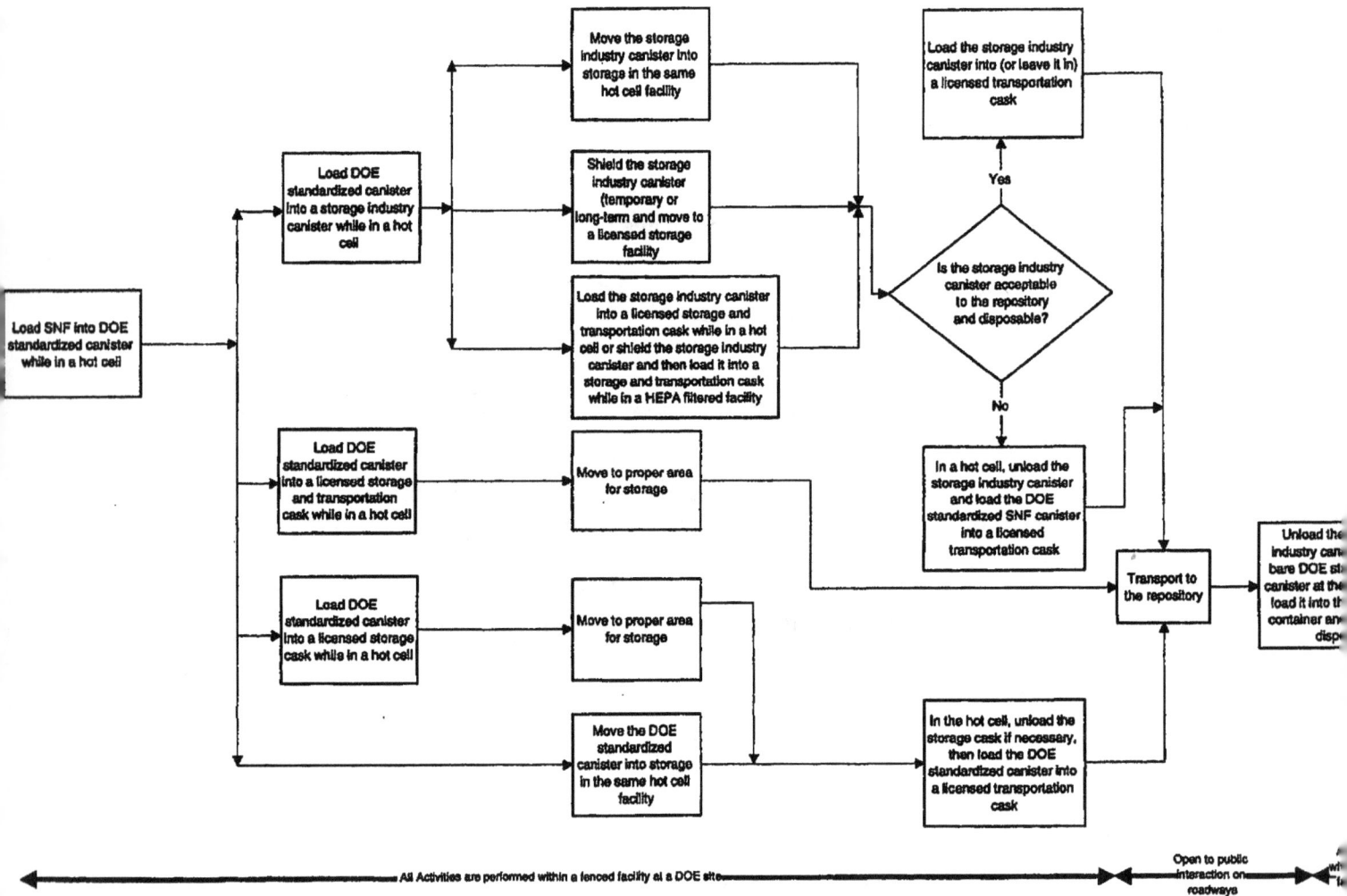

Fig. 2. Standardized DOE SNF canister usage (DOE 1999).

Assumptions for standardized DOE SNF canister design

The basic assumptions used for the functional performance of the standardized canister designs for DOE-owned SNF are:

1. The standardized canisters shall have an inherently robust design such that they can be readily incorporated into storage and transportation systems meeting 10 CFR Part 72 (USNRC 2003c) and 10 CFR Part 71 (USNRC 2003b) requirements respectively.
2. The standardized canisters shall be compatible with the currently-known repository requirements.
3. The design of any internal components (e.g., baskets, spacers, sleeves, dividers, and cans) necessary for the loading of the SNF and for the control of criticality must be constrained by the existing design and interior dimensions of the DOE SNF canisters.
4. Handling facilities will have to accommodate modification to their facilities if these canisters do not fit within the safety envelope of the design criteria for accidental drops and confinement for expected accidental release scenarios.
5. The SNF custodian shall select those canisters, or combinations of canisters, that provide an optimum packing configuration for the custodian's SNF.
6. The DOE SNF canisters shall be capable of accepting intact, failed, or damaged SNF—directly or canned.
7. The DOE SNF is loaded into a canister in a hot cell or shielded facility so that the fuel can be properly dried, and adequate radiation and temperature measurements can be taken to ensure compliance with applicable canister design limits. If necessary, the DOE SNF can be initially placed into a canister underwater (drain holes are provided in the canister design), but the fuel and canister must then be properly dried. Final loading and/or closure and testing would still occur in a hot cell or shielded facility.
8. The loading of the SNF into a canister will not cause significant localized thermal gradients in the pressure boundary, nor will it result in significant bowing concerns for the canister.
9. After SNF loading and radiation and temperature acceptance, the canisters shall be backfilled with an inert gas (e.g., helium). Canister sealing during storage is optional (depending on the storage system used) but seal-welded canisters (backfilled with an inert gas) are required for transportation and repository disposal.
10. The amount of added pressurization from the SNF during interim storage and transportation is assumed to be small such that the actual maximum pressure experienced by the canisters during their containment lifetime is less than the repository acceptance pressure of 151.7 kPa (22 psig).
11. After loading the canisters with SNF, all canisters will require either remote handling due to a lack of canister shielding or placement into another container or overpack which provides adequate shielding.
12. Inspections of the SNF after interim storage and prior to transportation to the repository are not anticipated at this time.

The SNFs being considered for movement to interim storage, transportation, and final disposal are those listed in the Spent Fuel Database, currently maintained by the NSNFP. The two standardized sized canisters, 457 mm (18 in) and 610 mm (24 in) nominal outer diameter (OD), each with two suggested lengths as listed in the specification, shall provide confinement for the listed DOE SNF, with minimal exceptions for larger-sized SNF that may or may not have the ability to be transported as bare SNF in transportation casks. Some SNF or other radioactive materials, such as the particulate waste described in 10 CFR 60.135(c)(2) (USNRC 2003a), may require an interior sealed container. An example of such a container is standard high-integrity can (SHIC) with approximately 5-in (12.5 cm) OD which provide for ease of packaging due to their smaller size. They also maintain structural integrity of the SNF or radioactive material; reduce the possibility of criticality configurations; reduce the possibility of occurrence of other dangerous scenarios; and reduce potential gas generation concerns. The SNF that is placed in the canister will be assumed to be adequately characterized for storage, transportation, and final disposal when sealed into the canister. This information includes radiation shielding; decay heat removal; corrosion; gas generation; criticality control; and other parameters necessary for adequate safety.

Prerequisites for the design specification for the DOE SNF canisters

This preliminary design specification establishes a common basis for all standardized DOE SNF canisters so that they can be accepted for disposal at the repository. The design specifications must consider that the DOE SNF canister final design must be integrated with the total system design, including storage and transportation requirements. Different groups of DOE SNF canisters may be stored in a variety of facilities, and these same groups of DOE SNF canisters may even be transported in a variety of transportation casks. Each unique group of DOE SNF canisters will require Design Specifications and Design Reports as specified by ASME (2001). The Design Reports must establish the design bases and the specified uses of the DOE SNF canister. The Design Reports may also have to be updated as necessary when design information changes. The preliminary design specification assumes that after the DOE SNF is placed in the canister, their use shall be limited to that indicated in four areas as follows:

1. **Initial Loading and Canister Handling:** Sealing of the DOE SNF canisters may be required for interim storage, depending on the type of storage system utilized. However, if the storage system permits it, the DOE SNF canisters may be sealed after interim storage. Incorporated into the DOE SNF canister design is the option of a threaded plug in the top and bottom head. However, prior to transportation and repository disposal, the DOE SNF canister must be seal-welded closed. The DOE SNF canisters shall be backfilled with an inert cover gas (e.g., helium) inside the canister at a pressure of 13.8 to 27.6 kPa (2 to 4 psig). The final closure weld shall implement a welding procedure that can be qualified to yield leak-tight welds.

2. **Interim storage:** For the purposes of this design specification, the SNF storage system must be designed to incorporate the DOE SNF canister into its system. Because the canister is assumed to provide a robust design, it is anticipated that incorporation of the DOE SNF canister into any storage system will not be difficult. Many design considerations, including retrievability, seismic loads, accidental drop loads, and environmental conditions must be adequately addressed by the storage system vendor in order to ensure the proper care and use of these DOE SNF canisters.

A major assumption made regarding the use of the DOE SNF canister in any SNF storage system is that the DOE SNF canister will either stay in the hot cell facility for interim storage purposes or be placed inside another storage container (either a storage industry canister; storage cask; or combination storage and transportation cask). Typically, this other container will be a containment vessel designed to withstand the anticipated operational and accidental loads identified for the SNF storage system.

3. **Transportation:** Although the DOE SNF canister is to be placed inside a licensed transportation cask, the precise details of how the canister is supported within the cask and associated cask details are not known at this time. Therefore, for the purposes of this design specification, the transportation cask must be designed to incorporate the DOE SNF canister into its system. Because the canister is assumed to provide a robust design, it is anticipated that incorporation of the DOE SNF canister into a transportation cask

system is achievable. Many design considerations, including the hypothetical accident conditions described in 10 CFR Part 71.73 (USNRC 2003b) must be adequately addressed by the entire transportation system (canister and cask) in order to ensure the proper care and use of these DOE SNF canisters.

4. **Disposal at the repository:** The loaded DOE SNF canisters will be handled at the repository during the unloading phase from the transportation cask, and during the placement of the canister into the repository waste package. All of these activities are assumed to occur within the confines of a HEPA-filtered building. Placing the canisters into the waste package means that the DOE SNF canisters must be able to conform to the material compatibility and criticality issues indicated in 10 CFR Part 60 (USNRC 2003a). There are additional canister design requirements for the DOE SNF canister but they are contained in various repository documents and cover allowable materials, pressures, weights, and other miscellaneous criteria including labeling.

The DOE SNF canisters are not currently listed on the Q-list for repository equipment, however, the repository has deemed the DOE canisters to be important to safety. The Waste Acceptance System Requirements Document (WASRD) (DOE 2002) has very specific requirements for the DOE canisters during surface facility drop events.

Besides the design requirements for safely lifting and handling the DOE SNF canisters, the repository has indicated that the only remaining requirements imposed on the DOE SNF canisters are those associated with accidental drops or tip-overs of the waste package. Because the canister is assumed to provide a robust design, it is anticipated that incorporation of the DOE SNF canister into the waste package design accidental drop scenarios (2-meter drop and tip-over) should not be a difficult task. The repository personnel has indicated a requirement to remove the DOE SNF canister (not the individual fuel pieces) from a damaged (due to tip-over or drop accident) waste package so that it can be reloaded into another undamaged waste package. Once the waste package has been placed into a drift, the repository personnel has indicated that there are no additional design requirements imposed on the DOE SNF canister. This includes any requirements associated with the retrieval of a waste package as described in 10 CFR Part 60.111(b) (USNRC 2003a). The repository personnel is only anticipating the ability to retrieve a waste package, and is not concerned with removing any contents from a waste package during any retrieval efforts.

Limitations of uses of the DOE SNF canisters

By definition, intact SNF is fuel with cladding that has no hairline cracks or pinholes. Failed SNF is fuel with cladding that has hairline cracks or pinhole leak defects. Damaged fuel, by definition, is fuel with cladding that has defects greater than hairline cracks or pinhole leaks. The DOE SNF canisters must be designed to directly accommodate the placement of intact, failed, or damaged SNF inside the canister. Using this approach, if the DOE SNF degrades during storage or transportation, the canister still satisfies the necessary requirements.

The DOE SNF canisters shall not be arbitrarily used for any other purposes beyond those indicated in the design specification without additional detailed analysis, evaluation, and testing in order to assure that other uses do not violate any of the limitations imposed by the design specification. Examples of use not considered herein include placing molten glass or molten HLW in these canisters, dropping the canister onto sharp projectiles, and dropping the canister from heights greater than that considered.

Canister barrier requirements

Because it has been assumed that damaged SNF will be loaded directly into the DOE SNF canister, the DOE SNF canister must perform the function of the cladding as the primary barrier. In this fashion, the storage system outer container (typically the storage industry canister) or the transportation cask will continue to provide the function of the secondary barrier. Redundant sealing is achieved with this design approach. The design specification assumes that the storage or transportation systems do not provide the double barriers themselves.

DEVELOPMENT OF A ROBUST DESIGN FOR THE CANISTER

Although details regarding SNF loading, interim storage, and transportation are not yet finalized, the NSNFP is establishing a design envelope for the DOE SNF canisters. This design must be robust defined to mean a design that has significant safety margins for the known or estimated loads. With this robust design, the NSNFP can proceed with the design and fabrication of the DOE SNF canisters. The storage system vendor and the transportation cask designer must incorporate the specified DOE SNF canister design into their respective system designs.

Accidental drop scenarios should not be a major design problem for the interim storage vendor or the transportation designer. This is based on the assumption that the smaller DOE SNF canister will be placed in a specifically-designed storage facility, inside another container consisting of the larger storage industry canister, or the transportation cask. The storage industry canisters have already been successfully designed to accommodate their applicable site-specific and system-specific accidental drop or tip-over loads for commercial fuel. Transportation casks must be designed to accommodate 9-m (30-ft) drops onto an unyielding surface and still be able to remove the SNF contents. Therefore, the DOE SNF canister, being inside the storage industry canister or transportation cask (and being robust), should present an easier design situation than deforming baskets. At most, impact limiters inside of the storage or transportation systems may be required for certain fuels with special criticality concerns.

The only remaining accidental drop concerns relate to the time when the canister is being handled by itself or it is inside the waste package at the repository. These anticipated drop scenarios defined to date are a 7.3-m (24-ft) accidental drop of a DOE SNF canister at the repository while being placed into a disposal container, a 2-m (6.56-ft) drop of the waste package with the DOE SNF canister inside, and a tip-over of the waste package with the DOE SNF canister inside. These accident events are listed in order to provide a clear understanding of the intended use of the canister with regard to potential accidental drop events that have been identified.

There are significant advantages to constructing a DOE SNF canister that has a robust design and is drop-resistant. By constructing a drop-resistant canister, more assurances are provided that the canister can be readily incorporated into storage and transportation systems. These systems must be designed for certain postulated drop accidents in addition to other dynamic events. Also, if a drop accident occurs even in a HEPA-filtered facility, recovery would be much simpler and quicker for a canister that maintains containment of the SNF than if contamination of the facility resulted. It is believed that a severe drop accident resulting in contamination of the repository could shutdown the receiving operations for a significant period of time and the ripple effect from such a shutdown would impose nationwide schedule delays.

The design of a usable DOE SNF canister that satisfies ASME (2001) Boiler and Pressure Vessel Code Section III (Subsection NB, Level D or Division 3, Subsection WB) hypothetical accident condition stress limits for an accidental drop scenario of significant height requires some form of energy absorption. Due to the high radiation levels involved, putting on and removing external impact limiters is not practical for the DOE SNF canisters. Consequently, having integral energy absorption capability is desirable. Given other constraints of size and usable volume, keeping the elastically-calculated stresses below the acceptable stress limits for the DOE SNF canister is extremely difficult even with self-contained energy absorption capabilities. Therefore, the NSNFP has designed the DOE SNF canister using a plastic strain approach. The canister is designed with symmetrical, skirted ends and a cylindrical shell sized to absorb a sufficient amount of energy during the impact event so as to limit pressure boundary material strains to acceptable levels. Permanent local deformation is expected as a result of accidental drops. However, low strain limits have been established (using a relatively severe SNF and basket load geometry) to reasonably assure that breech of the canister pressure boundary will not occur during impact for a variety of other SNF and internals. Actual drop testing from heights of 9 m (30 ft) with the canister and various internals in multiple essential orientations will be used to validate the containment capability of the DOE SNF canister design. The currently-unknown details make it impossible to accurately predict the resulting stress and strain levels in the DOE SNF canister material during impact. Therefore, the design of the specified DOE SNF canister was made robust in order to have margin between expected strain levels and those strains that could cause a loss of containment in the canister shell.

To support the specified DOE SNF canister design, the following actions should be considered when details and procedures affecting movement of the canister, design of the canister internals, and loading of the canister are finalized:

1. Movements should be planned to minimize the amount of handling of the canister and the number of lifts. Unnecessary lifting should be avoided. Movements should be made at minimum heights, as close to the floor as practical.
2. Movement paths should be defined and restricted to minimize lifts over walls, transport carts, machinery, and other obstructions that would require increased lift heights, and would impose sharp edges and other puncture threats to the canister.
3. To minimize high lifts over walls, facilities should be designed using a labyrinth arrangement for divider walls when shielding or other reasons for wall separation are required. Facilities and support equipment should be designed to eliminate sharp edges, corners, and protrusions. Tops of walls over which lifts may occur should be rounded and covered with replaceable energy absorbing impact structures.
4. All features of the related facility and support equipment design should give priority to protection of the canister and the SNF.

During accidental drop events, high stress and strain values are expected in local regions of the DOE SNF canister where internal structures contact the canister shell pressure boundary. Internal structures (baskets, spacers, sleeves, dividers, cans, etc.) and fuel elements may contain sharp, stiff elements that pose a puncture threat to the canister containment. Results of the canister drop test evaluations will be contingent on the internal structure designs. Canister internal structure designs must consider fuel safety, condition, stability, and criticality concerns for numerous unique fuel types. A generous volume has been allotted in the DOE SNF canister design to allow flexibility in meeting these demands. Canister protection shall also be considered in the internal structure design requirements. The DOE SNF-specified canister design includes two-in-thick internal end impact plates (part of the internals design) to protect the dished heads from internal impacts and punctures. Internal basket, sleeves, dividers, cans, and spacers shall also be designed to avoid sharp corners and protrusions; and provide large bearing surfaces and energy absorption features that minimize point loads on the containment shell pressure boundary and resulting high-localized regions of stress and strain.

Based on discussions with the repository personnel, once the DOE SNF canister is placed inside the waste package, there are no long-term performance characteristics expected of the DOE SNF canister itself. However, the waste package must satisfy certain design requirements for drop and tip-over accidents. A dropped or damaged waste package may no longer have the required long-term performance capability and may need to be replaced. In order to utilize a new waste package, the SNF must be removed from the damaged waste package. For commercial fuels, this means that the baskets inside a damaged waste package must limit deformations to such an extent that the commercial fuel can still be removed.

The consequences of a waste package drop or tip-over, however, are not expected to be as severe to the DOE SNF canister as they may be to commercial fuels. If a waste package is dropped or if it tips over, there is still the repository requirement to unload that waste package. Unloading the DOE SNF out of a damaged waste package does not require maintaining such a precise geometry as demanded for commercial fuels. All that is needed is to remove the DOE SNF canister (as well as the HLW canisters) so that they can be loaded into a new waste package. That means that the ease of removing the DOE SNF canister depends only on the removal of at least one of the HLW canisters. That one HLW canister could be the canister that was located on top of all the other canisters during the drop or tip-over accident. Being on top, it should be relatively undamaged. Once one HLW canister has been removed, there should be adequate room for removal of the remaining HLW canisters and the DOE SNF canister.

Additionally, the thick, internal end impact plates incorporated into all the DOE SNF canisters will provide support and resistance to crushing forces generated during the waste package drop or tip-over events. Other canister internal components (i.e., baskets, spacers, sleeves, cans, and dividers) can also be designed to protect the integrity of the canister as well as the fuel. This will help mitigate the consequences of these accident events. Therefore, the waste package drop or tip-over accident is not considered to be a design-controlling event for the DOE SNF canister itself.

All design concerns resulting from the materials (SNF and internals) placed inside the DOE SNF canisters must be addressed during normal and off-normal conditions of handling, interim storage, transportation, and disposal. The materials to be placed inside these sealed canisters must be controlled to the extent that adverse or excessive heat, internal pressures, corrosive conditions, and excessive radiation fields are not generated. Because a significant amount of the DOE SNF is highly enriched uranium (HEU) with ^{235}U enrichment levels greater than 20%, criticality concerns must also be carefully addressed when determining what SNF, how much SNF, and where the SNF is to be placed inside these canisters. Maintaining SNF placement may be crucial at times for certain fuels, requiring internal devices such as baskets and spacers to not deform under any specified canister loadings.

The design of the SNF, any internals, and the canister itself must be evaluated together. However, the design of the canisters is preceding the design of the canister internals. Internals required for the proper placement of SNF will have to be designed to conform to the design constraints imposed by the existing DOE SNF canisters. This includes the consideration of geometry as bounded by the inner diameter of the canisters, basket, spacer, or other internals deformations, canister deformations, and canister stresses from SNF or internals loads. Using the above approach, a lightweight, spacious and robust canister can be designed for DOE SNF that can meet the accident condition demands and satisfy the performance requirements for handling, storage, transportation, and disposal at the repository.

Regulatory guidance

There is little explicit regulatory guidance for design criteria for the standardized DOE SNF canisters. However, 10 CFR Part 72 (USNRC 2003c) regulations describe the design criteria necessary for an interim storage facility to be approved and licensed by the USNRC. The DOE SNF canisters will be part of the components comprising an interim storage facility, but it is not intended for the DOE SNF canisters to be an interim storage facility by themselves outside a hot cell facility. Furthermore, criteria necessary for transportation casks are provided in 10 CFR Part 71 (USNRC 2003b). The DOE SNF canisters by themselves are not transportation casks but will be placed into USNRC-licensed transportation casks. However, 10 CFR Part 71.63(b) (USNRC 2003b) does include double containment for certain SNF. The DOE SNF canisters will not be placed directly into the repository but will be placed into a disposal container (or waste package after the final barrier weld is accepted) before being officially disposed. There are generic requirements such as confinement or containment barriers for the SNF; material interactions and compatibilities; and the need for criticality control in the regulations. However, there are no complete and explicit design parameters for the DOE SNF canisters.

Design basis

With the assumption that the transportation cask does not provide double containment, the direct placement of damaged SNF into the DOE SNF canister imposes the requirement of using ASME (2001) Boiler and Pressure Vessel Code, Section III design criteria. The following changes have been made to the 2003 Addendum to Section III, Division 3 in order for the DOE SNF canisters to be fabricated as N-stamped vessels:

1. In order to complete construction (e.g., perform the final closure weld), Division 3 was changed to allow field operations at locations other than the Code shop.
2. Division 3 was changed to allow the actual N-stamping prior to the SNF being loaded into the DOE SNF canisters. In addition, clarifications that allow the use of ultrasonic examination for the final closure weld were made.
3. Due to the presence of the SNF after loading, the DOE SNF canisters shall not be required to satisfy pressure test requirements of Section III, Division 3, Subsection WB (ASME 2001) after loading the SNF and final closure welding. Changes permitting helium leak testing in lieu of pressure testing for low design pressure vessels are covered by code case 656 to Division 3 rules.

The DOE SNF canisters shall not contain any pressure relief devices. The DOE SNF canisters shall be considered N-stamped (with the Data Report signed) after the authorized nuclear inspector has accepted the final closure weld. The DOE SNF canisters shall be designed and evaluated to the acceptance criteria of ASME (2001) Section III, Division 3, Subsections WA and WB. Once storage and transportation details become available, the DOE SNF canisters shall also have the Design Reports incorporate storage and transportation system specifics so that the canisters can be utilized in those systems. The Design Report justifies how the DOE SNF canister can be used. As such, the design loads defined in 10 CFR Parts 71.71 and 71.73 (USNRC 2003b) shall be applied to the DOE SNF canister and the outer transportation cask as a combined system (canister and cask)—not individually. The same combined system design philosophy applies to the DOE SNF canister and the applicable storage system. Depending on how the DOE SNF canister is used for interim storage, there could exist the possible requirement to demonstrate DOE SNF canister structural adequacy. If this occurs, it may be necessary to demonstrate compliance with the requirements of the ASME (2001) Section III, Division 1, Subsection NB. However, this is achievable because the DOE SNF canisters are already Section III vessels.

Canister design limitations

Reasonable design criteria must be established to adequately address anticipated loadings and events. However, it is unreasonable to incorporate all potential site-specific or facility-specific circumstances into the design bases of these DOE SNF canisters. Hence, it becomes incumbent on all users of these canisters to understand their design limitations and to adjust their operations as necessary. One obvious example may be the potential internal pressure. If a specific user intends to load DOE SNF in such a manner that the internal pressure exceeds the maximum considered design pressure, then adjustments must be made to limit the amount of SNF; the amount of included water; or whatever variable that can be adjusted, such that the design pressure is not exceeded.

Another more subtle example of user awareness may be the potential for canister puncture during a possible drop event. From a design perspective, geometries of an unlimited number of potential targets that might puncture the canisters are not identifiable. Therefore, the DOE SNF canisters have not been explicitly designed for any specific puncture resistance, although the drop-resistant robust design instills a significant puncture resistance capability by itself. Hence, it will be up to the user to determine if any potential exists for canister puncture during any canister handling operations. If the potential for puncture exists, the user can either perform an analytical evaluation to determine if puncture is possible, or the user can pad the identified target to prevent puncture.

Establishing the acceptability of future use of canisters that may have been subjected to significant loadings imposed while empty or while filled with SNF (but before the top head is attached) becomes the responsibility of the user. Any significant loadings imposed on empty canisters (e.g., accidental drops) shall be evaluated as described in the final acceptance criteria specified by ASME (2001) Section III, Division 3, Subsections WA and WB (or possibly Section III, Division 1, Subsection NB if appropriate). Significant loadings inadvertently imposed on filled canisters (before the top head is attached) should be evaluated in the same fashion. If a filled canister (before the top head is attached) is "adversely" loaded or dropped, the facility must address the potential cleanup situation under its own operating requirements. Once the canister can be emptied and decontaminated (if necessary), then the user can perform the evaluations and inspections necessary to demonstrate the acceptability of future use or proper disposal of the nonconforming canister.

Although the criteria identified in the preliminary design specification will incorporate reasonable and appropriate design bases, the user must recognize that local conditions and situations must be addressed if they are expected to exceed the specified design criteria. Hence, as indicated before, it becomes incumbent on all users of these canisters to understand the design limitations of the canisters and be able to adjust their operations where necessary in order to achieve the expected functionality of these DOE SNF canisters. This includes the proper storage (e.g., dry storage without contamination by damaging environments) of empty DOE SNF canisters prior to loading.

Geometry and materials

All of the DOE SNF canisters shall be right circular cylinders that are able to stand vertically when placed on a flat surface after being loaded with SNF. The large canister shall have a nominal outer diameter (OD) of 610 mm (24 in) and a nominal wall thickness of 12.7 mm (0.5 in). The small canister shall have a nominal OD of 457 mm (18 in) and a nominal wall thickness of 9.53 mm (0.375 in). Both canisters shall be designed for a maximum overall length of either 3,000 mm (118.11 in) or 4,570 mm (179.92 in). Dimensional tolerances and fabrication processes (including weld grinding where necessary) shall be controlled so that the maximum dimensions are not exceeded.

Because the material specification for seamless and welded austenitic stainless steel pipes is SA-312, certain dimensional tolerances from material specification SA-530 in Section II of the ASME (2001) Code associated with the exterior canister shell are already specified. Outer diameter variations shall not exceed 3.2 mm (0.125 in) for the 610 mm (24.00 in) canister and 2.4 mm (0.093 in) for the 457 mm (18.00 in) canister. For ovality, the difference in extreme outside diameter readings in any one cross section shall not exceed 1.5% of the specified nominal outside diameter. Straightness tolerances for each canister shall not exceed 3.2 mm (0.125 in) maximum deviation for every 3.0 m (10 ft) of pipe length (both ends of a 3.0 m straightedge used for the measurement are in contact with the surface). In addition, provisions must be made for the as-welded condition of the final canister weld (attaching the top head to the canister shell). Therefore, an additional diameter increase of 4.76 mm (0.1875 in) shall be considered acceptable for the crown of the final canister weld. In order to ensure that certain contents can be placed inside these canisters and that the canister can be placed in the repository waste package, the dimensions listed in Table 2 shall be incorporated into the canister design.

Current repository criteria indicate that low-carbon austenitic stainless steel or stabilized austenitic stainless steel materials are acceptable. The National Spent Nuclear Fuel Program (NSNFP) is proceeding on the assumption that low-carbon austenitic stainless steel will be acceptable to the repository. Although canister-specific drop-tests have been performed that clearly demonstrate the acceptability of 304L, stainless steel 316L has better resistance to pitting corrosion and hydrogen embrittlement. Considering the variations in actual material properties of stainless steel, it is believed that the use of 316L would be an acceptable canister material, even including accidental drop events. Therefore, the DOE SNF canisters shall be made of SA-312, type 316L stainless steel for the shell; and SA-240, type 316L for all other parts including the heads, nameplates, and lifting rings. The optional plugs and plug thread plates shall be SA-479, type 316L stainless steel. All stainless steel materials shall be annealed and pickled. The DOE SNF and canister internals shall preclude chemical, electrochemical, or other reactions (e.g., internal corrosion) of the canister or waste package such that there will be no adverse effects.

Table 2. DOE SNF canister dimensions (DOE 1999).

Canister Size	Nominal Outer Diameter	Long Canister Max. External Length	Short Canister Max. External Length	Min. Internal Diameter	Long Canister Min. Internal Length	Short Canister Min. Internal Length
Large	610 mm (24.00 in.)	4,570 mm (179.92 in.)	3,000 mm (118.11 in.)	579 mm (22.80 in.)	4,038.6 mm (159 in.)	2,470.2 mm (97.25 in.)
Small	457 mm (18.00 in.)	4,570 mm (179.92 in.)	3,000 mm (118.11 in.)	430 mm (16.93 in.)	4,114.8 mm (162 in.)	2,540.0 mm (100 in.)

As part of the material selection process, important parameters to consider are the anticipated erosion and corrosion values expected during interim storage and transportation use. A total value of 1.27 mm (0.050 in) of pressure boundary wall thickness reduction has been established as the erosion and corrosion value to be used for canister design purposes. This corrosion/erosion value reflects the full design lifetime of 100 years. Therefore, prior to acceptance at the repository, the DOE SNF canisters shall be protected from adverse environmental conditions in such a manner as to prevent the total wall thickness corrosion/erosion limit from being exceeded. A 50-year interim storage and transportation interval shall be assumed for this specific wall thickness reduction evaluation. The assumption is made that once the DOE SNF canister is placed inside the waste package, insignificant corrosion or erosion will occur for the next 50-year interval.

The canisters will be subjected to a radiation environment. Because criticality is to be eliminated, the canisters should not be exposed to large neutron fluences. Long-term cumulative exposures to high neutron fluence (on the order of 10^{17} n/cm^2 and greater) has caused degradation in reactor vessels but this is not applicable to the DOE SNF canisters. Radiation fields (10^7 rad/h or less) are expected, but no significant material damage or degradation is anticipated for the stainless steel material.

Both the inner and outer surfaces of the canisters shall have a finished condition such that acceptable nondestructive examinations can be performed in order to satisfy ASME (2001) Code, Section III, Division 3, Subsection WB requirements. However, no specific surface finish is specified for the DOE SNF canisters. The repository has limits on outside surface contamination for the canister but has not imposed an associated surface finish in conjunction with the surface contamination requirement. The repository does require that any burrs, sharp edges, and weld edges shall not exceed 0.5 mm (0.0197 in). The interior surfaces shall be smooth enough to allow easy loading of any DOE SNF or internals (baskets, spacers, sleeves, dividers, cans, etc.) so as to not damage the SNF.

Contents

The contents of the canister including SNF and any applicable internals necessary for the safe placement and orientation of the DOE SNF, have not been specifically defined. However, it is assumed that the contents will not compromise the structural integrity of the DOE SNF canister. It is also assumed that the contents will satisfy all applicable regulations and requirements, especially those set forth by the repository. For example, the contents of the disposal canister shall contain no pyrophoric, combustible, explosive, or chemically-reactive materials in an amount that could compromise surface-facility, repository preclosure safety, or repository long-term performance. The DOE SNF canisters shall be designed for the total maximum allowable weights (canister plus contents) listed in the Table 3. These weight limits are equal to or less than the weight limits established in the repository's Interface Control Document (ICD).

Table 3. DOE SNF canister maximum total allowable weights (DOE 1999).

Canister Size	Nominal Outer Diameter	Long Canister Maximum Total Weight	Short Canister Maximum Total Weight
Large	610 mm (24.00 in.)	4,535 kg (10,000 lb$_f$)	4,080 kg (8,996 lb$_f$)
Small	457 mm (18.00 in.)	2,721 kg (6,000 lb$_f$)	2,270 kg (5,005 lb$_f$)

When loading the DOE SNF, the center-of-gravity of the entire contents (SNF, baskets, spacers, sleeves, dividers, etc.) shall be within 127.0 mm (5 in) of the canister centerline for the 457 mm (18.00 in) nominal OD canister, and within 203.2 mm (8 in) of the canister centerline for the 610 mm (24 in) nominal OD canister. These loading restrictions are to avoid excessive lop-sided loading situations and to limit resulting stresses in the lifting ring and adjacent skirt portion of the canister. When possible, the center-of-gravity of the loaded DOE SNF will be as close to the canister centerline as reasonably achievable. The axial location of the center-of-gravity of a loaded DOE SNF canister shall be within 609.6 mm (24 in) of the canister centroid. Sites performing SNF loading may make separate evaluations of center-of-gravity locations if the indicated center-of-gravity limitations are exceeded.

Canister sealing

Storage, transportation, and repository disposal criteria all indicate specific requirements associated with the safe and proper sealing of SNF containers. After loading the SNF, the DOE SNF canisters shall be covered with an inert gas (e.g., helium) inside the canister. Such a gas is intended to eliminate or significantly reduce SNF corrosion; provide more appropriate heat transfer conditions internally; and reduce combustion concerns.

Sealing the DOE SNF canisters may be required for interim storage, depending on the type of storage system utilized. However, if the storage system permits it, the DOE SNF canisters may be sealed after interim storage. Incorporated into the DOE SNF canister design is the option of a threaded plug in the top and bottom head. However, prior to transportation and repository disposal, the DOE SNF canister must be seal-welded closed. The DOE SNF canisters shall be backfilled with an inert cover gas (e.g., helium) inside the canister at a pressure of 13.8 to 27.6 kPa (2 to 4 psig). The final canister weld (attaching the top head to the canister shell) shall implement a welding procedure that can be qualified to yield leaktight welds. At a minimum, in order to demonstrate compliance with the ASME (2001) Code and obtain the Code stamp per the proposed Code changes, the DOE SNF canister shall be helium leak-tested to verify that no leakage is detected that exceeds the rate of 10^{-4} std cm^3/s. After it is closed, the DOE SNF canister shall not contain or generate free gases other than air; inert cover gas; and radiogenic gases with an immediate internal gas pressure not to exceed 151.7 kPa (22 psig). Therefore, the DOE SNF canister shall have a maximum allowable (design) pressure of 344.8 kPa (50 psig).

Incorporated into the DOE SNF canister design is the option of a threaded plug in the top and bottom head. These threaded plugs can be used, where necessary, for a number of functions including canister draining, degassing, and remote inspection. Installation or removal of the threaded plug(s) is expected to be performed while the DOE SNF canister is inside a hot cell because the containment feature of the canister depends upon the proper installation of the threaded plug. When using these threaded plugs, it is necessary to seal-weld the threaded plugs in order to establish an acceptable containment boundary per ASME (2001) Code, Section III requirements. The seal weld shall cover any exposed threads on the plug.

Shielding

For the purposes of the design specification, it is assumed that the DOE SNF canisters do not require additional shielding beyond that provided by the canisters themselves, or in conjunction with shielding provided by the facilities handling the DOE SNF canisters or the storage or transportation systems. It is assumed that any of the facilities handling the DOE SNF canisters or the repository will have adequate equipment necessary to handle the DOE SNF canisters by themselves. If additional shielding is required by local sites or facilities, it is up to the user to provide adequate shielding measures without adversely affecting the DOE SNF canisters.

Criticality

For the purposes of the design specification, it is assumed that adequate attention to the types and amounts (proper fissile limits) of SNF to be loaded into the canisters or proper configuration using properly-designed internals (baskets, spacers, sleeves, dividers, cans, etc.) will preclude any criticality concerns. Therefore, this design specification assumes no criticality events during the canister's design life. For criticality concerns, the DOE SNF canisters must be capable of maintaining reasonable geometric integrity only. The personnel responsible for the designs of internals necessary for criticality prevention must address any associated concerns should the DOE SNF canisters be loaded, dropped, or handled in such a fashion as to adversely load the SNF or internals.

Normal operating loads and environmental conditions

The weight considerations listed in Table 3 are for all the DOE SNF canister geometries. The DOE canisters shall not be horizontally or vertically stacked at any time with any other canisters without a proper evaluation of all possible consequences. Interim storage, transportation, and disposal scenarios are situations where the canisters are within other enclosures or facilities, and these canister placements shall be properly evaluated. The limitation of no vertical or horizontal stacking is imposed not from a strength concern, but rather from a safety viewpoint. Vertically stacking these long, slender canisters (either empty or filled) would prove to be a major safety concern for personnel due to a lack of stability. Horizontally stacking these canisters (like a cord of firewood) could be permissible in an empty condition, but that could also be unstable due to rolling concerns unless specific actions are taken to prevent it from happening. Horizontally stacking these canisters when filled with SNF appears to be undesirable from a rolling stability concern. Potential criticality implications and excessive heat-generation concerns may exist if the canisters were to be stacked in close proximity to each other. Therefore, due to a number of unknown implications and concerns, the preliminary design specification assumes that the canisters will not be vertically or horizontally stacked. If the canister user needs to stack the DOE SNF canisters, then the user must provide the justification, addressing all potential implications.

The lifting fixture is not part of the canister design specification. However, the design of the lifting fixture does affect the resulting design of the canister. The lifting fixture shall provide at least three locations to engage the canister lifting ring, with the capability to engage and disengage remotely. The lifting fixture shall be capable of engaging and disengaging while remaining within the projected perimeter of the DOE SNF canisters.

The DOE SNF canisters shall be designed to be vertically lifted with a lifting fixture that engages underneath the 12.7 mm (1/2 in) thick lifting ring. Material temperature limits for lifting the canisters shall be 148.9°C (300°F). Due to the symmetry of the specified DOE SNF canister design, either end shall be capable of being used to lift the canister. The Maximum Normal In-Plant Handling Pressure (MNIP) shall be considered to be acting coincidentally. With respect to recovering from an accidental canister drop or tip-over—regardless of severity—the canisters shall be designed to be picked up from both extreme ends or tilted back upright from a horizontal position. Stresses resulting from this action shall satisfy normal operating condition stress limits defined in the ASME (2001) Code, Section III, Division 3, Subsection WB. Worst case temperatures and pressures shall be considered to be acting coincidentally. The weight of the contents shall be assumed to be lumped at the centroid of the canister.

The Maximum Normal In-Plant Handling Pressure (MNIP) is the maximum pressure that would develop in a DOE SNF canister during initial handling; interim storage; transportation; initial repository handling; or loading of disposal container prior to actual emplacement in a repository drift under the most severe conditions of normal in-plant handling operations. The DOE SNF canister shall be designed for a MNIP not to exceed 344.8 kPa (50 psig) per the criteria of the ASME (2001) Code, Section III, Division 3, Subsection WB.

The Maximum Normal Operating Pressure (MNOP) is the maximum pressure that would develop in a DOE SNF canister during initial handling; interim storage; transportation; initial repository handling; or loading disposal container prior to actual emplacement in a repository drift without venting. The DOE SNF canister shall be designed for a MNOP not to exceed 151.7 kPa (22 psig) per the criteria of the ASME (2001) Code, Section III, Division 3, Subsection WB.

The Primary Service Temperature for a DOE SNF canister when it is not inside any other container is 176.7°C (350°F), and 343.3°C (650°F) after placement within another enclosed container (e.g., a storage industry canister for interim storage or a transportation cask), possibly with other heat-generating DOE SNF canisters or HLW canisters. The maximum operating temperature for a DOE SNF canister when it is not inside any other container is 148.9°C (300°F), and 315.5°C (600°F) after placement within another enclosed container (e.g., a storage industry canister for interim storage or a transportation cask), possibly with other heat-generating DOE SNF canisters or HLW canisters.

The DOE SNF canisters shall be designed for 20 full MNIP and temperature cycles of a canister achieving its maximum steady state operating temperature of 315.5°C (600°F) inside another container and then suddenly being exposed to an external calm air temperature environment of 10°C (50°F) while the canister simultaneously loses its internal pressure. The maximum thermal gradient associated with this event shall be evaluated per the criteria of the ASME (2001) Code, Section III, Division 3, Subsection WB. If the canisters are subjected to any other significant fatigue loads due to initial SNF loading, interim storage, transportation, or loading into a disposal container at the repository, a detailed fatigue analysis shall be performed per WB-3221.9(e). If necessary, cumulative usage factors from all uses (SNF loading, canister handling, storage, or transportation) shall be evaluated once these values are known.

The DOE SNF canisters shall be capable of maintaining containment in temperature environments that range from -40°C to 343.3°C (-40°F to 650°F), excluding accidental drop scenarios or other accidental events of a canister when being handled by itself or inside the waste package at the repository.

For situations requiring specific design evaluations where significant compressive stresses occur in the canister, the buckling stress shall be taken into account. Buckling situations need to be considered in terms of being able to remove the canister from the enclosing container (either the storage industry canister or the transportation cask).

The DOE SNF canisters shall be designed for any other normal operating condition loads resulting from initial handling, interim storage, transportation, or handling and loading into a disposal container at the repository once the loads and environments have been defined.

39

Hypothetical accident loads and environmental conditions

Anytime a loaded DOE SNF canister is being handled by itself, the canister shall be within a HEPA-filtered building or facility. This eliminates the requirement of specifically designing the DOE SNF canisters to any specific ASME (2001) Code stress limits for accidental drop events. However, when the DOE SNF canister is enclosed within a storage industry canister for interim storage purposes or within a transportation cask, the DOE SNF canister shall be designed in accordance with the criteria in the ASME (2001) Code, Section III, Division 3 stress limits identified in WB-3224 or Section III, Division 1, Subsection NB-3225 as required. For the repository waste package drop or tip-over event, the canister shall be considered adequate as developed since the only requirement is to remove the DOE SNF canister from the damaged waste package and place it into another undamaged waste package. The DOE SNF canister has a robust design because it was developed to maintain containment when subjected to a 9-meter (30-ft) drop onto an essentially unyielding surface. It is this robust design that makes the DOE SNF canister adequate to survive the waste package drop or tip-over event. For situations requiring specific design evaluations where significant compressive stresses occur in the canister, the buckling stress shall be taken into account. Buckling situations need to be considered only in terms of being able to remove the canister from the enclosing container (either the storage industry canister or the transportation cask).

Quality assurance

The designer and fabricator of the DOE SNF canisters shall establish, maintain, and execute a quality assurance program based on the criteria necessary to satisfy ASME (2001) Code, Section III, Division 3 construction criteria, 10 CFR Parts 71 (Subpart H), and 72 (Subpart G) (USNRC 2003b) quality assurance requirements.

Physical protection of SNF

Because the DOE SNF canisters are to be seal-welded for transportation and repository disposal, the DOE SNF canisters do not require the use of a USNRC-approved tamper-safe seal. Tamper indicating devices are only required on canisters containing strategic special nuclear materials that are not seal-welded. Depending on how the DOE SNF canister is being used within an interim storage system, a tamper-safe seal may be required if the DOE SNF canister is not seal-welded.

Labeling

The DOE SNF canisters shall be capable of being properly labeled as follows:

1. The labels shall be an integral part of the canister, engraved to a depth no greater than 0.8 mm (1/32 of an in) that can be reasonably expected to remain legible for 100 years at temperatures of 25°C (77°F) to 400°C (752°F).
2. The labels shall have a unique alphanumeric identifier.
3. The labels shall not impair the integrity of the canister.
4. The labels shall be chemically compatible with the canister material.
5. The top label shall be visible from the top of the canister with the lifting fixture engaged, with characters approximately 25.4 mm (1 in) in height.
6. The labels shall not cause the canister dimensional limits to be exceeded.

The DOE SNF canisters should labeled on the outermost surface of each lifting ring on the top and bottom ends of the canister (two places). The alphanumeric identifier shall be readable as if the remotely-operated cameras are on the outside of the canister looking inward toward the axial centerline of the canister. Placement of an alphanumeric label on the lifting ring shall not cause any interference or loading concerns.

Documentation

The designer of the DOE SNF canisters shall provide adequate documentation, reports, and design drawings in the proper form and format to satisfy the proper quality assurance requirements and record storage requirements. In addition, adequate documentation shall also be provided that:

1. Supports the acceptable design of the DOE SNF canisters.
2. Permits an independent review of all design procedures and calculations.
3. Identifies all software used in the design process.
4. Indicates appropriate validation and verification documentation of all software used for the design calculations.
5. Indicates the involved design personnel has the required experience, education, training, and proficiency.
6. Is legible and in a form suitable for reproduction, filing, and retrieval.

DROP TESTING RESULTS FOR THE 18-INCH STANDARDIZED SPENT NUCLEAR FUEL CANISTER

INTRODUCTION

The standardized canister design (DOE 1999) developed by the National Spent Nuclear Fuel Program (NSNFP) incorporates an energy-absorbing skirt that deforms on-impact during accidental drop events, providing a significant amount of protection to the actual pressure boundary or containment system of the canister. This deformed skirt can even be removed (cut off) if necessary without disrupting the canister containment, thereby enhancing the canister's ability to fit into other containers. The skirt helps to protect the canister containment system in virtually all accidental drop events excluding the horizontal (flat) impact orientation and various potential puncture events.

The goals and objectives of the drop-testing were developed early and were supplemented as results of various efforts became available as follows:

1. To demonstrate the canister's robust design by drop-testing a number of combinations of representative canisters (called test canisters) and contents (internals plus simulated SNF) that most significantly challenged the canisters from a containment viewpoint.
2. To determine the adequacy of the scientific prediction regarding the structural response of the test canisters and certain internals.
3. To demonstrate that the canisters could be fabricated as indicated on program drawings, making improvements where possible to improve the fabrication effort.
4. To demonstrate that the canister internals could be easily loaded into the canister, confirming that during actual use, the internals identified could be easily loaded using remote-handling techniques.
5. To gain insights into the use of ultrasonic examination for the final closure weld with a permanent backing ring.
6. To clearly demonstrate the magnitudes of the resulting deformations so that it could be determined if, after being dropped, the deformed canisters could be loaded into other containers. These other containers could include an interim storage canister; a transportation cask; a repository waste package; or a larger standardized DOE SNF canister.

The objective of the drop-tests was intended to demonstrate the canister's robust design by drop-testing a number of combinations of representative canisters (called test canisters) and contents (internals plus simulated SNF) that most significantly challenged the canisters from a containment viewpoint. These tests were performed at Sandia National Laboratories (SNL) as reported in Morton et al. (2000). Note that canister and internals deformations were not considered to be important with respect to SNF criticality for this drop-testing effort. Proposed internal geometries were reviewed to select the internals that would result in the worst possible damage on canister containment. For example, carbon steel material for the internals was utilized rather than ductile stainless steel to increase possible damage to the test canisters. Even the orientation of the internals within the canisters was specified in order to inflict maximum possible damage to the containment system of the test canisters. All possible situations were adjusted to inflict the most potential damage to the test canisters during their drop tests. Consequently, the test canisters contained a number of weld joints that did not receive any post-weld heat treatment. Similarly, the pipe used to fabricate the test

43

canister bodies and skirts was longitudinally welded pipe. It was assumed that if a problem were to develop, the weld joints would be a logical first location to check. If desired, seamless pipe can be used, but this test effort used welded pipe to demonstrate that the welds are adequate. Hence, the main focus of the drop-testing was the demonstration of maintaining a containment system, regardless of the impact orientation of the canister.

Using finite element methods and fully-plastic analyses, Morton et al. (2000) compared pre-test predictions with and post-test analysis. This effort not only provided validation of the unique computer models developed, but also allowed increased confidence in prediction of canister responses to situations not specifically tested.

DESIRED QUALIFIED DROP-TEST DATA RESULTS

The scope of work necessary to achieve the desired qualified drop-test data results considered a preliminary phase and six subsequent phases.

Preliminary phase

Snow and Rahl (1998; 1999) reported the results of both small- and full-scale drop-testing efforts. These initial drop tests were preliminary and scoping in nature, and were performed to give initial insights into the adequacy of the proposed canister design. Initial efforts included drop testing of full-scale simulated canisters from 30 ft (9 m) onto a hardened surface consisting of two-in thick steel plate placed on a thick concrete slab. Later on successful puncture testing was accomplished by dropping simulated canisters 40 in (1.02 m) onto an essentially rigid, six-in diameter bar. These preliminary tests performed at the Idaho National Engineering and Environmental Laboratory (INEEL), although limited in number and drop orientation, demonstrated that the proposed standardized DOE SNF canister design was robust.

At the end of this preliminary phase, the NSNFP initiated the fabrication, at the INEEL, of nine full-scale representative standardized DOE SNF canisters. These test canisters (loaded with carbon steel reinforcement bars to represent SNF) were drop-tested at SNL during the summer of 1999 and returned to the INEEL for post-drop test examinations and leak testing. With the available funding, only nine test canisters could be fabricated and drop-tested.

Phase I - material procurement

During this phase efforts were directed at procuring the materials necessary to fabricate the canisters, the internals, and the representative SNF. Basically, 316L stainless steel was purchased for the canisters, while carbon steel was utilized for the internals and reinforcement bar (hereafter called rebar). The only exception for the internals was the material obtained for the simulated High Integrity Containers (HICs) which utilized 316L stainless steel.

The stainless steel canister material that was purchased satisfied the ASME (2001) Boiler and Pressure Vessel (B&PV) Code, Section III, Division 3 requirements. Material certifications were obtained for the stainless steel materials. These followed Quality Level 3 requirements, of INEEL (LMITCO) Quality Assurance (QA) Program. Eighteen-in (457 mm) nominal diameter, schedule 40s, SA-312 pipe (longitudinally welded), 3/16 (4.8 mm) and ½ in (12.7 mm) thick SA-240 plate (including the flanged and dished pressure vessel heads), and SA-479 round bar material with 2- 5/8-in (66.6 mm) diameter was purchased for the canister fabrication effort. The SA-312 pipe was purchased using the normal material specification defaults. This included the requirement that each pipe length be hydro-tested to 600 psig (4.14 MPa). This was done as a proof test for the pipe and the longitudinal seam weld. In addition, 18 ASME flanged and dished heads were purchased, made from SA-240 plate such that the head thickness was a nominal 3/8 in (9.5 mm) thick. All of the stainless steel materials were required to be annealed and pickled. All of these items passed the receiving inspection performed by the INEEL's quality receiving group.

An A-36 carbon steel plate was obtained for fabricating the 2 in (50.8 mm) thick impact plates located inside of the canisters (considered part of the internals). For the remaining internals such as the spoked-wheel baskets, the sleeves, and the simulated Shippingport fuel bundles, A-36 plate and angle, A-500 Grade B structural tubing, and A-106 Grade B schedule 80 pipe were used. The rebar material was A-615, Grade 60 and was used to increase the test canister weight and represent the SNF. The simulated HICs were fabricated from 5-in (126.9 mm) nominal diameter, schedule 40s, welded SA-312, type 316L stainless steel pipe.

Phase II - fabrication and examination efforts

Test canisters: The fabrication and examination of the test canisters were performed at the INEEL under a Quality Level 3 effort with enhancements in order to approximate nuclear vessel construction. The actual fabrication of the test canisters was not subject to the NSNFP QA requirements. Although the INEEL does not have an "N-stamp", welders qualified to the INEEL Weld Program that follows ASME Section (2001) IX procedures were used. The qualification of the nondestructive examination personnel conformed to the requirements of SNTC-TC (1992). The welding techniques used are a combination of manual Tungsten Inert Gas (TIG) and manual pulse metal arc (wire feed). Material traceability was ensured through proper marking of the heat numbers on the specific components and assembly numbers on the finished components. All of the containment pressure boundary welds existing before loading the test canisters with internals were volumetrically examined using LMITCO (1996b) radiography testing (RT) and liquid penetrant (LP) examinations (LMITCO 1998) of the final pass.

During the fabrication and examination efforts, certain improvements were recognized that would make the canisters easier to fabricate and examine. Although limited, these changes were mainly associated with a slight geometry change of the plug thread plate and the elimination of RT for the welds attaching the skirt to the vessel head and the lifting ring to the skirt. Note that the ASME (2001) B&PV Code, Section III, Division 3 requires that these structural attachment welds be examined using liquid penetrant.

The test canisters fabricated were 18-in (457 mm) nominal diameter, and were either 10 or 15 ft long (nominally). The entire canister exterior was fabricated using 316L stainless steel. The test canisters themselves were fabricated mainly at the INEEL's Central Facilities Area (CFA) machine and weld shops. One item noticed during fabrication was that the full penetration weld attaching the lifting ring to the skirt caused the remaining skirt beyond or outboard of the lift ring to pull radially inward. This initial inward curving basically "controls" the deformation of the skirt during a drop accident event. Rather than deforming outward, the skirt will now deform inward during drop events, especially the 0° or vertical drop.

Internals: The function of most of the internals is to orient the SNF inside the canister and to prevent excessive SNF movement during canister movement situations. The desire is to load the SNF so that the total center-of-gravity is at, or near, the centroid of the canister. Internals consisted of both the upper and lower 2-in (50.8 mm) thick impact plates, a full cavity length 3/16-in (4.8 mm) thick sleeve (where used), the most potentially-damaging SNF basket referred to as the spoked-wheel basket (Fig. 3), and any spacer plates necessary to properly position the SNF or the two-in (50.8 mm) thick impact plates. All of these internals were made of A-36 carbon steel plate or A-106 Grade B schedule 80 carbon steel pipe.

The sleeve, employed on five of the canisters, was intended to separate the spoked-wheel basket and rebar from direct contact with the canister wail. It was suggested that during a drop event, contents with sharp edges or points might initiate a crack on the canister wall. If that crack were to propagate through the wall, then the containment would be breached. However, with a sleeve separating the contents from the actual canister wall, a potential crack could initiate in the sleeve but have no mechanism to propagate into the canister wall. It is expected that all canisters with contents having possible "sharp" edges or points would have an internal sleeve for this reason. The sleeve physically provides the margin of safety that is desired to be maintained in the canister design.

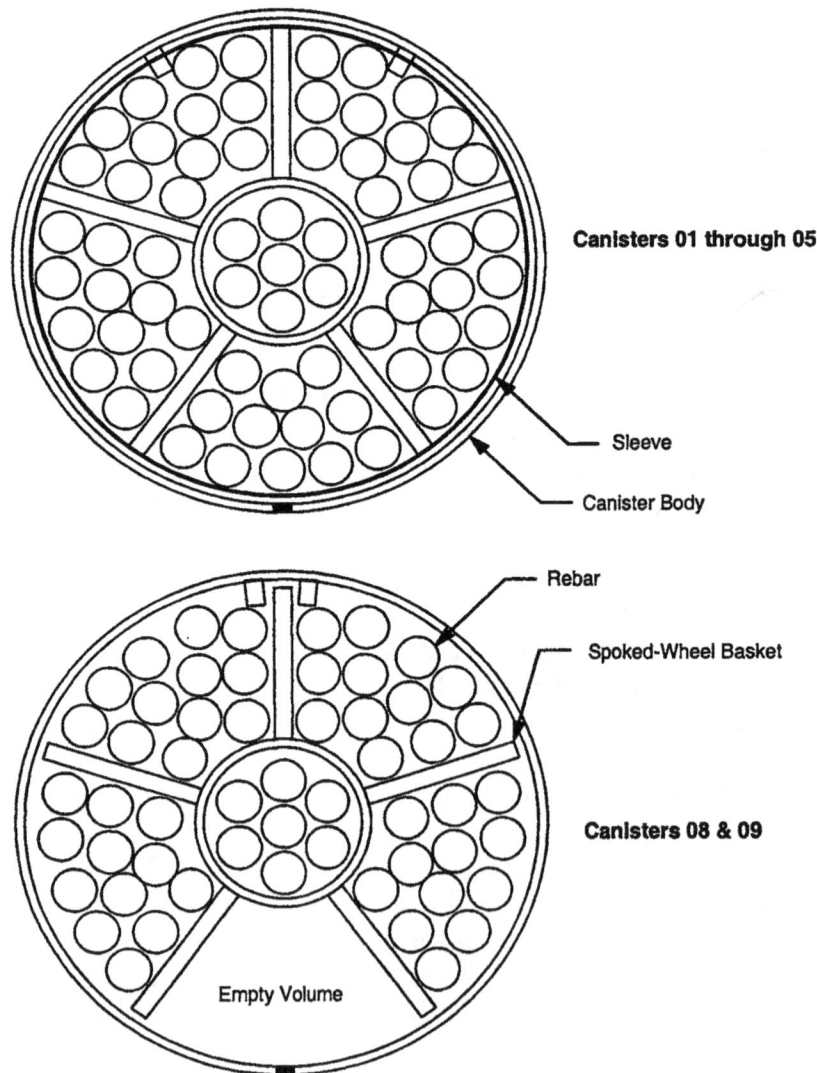

Fig. 3. Cross-section of test canister 01-05, 08 and 09 internal components (Morton et al. 2000).

For the two 10-ft (3 m) canisters, special internals were used (Fig. 4). One test canister contained two simulated Shippingport fuel bundles placed side-by-side into the canister without any sleeve. These simulated SNF elements were fabricated by welding four A-36 3 x 3 x 3/16 angles onto the outside corners of A-500 7 x 7 x 3/8 structural tubing. The other 10-ft (3 m) canister contained simulated HICs. The SA-312 welded stainless steel pipe was used to fabricate the HICs, including SA-240, 316L, ½-in (12.7 mm) thick endplates.

Because the internals have not been identified as having to perform any safety-related function (such as criticality spacing control), the techniques used to fabricate them were not as rigidly documented as for the canisters. However, qualified INEEL welders and inspectors were utilized. Appropriate INEEL Weld Program welding procedures were used for manual Tungsten Inert Gas (TIG) for the simulated stainless

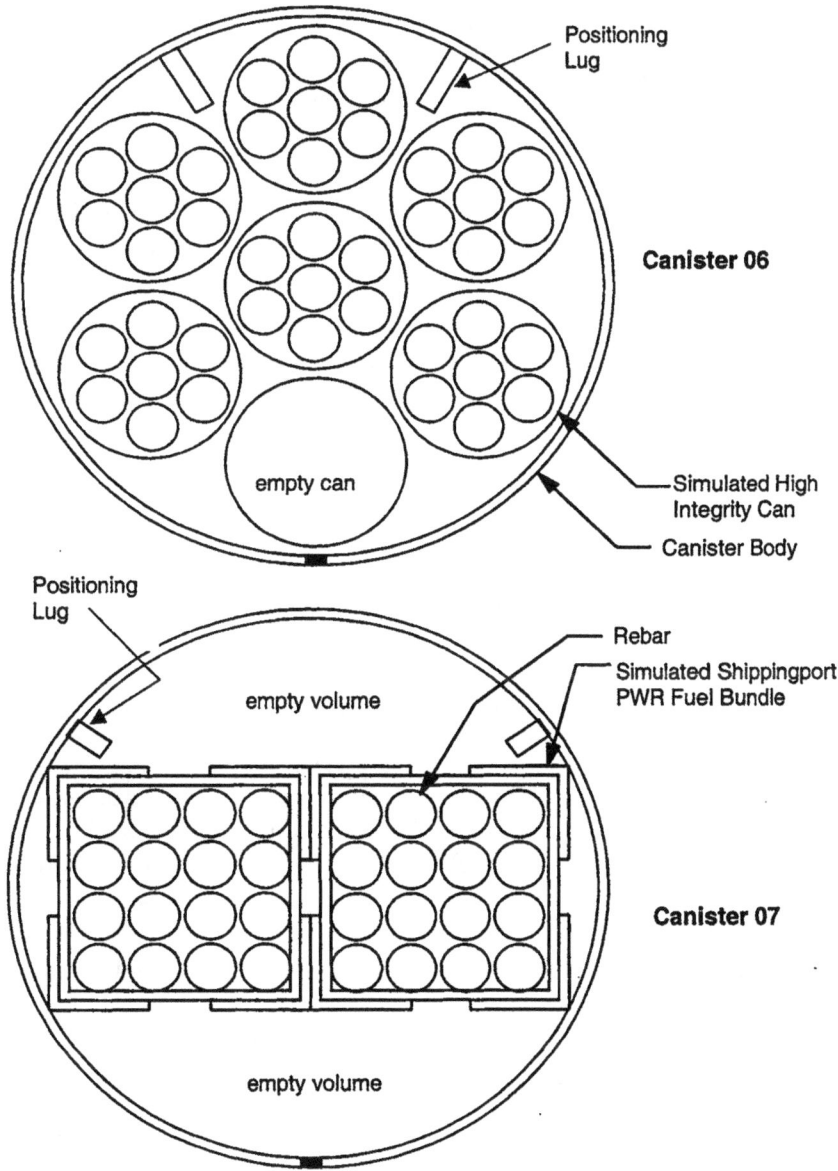

Fig. 4. Cross-section of test canisters 06 and 07 internal components (Morton et al. 2000).

steel HICs, and shielded metal arc welding techniques for the remaining carbon steel internals. All welds were inspected using liquid penetrant techniques on the final pass.

The 2-in (50.8 mm) thick impact plates that fit inside the canisters were machined at the INEEL'S Test Reactor Area (TRA). The plates were machined to a shape that corresponded to the interior dimensions of the top and bottom dished heads. The machining permitted a greater area of contact and better load distribution during potential accidental drop events while still having a flat area for the SNF and other internals to rest on during loading. INEEL Test Area North (TAN) personnel fabricated the remaining internals.

Phase III - canister assembly and drop-test preparations

Once the test canisters and internals were fabricated, all of these items were transported to the Water Reactor and Research Test Facility (WRRTF) at TAN. Where appropriate, still pictures and videotape of the test canisters and internals were taken, showing how the canisters were loaded, welded, tested, and measured.

Table 4 contains information regarding the number of test canisters; their unique identifiers; lengths; test weights; internals configurations; and the reasons for each specific test. This test matrix information was developed in order to achieve as much insight as possible into the structural response of the test canisters subjected to an accidental drop event. The major goal of this entire drop-test effort was to demonstrate (via the post-drop pressure and leak testing) that the test canisters could indeed maintain containment after being drop-tested.

Table 4. Canister information (Morton et al. 2000).

Canister Label	Normal Length	Desired Impact	Total Weight	Canister Intervals	Test Purpose
18-15-00-01	15 feet (4.5 m)	0°	6,033 lbs. (2,736 kg)	Sleeve and sprocket-wheel basket	Worst case internals with sleeve at multiple impact angles
18-15-06-02	15 feet (4.5 m)	6°	5,948 lbs. (2,697 kg)	Sleeve and sprocket-wheel basket	Worst case internals with sleeve at multiple impact angles
18–15-90-03	15 feet (4.5 m)	90°	5,995 lbs. (2,719 kg)	Sleeve and sprocket-wheel basket	Worst case internals with sleeve at multiple impact angles
18-15-45-04	15 feet (4.5 m)	45°	5,995 lbs. (2,719 kg)	Sleeve and sprocket-wheel basket	Worst case internals with sleeve at multiple impact angles
18-15-80-05	15 feet (4.5 m)	80°	5,965 lbs. (2,705 kg)	Sleeve and sprocket-wheel basket	Worst case internals with sleeve at multiple impact angles
18-15-90-06	10 feet (3 m)	90°	3,802 lbs. (1,724 kg)	Simulated High Integrity Cans	Round-shaped internals and simulated HICs response without sleeve
18-15-90-07	10 feet (3 m)	90°	2,997 lbs. (1,359 kg)	Simulated Shippingport Fuel Bundles	Margin test with sharp-edged internals directly impacting on canister interior without sleeve
18-15-PW-08	15 feet (4.5 m)	Initially 0° then tip over for puncture	5,972 lbs. (2,708 kg)	Sprocket-wheel basket	Determine actual response to proposed accidental drop during canister loading scenario
18-15-PP-09	15 feet (4.5 m)	Puncture	6,085 lbs. (2,760 kg)	Sprocket-wheel basket	Demonstrate puncture-resistance

Before the actual loading of the canisters began, the dimensional and weight measurements were recorded obtaining basic "as-built" information about each canister component. This information was recorded on data sheets that identified each canister by both the unique test canister identifier and the assembly numbers used during fabrication. Component and material traceability was maintained by noting which assembly was used for each test canister. The accuracy of measurements depended on the measuring device used. Measurements obtained using a tape measure had an estimated ±1/8 in (+3.2/-3.2 mm) accuracy. The tape measures were not calibrated. Micrometer and caliper measurements had a +0.010/-0.010 in (+0.25/-0.25 mm) accuracy. Weight measurements had an accuracy that depended on the load range involved. For lighter loads of less than 1,000 lbs. (453 kg), the accuracy was +/- 5 lbs (+/- 2.3 kg). For heavier loads, greater than or equal to 1,000 lbs. (453 kg), the accuracy was +/- 10 lbs (+/- 4.5 kg). Greater accuracy of all measurements was attained where possible. Measurement devices were calibrated at the INEEL and were tagged with unique identifying numbers.

The actual loading of the canister internals took place in a methodical fashion in order to obtain the necessary "as-built" information and to document the loading process. An overhead crane was used extensively during the loading of each canister. Typically (for the five canisters 18-15-00-01 through 18-15-80-05), the 2-in (50.8 mm) thick lower impact plate was first installed by lowering it into the bottom canister assembly. The threaded eye bolt and sling were removed using a long-handled tool developed specifically for that removal process. Note that the threaded hole for the eyebolt was drilled through the entire impact plate thickness. This permitted access to the canister interior for pressure and leak testing. During actual usage, this also permits water drainage and access to the interior. The design drawings also specify small grooves that can be machined into the curved surfaces of the impact plates. This can be done when desired to enhance water drainage out through the threaded plug opening.

Next, the 3/16-in (4.8 mm) thick sleeve was lowered into the canister. It rested on the fop flat surface of the lower impact plate. At this time, when required, positioning lugs were welded to the canister inside wall through the sleeve. By holding the sleeve in position, the spoked-wheel bask could also be held in the desired position for the drop-test. Then the spoked-wheel basket was lowered into the canister bottom assembly (inside the sleeve) and also rested on the top flat surface of the lower impact plate. Although the pipe material used for the canisters was not perfectly round (as expected during actual usage), the impact plates, sleeves, and spoked-wheel baskets all loaded with exceptional ease and did not require any force to position them. The dimensions specified on the engineering sketches were correctly specified to allow easy loading, yet still provide adequate space for the SNF. The remote loading of these internals can be performed with ease. The only recommended change would be to shorten the sleeve so that it does not overlap with the top head assembly backing ring. In this way, the backing ring would still protect the canister inside wall from impacting internals, and there would not be any interference between the backing ring and the sleeve while the top head assembly is being positioned for final welding.

For the remaining four canisters, the loading sequence was very similar. Canisters 18-15-PW-08 and 18-15-PP-09 were loaded identical to other 15-footers but without a sleeve. Because these canisters were being puncture-tested, it was decided to perform those tests without a sleeve. This was done in an effort to achieve a worst-case situation since the sleeve would more likely help the canister resist the effects of the puncture.

The two 10-ft (3 m) long canisters were also loaded without sleeves. However, the spoked-wheel baskets were not used on the ten-footers. Canister 18-10-90-06 had seven simulated HICs placed into the canister.

A sleeve was not considered necessary since the round shape of the HICs does not create a situation where large localized strains could occur, as would be the case with the spoked-wheel internals. For canister 18-10-90-07, the initial National Spent Nuclear Fuel Program (NSNFP) plan was to load two Shippingport fuel bundles side-by-side into a canister. The dimensions of the Shippingport fuel bundles would not allow a sleeve to be installed into the canister. However, the NSNFP later decided, based on criticality concerns, to load only one Shippingport fuel bundle per canister. Since the canisters and internals had already been fabricated when this change occurred, the decision was made to proceed with the test as originally planned. Instead of testing what was initially considered to be a unique loading scenario (because a sleeve would not fit), this test was treated as a demonstration of safety margins because the simulated Shippingport fuel bundles would bear directly on the test canister pressure boundary material, along two separate lines nearly the full length of the canister, during a horizontal impact orientation.

The rebar was loaded after the fabricated internals were loaded into the test canisters. Care was taken to properly orient the internals and rebar with respect to where the test canisters were designated to impact during the drop tests. Whenever the most significant impact occurred on a canister skirt, the canister internals were positioned to cause the most potential damage to the canister containment; and the canister was marked so that the initial impact occurred on the longitudinal weld seam of the skirt. Whenever the most significant impact occurred on the canister pressure boundary, the canister internals were positioned to cause the most potential damage to the canister; and the canister was marked so that the initial impact occurred on the longitudinal weld seam of the canister pressure boundary. For the two test canisters being subjected to puncture tests (18-15-PW-08 and 18-15-PP-09), rebar was omitted from that local puncture region in order to permit as much canister deformation as possible.

For canister 18-10-90-06 with the simulated HICs, the bottom simulated HIC (aligned with the canister longitudinal weld seam) was left empty, while all of the other simulated HICs were filled with rebar. This would produce more damage in the bottom simulated HIC. Finite element predictions were also made for these specific internals in order to ascertain how to best determine accurate internals deformations. After loading the rebar into the simulated HICs, ½-in (12.7 mm) thick endplates were welded into all seven of the simulated HICs, including the empty one. Except for the two canisters subjected to puncture-testing and the canister with the simulated HICs, the goal was to uniformly distribute the rebar placement across the canister.

After each test canister was loaded, the top impact plate and spacer plates (where necessary) were positioned. Next, the top head assembly was positioned onto the bottom canister assembly. After a quality examination checked for proper alignment, TAN personnel then completed the final closure weld that sealed each test canister. Manual TIG welding was utilized for the final closure welds.

Using a manually-adjustable lifting fixture, each test canister was then vertically lifted out from the scaffolding used for loading. This lifting fixture had two plates that extended out and engaged underneath the lifting ring on the test canister. The lifting ring functioned as intended, safely lifting each canister. The canisters were then positioned horizontally across large concrete blocks onto wooden cradles to prevent rolling. These 2 ft. x 2 ft. x 6 ft. (0.6 m x 0.6 m x 1.8 m) concrete blocks (weighing approximately 3,600 lbs. [1,631 kg] each) also provided a significant personnel safety feature while the loaded canisters were being worked-on, examined, and measured.

The final closure weld incorporates a permanent backing ring. The backing ring has four distinct purposes:

1. It provides a guide to aid in the installation and final alignment of the top head during final assembly.
2. The presence of this backing ring allows the full penetration butt weld to be made easier, especially since the weld will have to be made remotely.
3. The backing ring helps to protect the canister inside wall and final closure weld from impacting internals.
4. The presence of the permanent backing ring also helps protect the SNF inside the canister during the welding of the final closure weld.

Volumetric examinations of the final closure welds were planned using ultrasonic testing (UT) methods (LMITCO 1999). After a substantial number of indications were recorded by the UT examiner, a significant portion of the closure welds were ground-out on the first five test canisters (18-15-00-01 through 18-15-80-05). However, during this laborious grinding process, few if any actual defects were noticed. After re-welding these canisters, the UT examiner indicated problems in the exact same locations as before. This result was surprising, and the validity of the examination process being utilized was questioned. Additional efforts to determine the validity of the UT indications were made; however, no clear assurances were provided that the apparent UT indications were actually valid or significant. There was speculation that the backing ring was causing some sort of misinterpretation of the UT readings, especially as the calibration standard used did not include a backing ring. In addition, all of the initial UT examinations used just a 45° beam transducer. After consultations with another inspector with higher qualifications (Level III), both 45° and 60° beam transducers were tried to clarify the interpretations of the UT examinations.

Based on the known abilities of the welders being used and the many insights gained from the re-welding effort, the true capability of the welds to take the anticipated drop-test loads was not deemed to be a concern. Therefore, it was decided to "map" two separate test canister closure welds (canisters 18-15-80-05 and 18-10-90-06) and proceed with the drop-test preparations. The mapping consisted of determination and notation of the location of any significant UT indications around the full circumference of the weld. The final closure weld for canister 18-15-80-05 was deemed to satisfy the LMITCO (1999) TPR-4974 criteria, but the weld for canister 18-10-90-06 did not.

As an optional design feature, a threaded plug was incorporated into both the top and bottom head of each test canister. The threaded plugs allow for access to the canister interior. This design feature provided many optional uses. Fluids can be either added or released from the canister interior. Due to uncertainties regarding the generation of hydrogen gas during interim storage, the plugs can be either installed or left uninstalled, providing options to the user. However, the threaded plugs are expected to be installed prior to transportation to the repository. In addition, access is possible through these plugs for visual inspections using remote fiber cameras.

The reason for installing the bottom head threaded plug on the test canisters was to create a worst-case situation with respect to the deforming skirt potentially hitting the extended threaded plug assembly. The reason for installing the top head threaded plug assembly in the test canisters was to provide an opportunity to perform pressure and leak-testing. The bottom threaded plug was seal-welded in place, while the top head threaded plug was not seal-welded to maintain access to the canister interior.

At this stage, pneumatic pressure tests were performed to assure that all the test canisters provided containment prior to the drop-tests. The pressure tests followed these five steps:

1. After each test canister had been prepared, the canister was placed inside a facility so that the canister could attain a steady-state temperature. Once a canister achieved a steady-state temperature, the canister was pressurized to 50 psig (344.8 kPa).
2. The pressure test lasted at least one hour in duration after all connections were tightened to prevent leakage and a steady pressure had been achieved.
3. If leakage was indicated, attempts were made to eliminate all sources of leakage other than the canister itself. The goal was to eliminate leaking connections so that if any leakage were present, it would be attributable only to the canister.
4. During the hour-long pressure test, the canisters were not subjected to any significant temperature change that would affect the accuracy of the pressure test.
5. A pressure drop of no more than 0.5 psig (3.4 kPa) over the one-hour test duration was acceptable.

The results of the pressure tests clearly indicated that no measurable pressure loss was experienced by any of the canisters.

The canisters also needed to be marked in various locations in preparation of the drop-tests. These marks would permit before-drop-test and after-drop-test measurements to be taken at the same locations. Qualified NSNFP personnel utilized a variety of markers or tools to perform this task, including etching tools and permanent markers. Marking was based on tape measurement accuracy. Final total weight and "as-built" dimensional measurements were taken while the canisters were still positioned across the concrete blocks. The two 110-ft (3 m) long canisters (18-10-90-07 and -18-10-90-06 respectively) when loaded weighed approximately 3,000 lbs (1,359 kg) and 3,800 lbs (1,721 kg). All seven of the 15-ft (4.5 m) long canisters weighed approximately 6,000 lbs (2,718 kg) when loaded.

Final labeling of each canister was achieved by painting large black and yellow labels on each canister. This was done to make canister identification easier, and to provide labeling that could be read in the videotapes and still-pictures taken. Finally, as a backup to the large painted labeling, each top and bottom canister lifting ring was etched with the exact same label as that painted on the main canister body. Each canister was labeled using a unique sequence of alphanumeric characters with an AA-BB-CC-DD format. An AA represented the nominal diameter of the test canister in inches, which for this series of testing, was always 18. The BB characters reflected the nominal length in ft of the test canister (either 10 or 15 for these test canisters). The CC indicated the desired impact orientation in degrees, with 0 representing a vertical drop and 90 representing a flat or horizontal drop. In cases where there was not an impact angle but a puncture type test, the CC represented alpha characters that indicted the type of puncture test: PP represented the 40-in (1.02 m) drop onto a 6-in (152.3 mm) diameter puncture post; and PW represented the potential scenario of dropping a canister while loading the canister into a repository waste package or transportation cask or other similar larger container. Finally, DD was an additional numerical identifier that was necessary to achieve a unique canister number. For this series of test canisters, the DD was simply numbers 01 through 09. After final labeling, the test canisters were loaded onto a flatbed trailer and trucked to SNL.

Material testing was also performed during this phase. Although material certifications were obtained for the materials used to fabricate the canisters, the actual stress-strain relationship was not accurately known. Therefore, limited material testing was completed (using a calibrated tensile testing machine) to define appropriate strain behavior for the canister materials utilized.

Phase IV - Drop and pressure testing at Sandia National Laboratories

Sandia National Laboratories (SNL), operating under a Quality Assurance (QA) program based on ASME (1986) NQA-1, has an ongoing, qualified drop-testing program in place that has been utilized by numerous organizations, including the U.S. Department of Defense; the U.S. Nuclear Regulatory Commission (USNRC); and the DOE. This facility contains an essentially unyielding flat surface, capable of dropping very large test specimens up to approximately 80,000 pounds (362,400 kg) from heights up to 100 ft (30 m). Smaller items can be raised to almost a 700-ft (210 m) drop height. The SNL mobile instrumentation data acquisition system (MDAS) is a self-contained data acquisition facility that can produce fully-qualified data documentation. Records of equipment parameters and performance can be produced, providing a computer-generated audit trail.

The SNL received the nine test canisters on May 4, 1999. In order to attribute any damage received by the canisters solely to the drop-testing, canister movement activities were performed to prevent excessive or undue harsh treatment of the canisters. At SNL part of May and June of 1999 were devoted to pre-test measurements, and the actual drop-testing began on June 23 and ended on June 30, 1999. Post-drop pressure testing was performed on July 1, 1999, and post-drop measurements began later in July. The canisters were loaded onto a flat bed trailer July 20, 1999, and trucked back to the INEEL.

Although SNL executed the intended test plan and provided extremely valuable results, these tests were unable to precisely hit the intended target point on a number of canisters. The main cause was the excessive tension in the tag lines used to align some of the test canisters before the actual drop. Table 5 lists the canisters; if the intended target was achieved; and the magnitude of potential discrepancy.

Table 5. Accuracy of canister impact (Morton et al. 2000).

Canister Label	Target Achieved	Discrepancy
18-15-00-01	Very Close	About ½° from vertical
18-15-06-02	Yes	---
18–15-90-03	No	About 24° rotation about the canister longitudinal axis
18-15-45-04	Yes	---
18-15-80-05	Yes	---
18-15-90-06	No	About 24° rotation about the canister longitudinal axis
18-15-90-07	No	About 19° rotation about the canister longitudinal axis
18-15-PW-08	No*	About -1° from vertical and about 15° rotation about the canister longitudinal axis
18-15-PP-09	No	About ½-inch (1.2 cm) towards the lower head and about 1 1/4-inch (3.2 cm) off the longitudinal weld seam

* An accurate secondary impact was very difficult to achieve for the specific test.

At SNL, seven out of nine test canisters were dropped from 30 ft (9 m) onto an essentially unyielding flat surface. One test canister was dropped 40 in (1.02 m) onto an essentially rigid 6-in (152 mm) diameter, 24-in (609 mm) high bar.

The last remaining canister was subjected to a test to demonstrate what could occur if an actual standardized DOE SNF canister were accidentally dropped while being loaded into a repository waste package, a transportation cask, or interim storage canister. Consequently, this last remaining test canister was dropped 2 ft (51 mm) onto a 2-in (50.8 mm) thick plate. This plate was 18 in (457 mm) high and 36 in (914 mm) long and was positioned vertically for an initial impact. Subsequently, the canister was left to tip over and impact another 2-in (50.8 mm) thick plate with dimensions of 12 in (304.7 mm) high by 96 in (2.44 m) long positioned vertically, approximately 2 m (78 in center-to-center) away from the first plate.

All nine of the test canisters adequately survived the drop-tests from a deformation standpoint. They were able to pass the pressure test, holding 50 psig (344.8 kPa) of air pressure steady for one hour without any visible loss of pressure. These tests provided a few highlights that occurred during the drop testing at SNL as follows:

1. Canister 18-15-00-01 dropped 30 ft (9 m), bounced a couple of times, and then remained standing. It reflected not only the accuracy with which SNL dropped this particular canister but also how uniformly the canister was loaded at the INEEL.
2. The puncture post canister, 18-15-PP-08, impacted the 6-in (152.3 mm) diameter bar twice before it rolled off the bar. However, the 24-in (609.3 mm) long bar was welded to a larger plate with large eyebolts welded to the plate. These eyebolts were used to lift the puncture bar into position. After the test canister rolled off the bar, the canister impacted one of the eyebolts, resulting in a second (but unintentional) puncture test. The result was another significant "dent" but no containment system concern.
3. Finally, test canister 18-15-80-05 was dropped from an 80-degree orientation in order to achieve the slap-down effect. The videos taken by SNL clearly show an increased rotational acceleration of the top portion of the canister just prior to top head impact. Slap-down was achieved.

Phase V - post-drop test activities

The nine test canisters arrived back at the INEEL for post-drop examination and testing on July 22, 1999. As with the initial loading and unloading activities prior to the drop-tests, the loading and unloading activities after the drop-tests were intended to prevent excessive or undue harsh treatment of the canisters such that any damage received by the canisters could be attributed to the drop-tests only. The post-drop examination and testing activities included the following actions:

1. Helium leak-testing of the four worst damaged test canisters
2. Detailed measuring of all the deformed test canisters and recording the information onto data record sheets
3. Cutting open all the test canisters
4. Brief visual observations of the internals and inside surfaces of the test canisters.

Once the test canisters were back at the INEEL, the canisters were placed horizontally across the same concrete blocks in an effort to duplicate the conditions when the pre-drop test measurements were recorded. Various examinations were made to better understand the structural response of each test canister during its drop-test, and how the test canister geometry changed. Many of the measurements taken were identical to those taken prior to the drop-tests. As with the pre-drop measurements, these post-drop measurements were taken using calibrated measuring devices (except the measuring tape), using the same measuring tolerances, and recorded on similar data sheets. These examinations and measurements were typically canister-specific due to the varying structural responses.

None of the deforming skirts touched the pressure boundary or containment system of the test canisters during their drop-testing, including the extended threaded plugs. Post-drop inspections clearly indicated that the deforming skirts provided their intended energy-absorbing function. Post-drop non-destructive examination of certain skirts did not reveal any cracks or material tearing, even on skirts that were subjected to significant deformation. The pressure boundary or containment system of all nine canisters did not experience extreme deformations. The most significant change in an outer diameter of the pressure boundary was on test canister 18-15-90-03. That canister's maximum deformed diameter after dropping was slightly less than 19 in (482 mm).

The canisters that had the most significant damage were chosen to be helium leak-tested. At first, two canisters were thought to have been adequately damaged to undergo helium leak testing. However, after second consideration, it was decided to test the four most damaged test canisters. This would not only minimize concerns over a very limited number of canisters being leak-tested, but more canisters would cover a wider range of canister impact angles. The canisters chosen for helium leak testing were 18-15-00-01, 18-15-45-04, 18-15-80-05, and 18-15-PP-09.

The leak-testing effort utilized procedure TPR-4976 (LMITCO 1996a) which ties back to the ASME (2001) B&PV Code, Section V, Article 10. A full vacuum was pulled inside the test canister, and the outside surface was "bagged" in order to permit a 99% pure helium environment to exist on the outside surface of the test canister. The results of the leak-testing indicate that a helium leak rate of less than 10^{-7} std cm³/s was achieved. Note that the ANSI (1987) N14.5 standard contains acceptance criteria of leaktight or 10^{-7} std cm³/s leak rate; and USNRC recognizes leak rates of this magnitude to reflect a leaktight containment system.

After the post-drop measurements and the helium leak-testing were completed, the test canisters were cut open in order to examine the condition of the internals, the rebar, and the interior surfaces of the test canisters. After the canisters were cut apart, the two final closure welds previously ultrasonic testing (UT) examined and "mapped" (canister 18-15-80-05 was accepted and canister 18-10-90-06 was rejected) were UT-examined again by the same person. No changes in the results of the UT examinations were noted. These welds were then radiographic test (RT)-examined in order to gain more insights on UT versus RT capabilities. The results indicate that both of the final closure welds (including canister 18-10-90-06) were acceptable per TPR-4970 (LMITCO 1996b) criteria. Although canister 18-15-80-05 was difficult to interpret due to the backing ring, and "info only" was indicated for weld acceptance, the examiner could not see any indications in the radiograph. This suggests that the UT methodology used for these closure welds needs improvement. The longitudinal weld seam that was previously RT-examined during canister fabrication was RT-examined after the drop-test, and no changes were observed. Finally, the butt weld adjacent to the longitudinal weld was also RT-examined after the drop-test. This butt weld had been previously RT-examined during canister fabrication and had obviously passed. With this most recent RT examination, the weld was still acceptable. From this small amount of data, it appears that the drop-tests had little degradation effects on the stainless steel welds.

Material used for the vessel heads was not previously tested, and the only material available was obtained from one of the heads used in the test canisters. The material from the bottom head in canister 18-15-PW-08 was used because it was not significantly strained during its drop-test. This material-testing was performed using a calibrated tensile testing machine. Although material certifications were obtained for the head materials, the actual stress-strain relationship was not accurately known.

It is important to differentiate between damage to the canister skirt and damage to the canister pressure boundary (or containment system). The skirt was incorporated into the canister design to act as an energy-absorption device, and therefore significant deformation is expected. Yet skirt deformation does not necessarily affect the containment function of the canister. More important is deformation of the canister pressure boundary. This directly affects the containment system function of the canister.

Canister 18-15-00-01: This test canister was dropped from a vertical orientation 30 ft (9 m) onto the essentially unyielding surface. The skirt protected the canister pressure boundary as expected. The top portion of this canister was not damaged since the test canister remained standing vertical after the drop-test. The internals showed no recognizable damage.

Canister 18-15-06-02: This test canister was dropped from a "center-of-gravity-over-the-corner" orientation (approximately 6 degrees from vertical) 30 ft (9 m) onto the essentially unyielding surface. The skirt protected the pressure boundary of the test canister as expected. The top portion of the canister was only slightly deformed as it fell-over after the initial impact. The internals showed no recognizable damage.

Canister 18-15-90-03: This test canister was dropped from a horizontal orientation 30 ft (9 m) onto the essentially unyielding surface. The skirt ends deformed very little. Most of the deformation occurred in the mid-section where the canister body flattened along the point of impact, and the body ovalized to a maximum diameter of slightly less than 19 in (482.4 mm). The internals were somewhat deformed, but could easily still carry-out their intended function of separating the SNF and keeping it adequately positioned. The drop-test missed impacting the longitudinal weld seam on the test canister body.

Canister 18-15-45-04: This test canister was dropped from a 45-degree orientation 30 ft (9 m) onto the essentially unyielding surface. Both the top and bottom skirts deformed—especially the bottom skirt. However, the skirt still absorbed much of the impact energy and significantly reduced the damage that could have potentially occurred to the canister pressure boundary. The test canister had a noticeable bow over the entire canister length after the drop-test. The internals showed no recognizable damage with the exception of the 2-in (50.8 mm) thick impact plates that were slightly deformed in the local area of the impact point. However, they could still perform their intended function.

Canister 18-15-80-05: This test canister was dropped from an 80-degree (near horizontal) orientation 30 ft (9 m) onto the essentially unyielding surface. Since the slap-down effect was achieved, the top portion of the canister pressure boundary was more damaged than the bottom. This test canister experienced the most damage to the canister pressure boundary, especially for the welds made at the INEEL. The top head to canister body weld was significantly flattened during slap-down, and the dished portion of the head had a noticeable bulge at the slap-down impact location near where the skirt was welded onto the head. The spoked-wheel basket internal for this test canister showed the most significant damage of all the canisters. However, that damage was not that significant: Three of the spokes showed some deformation due to the rebar impacting the spokes. The 2-in (51 mm) thick impact plates were slightly deformed at the point of impact, but were still able to perform their intended function.

Canister 18-10-90-06: This test canister was dropped from a horizontal orientation 30 ft (9 m) onto the essentially unyielding surface. The most damage occurred in the middle section of the canister, where the canister body flattened along the point of impact, and the body ovalized to a maximum diameter of approximately 18-5/8 in (473 mm). The drop-test missed impacting the longitudinal weld seam on the test canister body. The simulated high integrity cans (HICs) were inside this test canister. All of the simulated HICs had a slight bow over their entire length due to the 90-degree impact. The one simulated HIC that was left empty was flattened. The mid-section maximum measured diameter on this empty simulated HIC was 6-11/16 in (170 mm), and the minimum diameter was measured as 3-15/16 in (100 mm).

Canister 18-10-90-07: This test canister was dropped from a horizontal orientation 30 ft (9 m) onto the essentially unyielding surface. This test canister experienced even less apparent damage than canister 18-10-90-06. The most damage occurred in the middle section of the canister, where the canister body flattened along the point of impact, and the body ovalized to a maximum diameter of approximately 18-1/2 in (470 mm). This canister deformed a smaller amount than canister 18-10-90-06 due to the different internals configuration and the lighter total weight of the test canister. The drop-test missed impacting the longitudinal weld seam on the test canister body. The internals for this test canister were the simulated Shippingport fuel bundles. The corners were ground down during canister loading to allow the backing ring on the top head assembly to fit properly. Because these internals represented a specific fuel bundle and not a basket design, the deformations were not a major concern to this drop-test effort.

Canister 18-15-PW-08: This test canister was dropped from a vertical orientation 2 ft onto a rigid and vertically-oriented 2-in (50.8 mm) thick plate (representing an edge of a repository waste package or other similar component). The test canister was then allowed to tip over onto another essentially rigid and vertically-oriented 2-in (51 mm) thick plate (representing the other edge of the waste package or other component). The skirt was indented approximately ½ in (13 mm). The damage done to the test canister body due to the secondary impact was noticeable, but not very significant. The minimal diameter at this secondary impact point was measured as approximately 16-5/8 in (422 mm), resulting in an indent of approximately 1-3/8 in (34.9 mm). The edges on both the 2-in (51 mm) thick plates were relatively sharp since these edges were not ground-down or rounded but were "as-received." The top portion of this test canister impacted the essentially unyielding surface at the position where a welded-on lifting lug had been placed. This third main impact resulted in the localized deformation of the top skirt. The uniqueness of this deformation was due to a "pad eye" (welded-on lifting lug) attached to the skirt at the point of the third impact. This test canister had loaded into it a spoked-wheel basket without a sleeve, and rebar was not placed adjacent to the desired impact location. The spoked-wheel basket received only minor localized deformation to one of the spokes adjacent to the impact location.

Canister 18-15-PP-09: This test canister was dropped from 40 in (1.02 m) onto a six-in (152 mm) diameter puncture bar. The impact target was the center-of-gravity of the test canister. This orientation would produce the most deformation to the canister body. However, the drop-test actuary impacted about ½ in (13 mm) more toward the lower head and about 1-1/4 in (32 mm) off the longitudinal weld seam. Damage to the test canister body was significant. The post-drop measurements indicated that the resulting puncture deformation was approximately 2-3/4 in (70 mm) into the canister. Both the top and bottom portions of the canister were virtually undamaged due to the short fall to the essentially flat unyielding surface of the drop pad. However, after impact, the test canister rolled off the puncture bar, impacting an eye bolt used to lift the puncture bar assembly. This test canister was loaded with a spoked-wheel basket only (no sleeve) and with no rebar adjacent to the impact point. The resulting damage to the spoked-wheel basket was minimal. The edge of the one spoke in the vicinity of the puncture impact had a slight mark that resembled a rub mark. The damage was so slight that it was very hard to see. The interior surface of the canister where the puncture occurred also shows little if any noticeable damage, except in the immediate area of the puncture bar deformation.

Phase VI - final report and documentation packages

The last phase of this effort consisted of a final report by Morton et al. (2000) that addressed the associated activities, especially the computer prediction efforts. The computer code ABAQUS/Explicit Release 5.8-1 (ABAQUS 1998) was used for the finite element modeling.

CONCLUSIONS

The results of these tests show that the design of the standardized DOE SNF canister is robust, and that its containment system remains functional after an accidental drop event. In addition, these nine test results provide validation of the capability of computer analyses to predict the structural response of these canisters under a variety of situations not necessarily tested.

ANALYTICAL EVALUATION OF THE IDAHO SPENT FUEL PROJECT CANISTER SUBSEQUENT TO AN ACCIDENTAL DROP EVENT

INTRODUCTION

Prior to the 1999 full-scale drop-testing effort, only 18-in (457 mm) diameter standardized canisters were expected to be used by the U.S. Department of Energy (DOE) spent fuel sites. Subsequently, in order to proceed with interim storage of DOE spent nuclear fuel (SNF) at the Idaho National Engineering and Environmental Laboratory (INEEL), private industry was contracted to construct a dry storage facility that incorporated the standardized canister. This facility, the Idaho Spent Fuel Project (ISFP) incorporated both the 18-in and the 24-in standardized canisters but modified their design. Snow and Morton (2003) investigated the significance of the proposed Foster Wheeler Environmental Corporation (FWENC) modifications on the robust design of the standardized canister using both 18-in (457 mm) and 24-in (610 mm) diameter standardized canisters, including the ability to maintain a leaktight containment during and after an accidental drop event.

In order to determine the significance of the FWENC modifications, analytical evaluations of 18-in (457 mm) standardized canisters during the 1999 full-scale drop-testing effort must be compared with 24-in (610 mm) diameter standardized canisters. As the actual deformations matched well with the analytical predictions, this comparison effort should provide valid insights into ISFP canister drop performance. Blandford (2003) performed an analytical evaluation for the 24-in (610 mm) diameter standardized canister. Snow and Morton (2003) used the same methodology used for the 18-in (457 mm) diameter canister. Therefore, comparison of the 24-in (610 mm) diameter ISFP canister against the 24-in (610 mm) diameter standardized canister is also expected to provide valid insights.

Future uses for the standardized canisters are not fully known. Therefore, it is important to judge the adequacy of the ISFP canisters not only for drop events specifically applicable to the ISFP facility, but for a common basis that was used to judge the viability of the original standardized canister design. Snow and Morton (2003) considered the drop events from the 1999 drop-testing as well as the expected performance for the two repository drop events defined in the WASRD (DOE 2002). The repository considered the performance of the standardized canister during the 30-ft (9 m) drops and the 40-in (1.02 m) puncture event when it was completing its performance allocation study of all the canisters expected to be received. Therefore, evaluating the ISFP canisters for these same drop events provides a common basis to evaluate canister drop performance.

Snow and Morton (2003) applied for all ISFP canister evaluations—room temperature and beginning of life material properties—as used for the 18-in (457 mm) diameter standardized canister analytical evaluations. The ABAQUS/Explicit, Version 6.3-3, was used to determine the structural response. This is an updated version of the software used to evaluate the 18-in (457 mm) diameter standardized canister (Version 5.8-1) and is the current National Spent Nuclear Fuel Program (NSNFP) validated and verified (V&V) version (DOE 2003b). The modeling methodology adopted is the same one previously used in the 1999 full-scale 18-in (457 mm) diameter standardized canister effort, except where changes are necessary to comply with ABAQUS/Explicit, Version 6.3-3 and computer program V&V requirements.

The 1999 drop tests included edge-impact and puncture post-drop events. Snow and Morton (2003) did not investigate these drop events on the ISFP canister. However, the differences between the standardized canisters and the ISFP canisters are not considered significant under these load conditions.

Table 6 identifies the drop events evaluated for the ISFP canisters. All drops were onto a rigid, flat surface. A solid model is first developed using appropriate software. The actual ABAQUS/Explicit FE model is generated and then subjected to rigorous checks to assess adequacy before actual analysis is performed. This rigorous checking process eliminates the need to control or validate the solid modeling software.

IDAHO SPENT NUCLEAR FUEL PROJECT CANISTER DESIGN

The ISFP canister design includes the following configurations:

1. An 18-in (457 mm) outer diameter canister, 15 ft (4.57 m) in overall length, with three distinct internal component configurations
2. An 18-in (457 mm) outer diameter canister, 10 ft (3.0 m) in overall length, with one internal component configuration
3. A 24-in (610 mm) outer diameter canister, 15 ft (4.57 m) in overall length, with three distinct internal component configurations

Figure 5 shows the basic canister configuration, and Fig. 6 shows an end close-up.

ISFP canister main components

Basic features of the 18-in (457 mm) diameter ISFP canisters are as follows:

1. Canister body and skirts ("impact limiters") made of 18-in (457 mm) nominal outer diameter pipe, 3/8-in (9.5 mm) nominal thickness (SA-312 type 316L SST)
2. Canister heads are ASME flanged and dished, 5/8-in (16 mm) nominal thickness, with a 2-in (51 mm) long straight flange, later machined to 3/8-in (10 mm) thickness after skirt is attached, (SA-240 type 316L SST)
3. Canister lifting rings with a 17-1/8-in (435 mm) outer diameter by 14-7/8-in (378 mm) inner diameter made of 5/8-in (16 mm) thick plate, later machined to ½-in (51 mm) thick (SA-240 type 316L SST)
4. Canister interior impact plates made of 2-in (51 mm) plate (SA-240 type 316L, or SA-351 type CF3M or CF3MN), flat on one side for the internal components to rest-on, and contoured on the other side to match the geometry of the inside surface of the head.

Similarly, the basic features of the 24-in (610 mm) ISFP canisters are as follows:

1. Canister body and skirts made of 24-in (610 mm) nominal outer diameter pipe, 1/2-in (13 mm) nominal thickness (SA-312 type 316L SST)
2. Canister heads are ASME flanged and dished, 3/4-in (19 mm) nominal thickness, with a 2-in (50.8 mm) long straight flange, later machined to 1/2-in (13 mm) thickness after skirt is attached (SA-240 type 316L SST)
3. Canister lifting rings with a 22-7/8-in (581 mm) outer diameter by 20-3/4-in (527 mm) inner diameter made of 5/8-in (16 mm) thick plate, later machined to ½-in (13 mm) thick (SA-240 type 316L SST)
4. Canister interior impact plates made of 2-in (51 mm) plate (SA-240 type 316L, or SA-351 type CF3M or CF3MN), flat on one side for the contents to rest-on, and contoured on the other side to match the geometry of the inside surface of the head.

Table 6. Drop events evaluated for the ISFP canister (Snow and Morton 2003).

Criteria	Drop Height		Orientation[1]
	(ft)	(m)	
10 CFR 71	30	9	Vertical (0-degree or near)
10 CFR 71	30	9	Center-of-gravity over corner (CGOC)
10 CFR 71	30	9	Horizontal (90-degree)
10 CFR 71	30	9	45-degree
10 CFR 71	30	9	Slapdown
WASRD	23	6.9	Vertical (0-degree)
WASRD	2	0.6	Worst Orientation

[1] Orientation angle in degrees with respect to vertical.

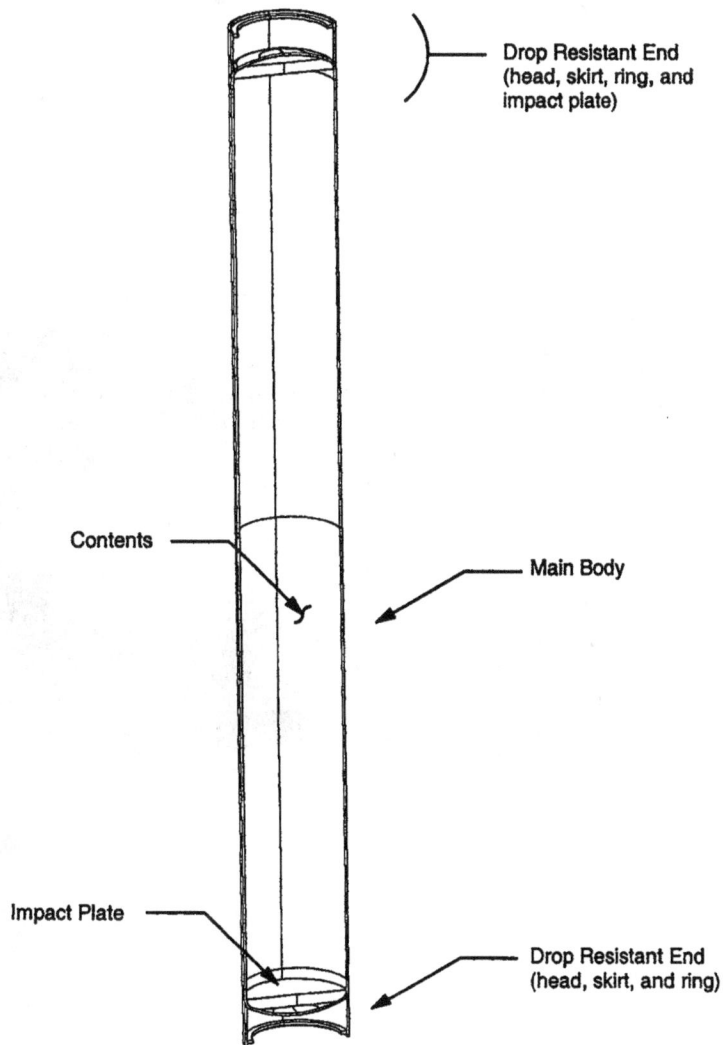

Fig. 5. ISFP canister design (Snow and Morton 2003).

61

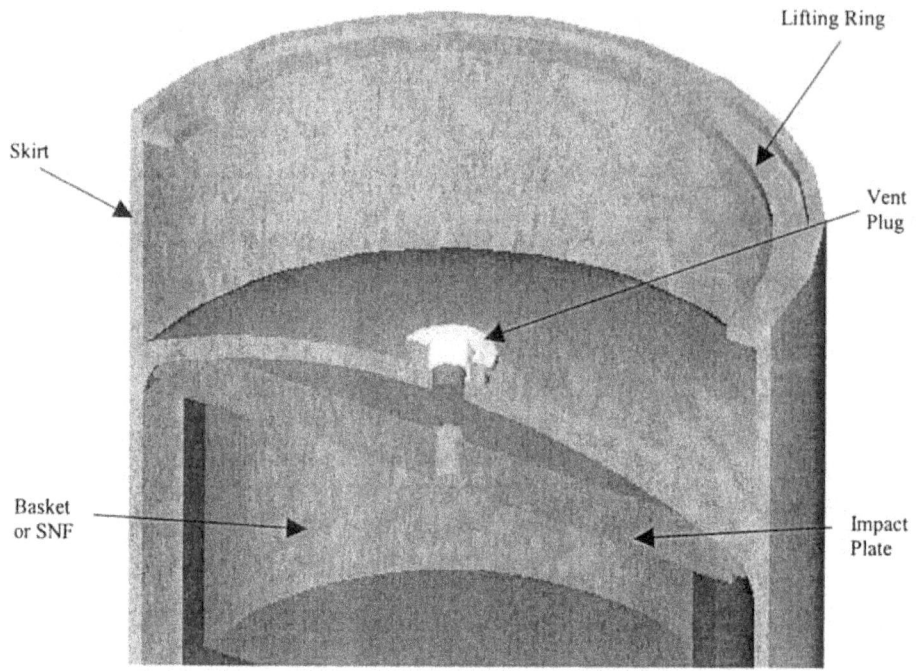

Top End

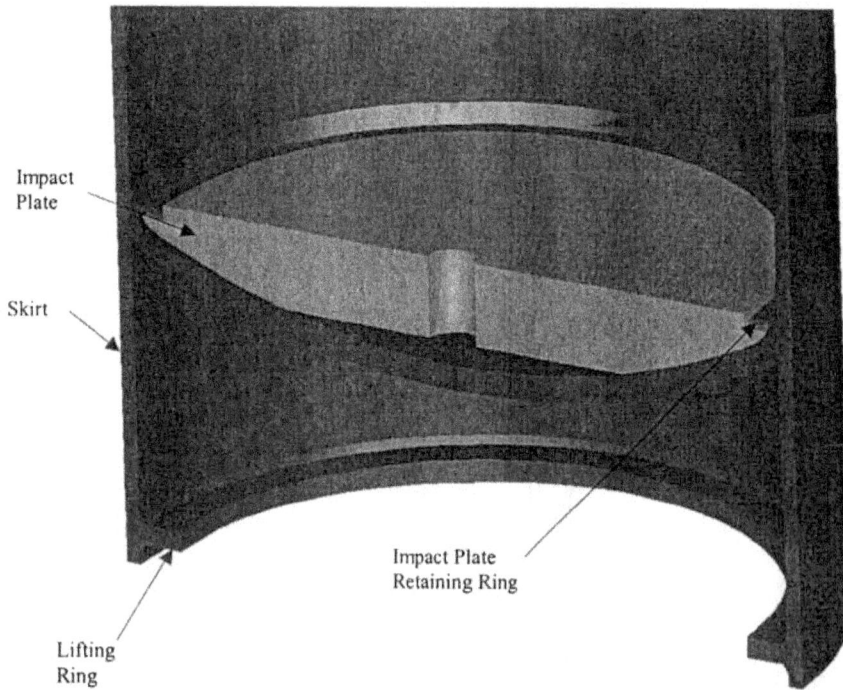

Bottom End

Fig. 6. Close-up of typical canister ends (Snow and Morton 2003).

62

ISFP canister internal components

The varied internal configurations for these ISFP canisters have several common components:

1. Each canister has an impact plate located next to a head and held in place by a retaining ring.
2. Each canister has a fuel basket with loaded fuel.
3. Each canister has a shield plug that is placed above the fuel basket.

Two of the 24-in (610 mm) canisters (those that hold Shippingport Reflector IV and V baskets) also have a ring welded to the inside of the canister on which the shield plug rests. In all other canisters, the shield plug rests on the fuel basket.

Differences between the standardized and the ISFP canisters

The main differences between the standardized canisters and the ISFP canisters are summarized as follows:

1. The standardized canisters used heads that were nominally 3/8 in (9.5 mm) thick for the 18-in (457 mm) size, and ½ in (13 mm) thick for the 24-in (610 mm) size; whereas the ISFP canisters use heads that are ¼ in (6.3 mm) thicker everywhere except in the straight flange. These are 5/8-in (15.9 mm) thick for the 18-in (457 mm) size, and ¾ in (19 mm) thick for the 24-in (609 mm) size. The standardized canisters avoided thicker walls to minimize potential thermal gradient effects and to keep costs down. Extra machining is required to achieve a match in the head and shell diameters for the ISFP canisters.
2. The ISFP canisters use a retaining ring that is welded to the inside of a head to hold the internal impact plates in place; whereas the standardized canisters do not use a retaining ring of any kind, but use the internals (baskets or spacers) to maintain the proper positioning of the internal impact plates.
3. The ISFP canisters use an internal shield plug, whereas the standardized canister design assumed typically the use of remote welding techniques for the closure weld.
4. Two of the ISFP canisters use a shield plug support ring that is welded to the inside of the canister body on which the shield plug rests, whereas the standardized canisters do not use any support rings (or any other component) welded to the inside of the canister containment body.
5. The standardized canister assumes an internal sleeve to prevent sharp internal components from bearing directly on the containment boundary (reducing localized strains) thereby increasing margins of safety during accidental drop events. These internal sleeves also provide a barrier between the DOE SNF or internals and the canister containment to reduce galvanic corrosion concerns. The ISFP canisters do not use any internal sleeves.
6. The ISFP canister and the assumed standardized canister fuel basket designs used for the analyses are significantly different.
7. The standardized canisters use heads with an optional small threaded plug (top or bottom). In contrast, the ISFP canisters employ a larger vent plug on the top head only. The ISFP computer models did include this feature.

Drop orientations and events

The main purpose of the evaluation performed by Snow and Morton (2003) was to determine the effect of the changes that were made to the design of the standardized canisters to arrive at the ISFP canister configurations. This was done by evaluating two categories of drop events:

1. **Drop Events:** The 30-ft (9 m) drop conditions evaluated in the 1999 standardized canister tests were evaluated for the ISFP canisters. This consisted of a drop onto a flat, rigid surface by an ISFP canister from the following orientations:

1.1. Vertical
1.2. Center-of-gravity-over-corner; 6 degrees off-vertical for the 18-in (457 mm) size, 7 degrees off-vertical for the 24-in (610 mm) size
1.3. 45-degree
1.4. slap-down consisting of 80 degrees off-vertical for the 18-in (457 mm) size, 70 degrees off-vertical for the 24-in (610 mm) size
1.5. Horizontal.

2. **Repository Drop Events:** The repository-defined drops consist of a 23-ft (6.9 m) vertical drop and a 2-ft (0.6 m) worst-orientation drop onto a flat, rigid surface by an ISFP canister.

Analytical modeling of ISFP canisters

Snow and Morton (2003) evaluated three configurations that were considered to envelop the response of the seven canisters under the prescribed drop events. Maximum canister design weights as well as actual loaded weights for certain configurations were evaluated.

Finite elements: The ISFP canisters were modeled using linear quadrilateral shell elements (element type S4R) for the canister body; upper and lower heads; and skirts. Shell elements were located on the geometry mid-surface. The internal impact plates and the lifting rings were simulated using solid linear brick elements (element type C3D8R) and wedge elements (element type C3D6). The head-to-body joints and the lifting ring-to-skirt connections all consisted of full penetration welds and were represented by using common nodes. The skirt-to-head welds were also full penetration, but were modeled using common nodes and multipoint constraints (element type MPC BEAM and LINK). Because all shell elements were placed at the mid-surface of the component, the skirts required a tie-back to the heads in the area of the attachment weld. Otherwise, the skirts would have appeared longer in the finite element (FE) model than in reality (which would affect their stiffness and buckling responses).

The internal components (baskets, fuels, shield plugs, etc.) were also modeled with shell and solid elements to represent their geometry, stiffness, and weight. This evaluation did not determine the condition of the fuel during and after a drop event. Therefore, the modeling of fuel was only sufficient to represent its effect on the basket and canister structure. Only half a canister was explicitly modeled due to the symmetry in geometry, loading, and response during the drop event. Symmetry boundary conditions were applied to ensure that the half-model responded exactly like the full canister.

Mesh size: The element sizes for the canister models were chosen based on the type of event being simulated and the expected response. Because large plastic deformations were expected, the element sizes could not be too small or they would distort excessively (causing the calculation to terminate) before the event was completed. At the other extreme, elements that were too large would not respond properly (e.g., a bulge in the canister would be shown as a sharp edge instead of a smooth curve) and the results would be in question. This was particularly important in areas where significant deformations would occur. When the number of elements for a particular component is mentioned, that number applies to the half-model. The element sizes used in these ISFP canister models were comparable to those used in the 1999 evaluations.

Model configuration #1: The three 18-in (457 mm) diameter, 15 ft (4.5 m) long ISFP canisters all have internal baskets that are similar. Therefore, the canister with the Peach Bottom 1 Basket was selected to represent the response of these three configurations. In order to compare the results of these evaluations with those from the 1999 tests, the weight of the internal components has been adjusted to achieve the maximum design weight of 6,000 pounds (2,718 kg). This 6,000-pound (2,718 kg) model has been evaluated under the 30-ft (9 m) and repository drop events. The maximum listed loaded weight of these three canisters is 4,090 pounds (1,853 kg) with the lowest loaded weight being 3,807 pounds (1,725 kg). This model was also evaluated under the defined repository drop events at a weight of 4,090 pounds (1,853 kg). The component meshes were as follows:

1. **Skirts:** The lower and upper skirts each used 1,776 shell elements each, sized at about 1/4-in (6.3 mm) x 6/10 in (15.2 mm) (longitudinal x circumferential dimension). This gave a fine mesh to represent these components as they deform during impact, absorbing the majority of the drop energy for all end-drops.

2. **Lifting Rings:** The lifting rings used the same circumferential spacing as the skirts. The brick mesh used four elements through the width and three brick elements through the thickness, with wedges on the interface with the skirt to represent the expected condition of the welds. A total of 672 elements were used to represent each lifting ring.

3. **Heads:** The lower head used 2,160 shell elements to represent the head and the integral straight flange. Elements were sized at about 1/4 in (6 mm) x 6/10 in (15 mm) (longitudinal x circumferential) in the straight flange and all but the center of the head. At the center of the head the element sizes were smaller. This provided a fine mesh throughout. The upper head used 2,064 elements, with the same mesh as the lower head, but with the center portion replaced with the vent plug.

4. **Canister Body:** Because the more significant deformations were expected to occur in the skirts and heads of these canisters, the shell element size for the main body began at about 1/4-in (6.3 mm) x 6/10-in (15.2 mm) (longitudinal x circumferential) at the heads and then grew to 1-1/2 in (38.1 mm) x 6/10 in (15.2 mm) through the remaining canister body. A total of 5,040 shell elements was employed.

5. **Impact Plates:** Each interior impact plate used a total of 4,848 brick elements, with five elements through the thickness. This allowed the impact plate to bend correctly due to the internal basket-load from above and the impact-load from below.

6. **Impact Plate Retaining Ring:** The impact plate retaining rings were modeled with four shell elements through the width, with a circumferential mesh that matched the heads (6/10 in [15.2 mm]). These rings were welded to the head knuckle (dish-to-straight flange transition) at about 30-degree segments (30 degrees of weld, 30 degrees no weld, and so on). Because the worst strains in the canister would be produced if impact were on one of these welds, the rings were modeled with a continuous weld (modeled using common nodes between the retaining ring and the head).

7. **Shield Plug:** The shield plug is basically a large mass of steel. It was not expected to deform much, if at all, during the defined drop events. The shield plug only needed to be modeled to simulate its impact on the adjacent components (upper impact plate, canister wall, internal basket, etc.) during the drop event. Therefore, the shield plug was modeled with 696 brick elements.

8. **Peach Bottom 1 Basket:** The bottom end spacer plate is 1-1/4 in (31.7 mm) thick and just under 17 in (431.6 mm) in outer diameter (O.D.) (SA-193 Type 630 17-4 PH). Seven intermediate spacer plates, and one top end spacer plate spaced at just under 17 in (431.6 mm) apart, were tied together (with the bottom plate) with threaded tie bars (1-1/2 in or 38 mm diameter solid bars, SA-564 Type 630 17-4 PH). A total of 10 storage tubes (4-in or 102 mm O.D. x 0.064 in or 1.6 mm wall, SA-213 316L SST) for holding fuels fit between the bottom end spacer plate and the top end spacer plate, passing through the intermediate plates. Four tubes filled with gadolinium were also located between the top and bottom end spacer plates. A base plate and top plate were attached with threaded fasteners to close the basket ends. Other components are not significant to this evaluation. Because the exact condition of the basket and contained fuels during and after the drop events was not needed or desired, the basket was modeled in a simplified manner. A total of 2,870 brick elements represented the plates, rods, and gadolinium tubes. A total of 3,650 shell elements simulated the storage tubes. The density of the basket components was adjusted to reach the desired loaded weight (this eliminated the needed to explicitly model the fuel). The mesh size was fairly coarse—if anything the basket was a little more rigid (which is conservative) than in reality—but was considered adequate in representing the effects of the basket on the canister.

Model configuration #2: The response of the 18 in (457 mm) diameter, 10 ft (3 m) long ISFP canister with the TRIGA Basket to the 30 ft (9 m) and repository drop events is considered to be enveloped by the 15-ft (4.5 m) long canister (model configuration #1 discussed above) for all but one orientation. That orientation is the slap-down event. This model has been evaluated for the 30-ft (9 m) slap-down (80 degrees off-vertical) event at its listed loaded weight of 3,648 pounds (1,653 kg). The model configuration canister contained two TRIGA baskets. The FE meshes discussed for model configuration #1 were identical to those for model #2 for the skirts, heads, lifting rings, impact plates, impact plate retaining rings, and shield plugs. The canister body was only different in that the center portion was removed to shorten the canister to the 10 ft (3 m) total length.

Two TRIGA baskets are held within this canister—one above the other. Each basket has a 1-1/4-in (32 mm) thick and just under 17 in (432 mm) in O.D. base plate (SA-693 Type 630 17-4 PH). A 1-3/4-in (44 mm) thick and also just under 17 in (432 mm) in O.D. lid (SA-240 316L SST) is attached to the base plate with three 1-1/2-in (38 mm) diameter tie-bars (SA-564 Type 630 17-4 PH). Between the lid and base plate are held fifty-four 1-3/4-in (44 mm) x 0.064-in (1.6 mm) wall storage tubes (SA-213 316L SST) that hold the fuel elements. Below the base plate is a 10-1/2-in (267 mm) long spacer pipe (14-in or 355 mm pipe, Schedule 80, SA-312 316L SST). Other components are not significant to this evaluation. Because the exact condition of the basket and contained fuels during and after the drop events was not needed or desired, the basket was modeled in a simplified manner. A total of 1,088 brick elements represented the base plate, lid, tie-bars, storage tubes, and fuel elements. A total of 144 shell elements simulated the spacer pipe. The density of the basket elements was adjusted to reach the desired loaded weight. The mesh size was fairly coarse, providing a basket that was more rigid than in reality—but was considered adequate in representing the effects of the basket on the canister.

Model configuration #3: The three 24-in (610 mm) diameter, 15-ft (4.5 m) long ISFP canisters have two distinct internal component configurations. In order to envelop the response of these three canister internal configurations with one model, the analysis included the components that are deemed to cause the greatest damage to the canister containment boundary in these drop events as follows:

1. **The Shippingport Reflector Rod (Loose Rods) Basket:** This component configuration was chosen because it has several thick plates that would impact the canister containment body wall in a drop event. All three canisters have comparable shield plugs.

2. **The Reflector IV and V:** These component configurations also include a shield plug support ring welded to the canister body inside wall. Impact on such a ring during a drop event would cause increased straining on the canister body, and was therefore included in this composite model.

As with model configuration #1, this configuration has been evaluated for the 30-ft (9 m) and repository drop events at its maximum design weight of 10,000 pounds (4,530 kg). It has also been evaluated for the repository drop events under the lowest listed loaded canister weight of 8,280 pounds (3,751 kg). The component meshes were as follows:

1. **Skirts:** The lower and upper skirts each used 2,624 shell elements, sized at about 1/4 in (6.3 mm) x 6/10 in (15.2 mm) (longitudinal x circumferential dimension). This gave a fine mesh to represent these components as they deform during impact, absorbing the majority of the drop energy for all end-drops.

2. **Lifting Rings:** The lifting rings used the same circumferential spacing as the skirts. The brick mesh used four elements through the width and three brick elements through the thickness, with wedges on the interface with the skirt to represent the fillet welds. A total of 896 elements were used to represent each lifting ring.

3. **Heads:** The lower head used 2,351 shell elements to represent the head and the integral straight flange. Elements were sized at about 1/4 in (6 mm) x 6/10 in (15 mm) (longitudinal x circumferential) in the straight flange and all but the center of the head. At the center of the head the element sizes were smaller. This provided a fine mesh throughout. The upper head used 2,385 elements, with the same mesh as the lower head, but with the center portion replaced with the vent plug.

4. **Canister Body:** Because the more significant deformations were expected to occur in the skirts and heads of these canisters, the shell element size for the main body began at about 6/10 in (15 mm) x 6/10 in (15 mm) (longitudinal x circumferential) at the heads and then grew to 2-3/10 in (58 mm) x 6/10 in (15.2 mm) through the remaining canister body. A total of 7,680 shell elements were employed.

5. **Impact Plates:** Each interior impact plate used a total of 3,668 brick elements, with four elements through the thickness. This allowed the impact plate to bend correctly due to the internal basket load from above, and the impact load from below.

6. **Impact Plate Retaining Ring:** The impact plate retaining rings were modeled with four shell elements through the width, with a circumferential mesh that matched the heads (6/10 in or 15 mm). These rings were welded to the head knuckle (dish-to-straight flange transition) at about 30-degree segments (30 degrees of weld, 30 degrees no weld, and so on). Because the worst strains in the canister would be produced if the impact were on one of these welds, the rings were modeled with a continuous weld (modeled using common nodes between the retaining ring and the head).

7. **Shield Plug:** The shield plug is basically a large mass of steel. It was not expected to deform much, if at all, during the defined drop events. The shield plug only needed to be modeled to simulate its impact on the adjacent components (upper impact plate, canister wall, internal basket, etc.) during the drop event. Therefore, the shield plug was modeled with 1,152 brick elements.

8. **Shield Plug Support Ring:** The shield plug support ring is a 1-in (25 mm) wide and ¾-in (19 mm) tall 316L stainless steel ring that is welded to the inside of the canister body. This ring was modeled with three

67

shell elements wide, matching the mesh size of the canister body around the circumference. This ring was connected to the body using common nodes to represent the full-penetration attachment weld.

9. Shippingport Reflector Rod (Loose Rods) Basket: The bottom end spacer plate is 1-1/4 in (31.7 mm) thick and just under 22-1/2 in (571 mm) in O.D. (SA-193 Type 630 17-4 PH). Four intermediate spacer plates, and one top-end spacer plate spaced at just under 20-1/2 in (520.5 mm) apart, were tied together (with the bottom plate) with threaded tie bars (1-1/2 in = 38 mm diameter solid bars, SA-564 Type 630 17-4 PH). A total of 127 storage tubes (1-1/4 in = 32 mm O.D. x 0.064-in or 1.6 mm wall, SA-213 316L SST) for holding fuel rods fit between the bottom end spacer plate and the top end spacer plate, passing through the intermediate plates. A base plate, a top plate, a lid, and a lid locking plate were attached with threaded fasteners to close the basket ends. Other components are not significant to this evaluation. Because the exact condition of the basket and contained fuel rods during and after the drop events was not needed or desired, the basket was modeled in a simplified manner. A total of 560 brick elements represented the basket structure (plates, rods, storage tubes, etc.). The density of the basket elements was adjusted to reach the desired loaded weight. The mesh size was fairly coarse—making the basket more rigid than in reality—but was considered adequate in representing the effects of the basket on the canister structure. A 26-in (660 mm) long spacer takes up the remaining space in this canister. The spacer consists of a top and bottom 1-in (25 mm) thick plate approximately 22 in (557 mm) in diameter (SA-240 316L SST) and a 20-in (508 mm) Schedule 60 pipe body (SA-312 316L SST). This spacer was simulated with 1,088 brick elements for the plates, and 384 shell elements for the pipe.

Component thickness: Because this evaluation effort compares the response of the ISFP canisters to those of the standardized canisters tested in 1999, the actual average thickness of several components from the 1999 tests was used for the 18-in (457 mm) diameter ISFP canisters. Those components were the canister heads, body, and skirts. The ISFP canister models use a 0.350-in (9 mm) thickness for the head-straight flanges, body, and skirts, instead of the nominal 3/8 in (10 mm). A thickness of 0.471 in (12 mm) was used for those same components on the 24-in (610 mm) size ISFP canisters. All other components used nominal dimensions.

Material density: The basic density of 316L stainless steel is 0.283 pounds per cubic in (7,833 kg/m³). However, this value was modified where necessary in the FE models in order to obtain the listed weights for the canister body, lid, and impact plates. Material densities for baskets with fuel were adjusted to achieve the correct listed weights as well. Attempts were made to obtain specific center-of-gravity information for each fuel, but this information was not readily available. Therefore, uniformly distributed fuel mass was assumed over the basket length. This was believed to be a reasonable approximation.

Contact modeling: Contact between components was simulated using the ABAQUS General Contact option supplemented by the Contact Pairs option in areas of interest (impact locations). This is one of the approved methods detailed in ABAQUS Software Report. These contact options employ penalty contact stiffness. Preliminary evaluations increased the default stiffness calculated within ABAQUS/Explicit Version 6.3-3 by a factor of 10. The results were the same as those obtained using the default stiffness values. This indicated that the default penalty stiffness calculated within ABAQUS was adequately stiff to simulate a "hard impact" for these ISFP canister evaluations.

Flat, rigid impact surface: The flat, rigid impact surface was modeled using one large rigid quadrilateral element (element type R3D4) that was fixed in space.

Friction: The 1999 standardized canister evaluation used a coefficient of friction value of 0.3 for all post-drop model analyses. This friction value was used for contact between the canister and the rigid surface, and contacts between canister internal components and the canister itself. Snow et al. (2001) have shown that the most valid coefficient of friction value is dependent on the drop angle. However, for all drop angles evaluated herein, a coefficient value of 0.3 is appropriate (in order to match that used in the 1999 efforts), yielding a valid comparison for the ISFP canisters.

Initial conditions: The FE models began the drop event by locating the canister model just above the rigid surface and applying a gravitational acceleration and an initial velocity. This allowed the elimination of calculations while the canister was freely falling through air. The initial velocity was calculated by equating the potential energy of the canister at the beginning of the drop (mass · gravity · drop height) to the kinetic energy just before impact

Model solution termination: Model solution was terminated when the canister model had progressed through the first impact and bounced off the rigid surface. This applied to all models except for the slap-down event, which continued the calculation through the impact of the top-end.

ISFP canister material properties: The ISFP canisters use 316L stainless steel for all skirts, lifting rings, heads, bodies, impact plates, impact plate retaining rings, shield plugs, and shield plug support rings. Impact plates can also be made of SA-351 CF3M, which is comparable in strength and ductility to 316L stainless steal (SST). Minimum material properties are specified by ASME (2001) for these 316L materials, but actual material properties (such as yield strength, ultimate strength, elongation, etc.) are usually much higher than the minimum specified values. However, the purpose of this evaluation is to compare the ISFP canisters to the standardized canisters under the same drop events. Therefore, the material properties used in the 1999 evaluations of the standardized canisters were used in this evaluation of the ISFP canisters.

The 1999 drop-test evaluations of the standardized canisters used actual material properties obtained from quasi-static material tensile tests on the several material "heats" used in the canister fabrications. The stresses on the stress-strain curves from these tensile tests were increased by 20% to account for dynamic strengthening due to the strain rate occurring during the drop events (Morton et al. 2000).

Even though there were several different 316L material heats used in the 1999 testing, the one most commonly used (Y2613) is considered representative of the other heats, and is the only one used on the ISFP canister evaluations for 316L SST components. Those properties at 70°F (21°C) are as follows:

1. Modulus of Elasticity (E) = 28.3 x 10^6 psi (195 x 10^6 kPa)

2. Poisson's Ratio (μ) = 0.29

3. Average Yield Strength ($\sigma_{nominal\ yield}$) = 37.8 ksi (0.26 x 10^6 kPa) (for Y2613)

4. Average Ultimate Strength ($\sigma_{nominal\ ultimate}$) = 76 ksi (0.52 x 10^6 kPa) (for Y2613)

ABAQUS/Explicit requires that material stress and strain data be given as true values, not engineering values. Table 7 lists the average engineering (nominal) stress-strain curve data and their true values using the following conversion equations:

1. True Stress $(\sigma_{true}) = \sigma_{nominal} \times [1 + \varepsilon_{nominal}]$

2. True Strain $(\varepsilon_{true}) = \ln [1 + \varepsilon_{nominal}] - \sigma_{true} / E$

Internal basket and spacer material properties: Many of the internal basket and spacer components use 316L SST. The properties from the previous section have also been used for these components. Other basket components used SA-193; SA-564; SA-693; or all Type 630 17-4 PH. This is a very high strength material (ASME 2001) with a room temperature yield strength at 105 ksi (0.72 x 10⁶ kPa); an ultimate strength at 135 ksi (0.93 x 10⁶ kPa); an elongation of 16%; and an area reduction of between 40% and 50%. Therefore, the material properties for these Type 630 17-4 PH components at 70°F (21°C) used in the following evaluations were:

1. Modulus of Elasticity (E) = 28.3 x 10⁶ psi (195 x 10⁶ kPa)

2. Poisson's Ratio (μ) = 0.29

3. True Yield Strength $(\sigma_{true\ yield})$ = 105 ksi (0.72 x 10⁶ kPa)

4. True Ultimate Strength $(\sigma_{true\ ultimate})$ = 160 ksi (1.1 x 10⁶ kPa) (assumed at 70% true strain). The true ultimate strength was likely much larger than 160 ksi (1.1 x 10⁶ kPa) @ 70% strain. However, the basket structures and their simplified modeling were such that not much energy absorption was expected in the baskets due to these drop events. Therefore, a true ultimate strength of 160 ksi (1.1 x 10⁶ kPa) @ 70% strain was considered adequate.

Table 7. Stress-Strain curves for canister body & skirt 316L SST, Heat No. Y2613 (Snow and Morton 2003).

Nominal Stress		Nominal Strain	True Stress[1]		True Strain
(ksi)	(MPa)		(ksi)	(MPa)	
37.8	261	0.002[2]	45.5	314	0
45.4	313	0.10	59.9	413	0.094
49.4	341	0.15	68.2	470	0.138
53.3	367	0.20	76.8	530	0.180
61.2	422	0.30	95.5	658	0.260
69.1	476	0.40	116.0	800	0.333
76.0	524	0.488[3]	135.7	936	0.393

[1] Average value increased by 20% to account for dynamic strengthening.
[2] By definition, strain at the yield stress.
[3] Average total elongation from testing.

Fuel material properties: Fuels within the canister baskets were not explicitly modeled. The exact condition of the baskets and contained fuels during and after the drop events was not needed or desired in these evaluations. Therefore, the fuels were only represented by an increase in density (mass) of the basket materials. If the fuels were quite strong/rigid (undeforming), then this would be an accurate way to represent them in the models. If the fuels actually deform significantly (absorbing energy) during the drop event, then modeling the fuels as mass only will be conservative with respect to the canister structure (canisters would be calculated to deform more than would actually occur). This is considered acceptable.

Analytical Calculations

The I-DEAS Master Series Version 9 m2 computer program was used to create the finite element models of the ISFP canisters. A solid model of a canister was created first, then used to generate the finite element model. The I-DEAS software was not used for the calculations—only for modeling purposes. The models were checked in the calculation software for accuracy.

The computer program ABAQUS/Explicit Version 6.3-3, a nonlinear FE analysis software package that is widely used in many industries, was employed to calculate the response of the ISFP canisters to the specified drop events. Extensive validation and verification (DOE 2003b) has been performed by the NSNFP on this software, approving it for such canister drop evaluations.

ISFP canister model #1: The ISFP canister model #1, at a maximum design weight of 6,000 pounds (2,718 kg), was evaluated for a 3-ft drop (9 m) onto a flat, rigid surface at five drop angles: 1) ½ degrees off-vertical (to match the actual 1999 drop test condition); 2) 6 degrees off-vertical (CGOC); 3) 45 degrees off-vertical; 4) 80 degrees off-vertical; and 5) 90 degrees off-vertical (horizontal). This model was also evaluated for a 23-ft (6.9 m) repository vertical drop and a 2-ft (0.6 m) worst angle (80 degrees off-vertical) repository drop. Additionally, the model was evaluated at a maximum actual loaded weight of 4,090 pounds (1,853 kg) for the 30-ft (9 m), 6 degrees off-vertical and 30-ft (9 m), 80 degrees off-vertical drops as well. This is shown in Table 8.

Table 8. ISFP canister model #1 drop conditions (Snow and Morton 2003).

Model File Name	Drop Height		Drop Angle from Vertical (Degrees)
	(ft)	(m)	
6,000-Pound (2,718-kg) Canister Weight			
model1_30ft_05deg	30	9	½
model1_30ft_6deg	30	9	6
model1_30ft_45deg	30	9	45
model1_30ft_80deg	30	9	80
model1_30ft_90deg	30	9	90
model1_23ft_vertical	23	6.9	0
model1_2ft_80deg	2	0.6	80
4,090-Pound (1,853-kg) Canister Weight			
model1_30ft_6deg_4090lbs	30	9	6
model1_30ft_80deg_4090lbs	30	9	80

The information reported from these drop evaluations is limited to: model deformations; model energy plots; and canister component strains. The condition of the internal components was not of interest in this evaluation. Therefore, the reported results include only those that relate to the containment boundary components and the canister skirts.

Figures 7 through 19 show the deformed shape of the canisters as they were rebounding off of the rigid surface. These deformed shapes are consistent with those obtained in the 1999 drop testing efforts (Morton et al. 2000).

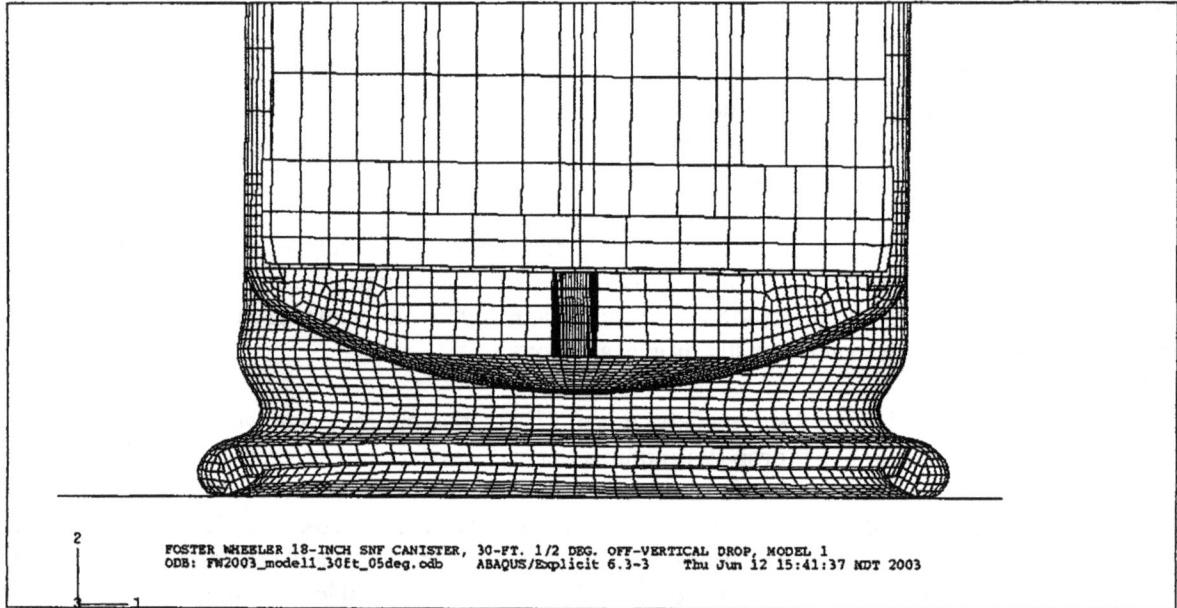

FOSTER WHEELER 18-INCH SNF CANISTER, 30-FT. 1/2 DEG. OFF-VERTICAL DROP, MODEL 1
ODB: FW2003_model1_30ft_05deg.odb   ABAQUS/Explicit 6.3-3   Thu Jun 12 15:41:37 MDT 2003

Fig. 7. Model #1 canister deformations, 1/2 degrees off-vertical, 30-foot (9-m) drop, 6000 pounds (2718 kg) (Snow and Morton 2003).

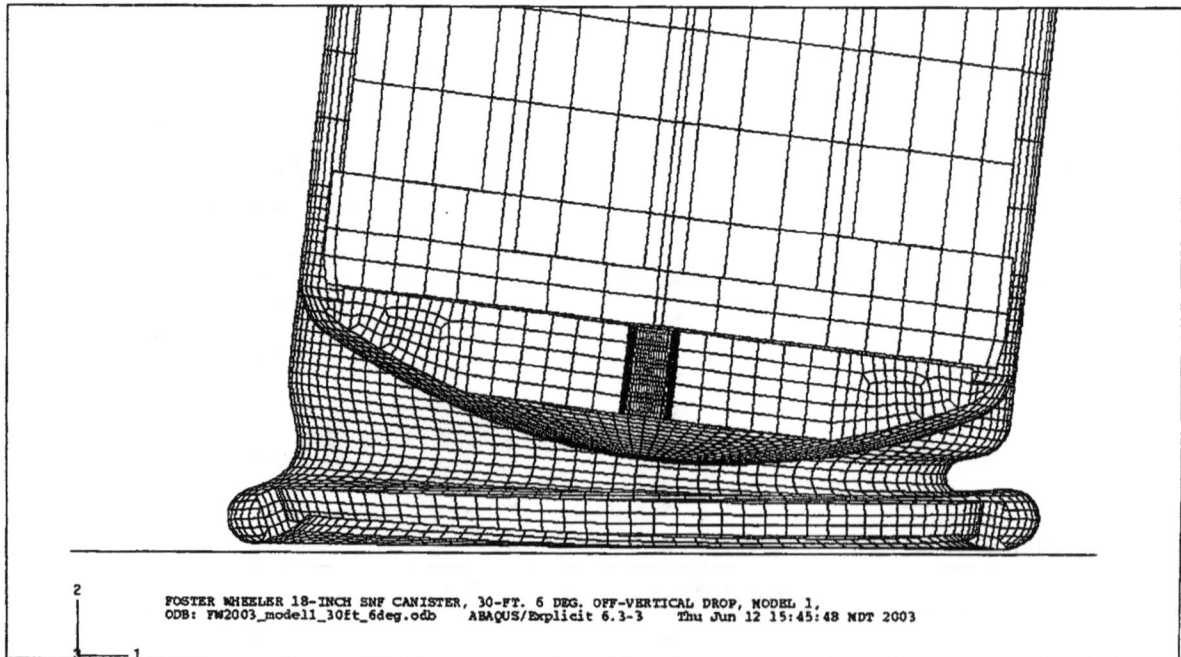

FOSTER WHEELER 18-INCH SNF CANISTER, 30-FT. 6 DEG. OFF-VERTICAL DROP, MODEL 1,
ODB: FW2003_model1_30ft_6deg.odb   ABAQUS/Explicit 6.3-3   Thu Jun 12 15:45:48 MDT 2003

Fig. 8. Model #1 canister deformations, 6 degrees off-vertical, 30-foot (9-m) drop, 6000 pounds (2718 kg) (Snow and Morton 2003).

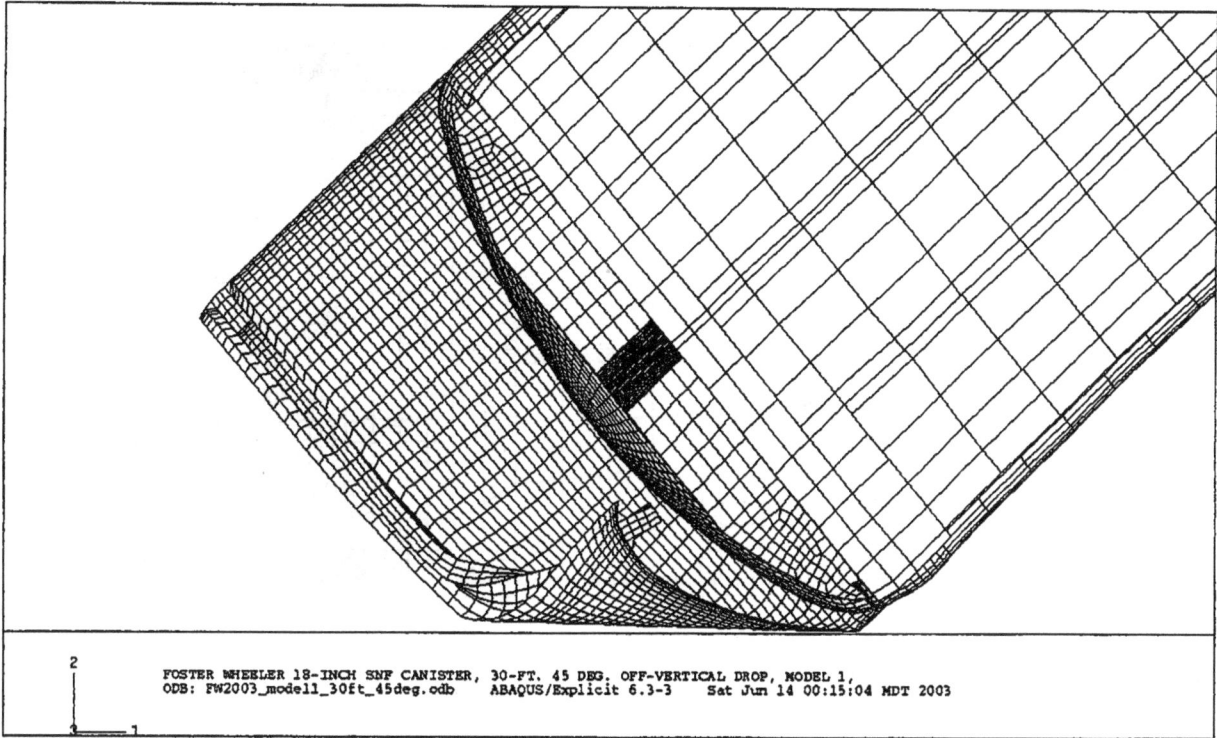

Fig. 9. Model #1 canister deformations, 45 degrees off-vertical, 30-foot (9-m) drop, 6000 pounds (2718 kg) (Snow and Morton 2003).

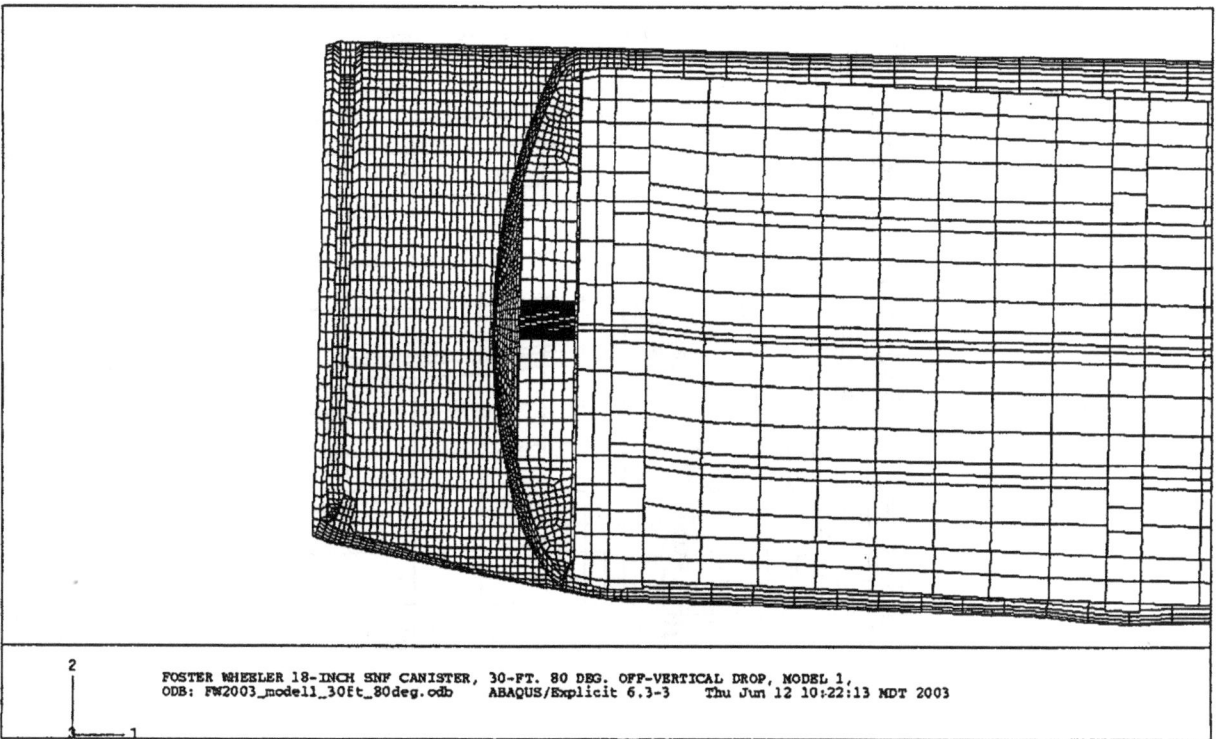

Fig. 10. Model #1 canister bottom end deformations, 80 degrees off-vertical, 30-foot (9-m) drop, 6000 pounds (2718 kg) (Snow and Morton 2003).

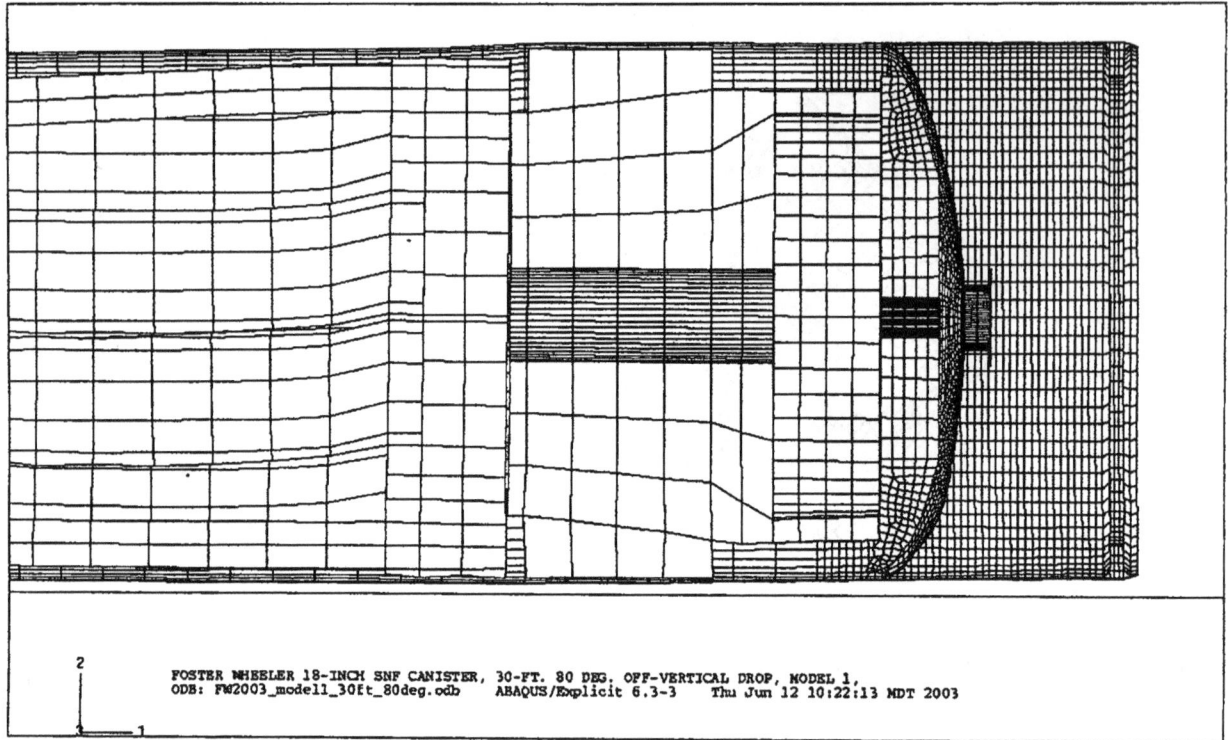

FOSTER WHEELER 18-INCH SNF CANISTER, 30-FT. 80 DEG. OFF-VERTICAL DROP, MODEL 1,
ODB: FW2003_model1_30ft_80deg.odb   ABAQUS/Explicit 6.3-3   Thu Jun 12 10:22:13 MDT 2003

Fig. 11. Model #1 canister top end deformations, 80 degrees off-vertical, 30-foot (9-m) drop, 6000 pounds (2718 kg) (Snow and Morton 2003).

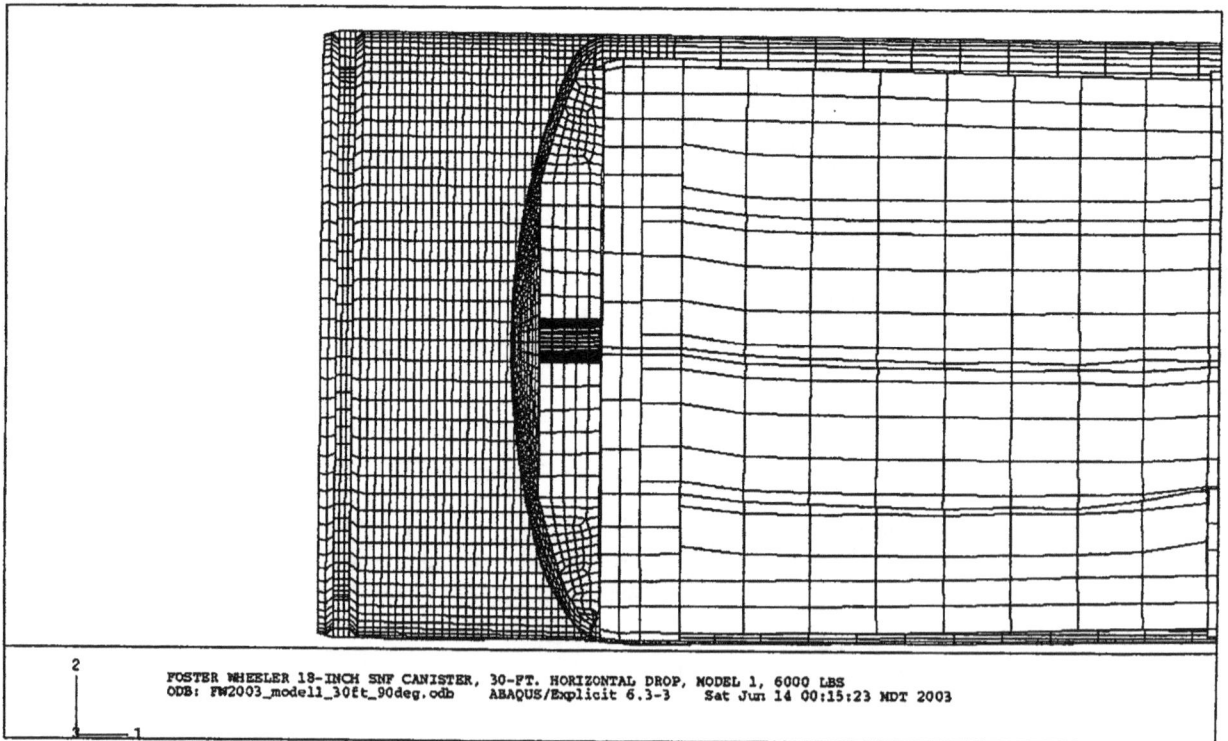

FOSTER WHEELER 18-INCH SNF CANISTER, 30-FT. HORIZONTAL DROP, MODEL 1, 6000 LBS
ODB: FW2003_model1_30ft_90deg.odb   ABAQUS/Explicit 6.3-3   Sat Jun 14 00:15:23 MDT 2003

Fig. 12. Model #1 canister bottom end deformations, 90 degrees off-vertical, 30-foot (9-m) drop, 6000 pounds (2718 kg) (Snow and Morton 2003).

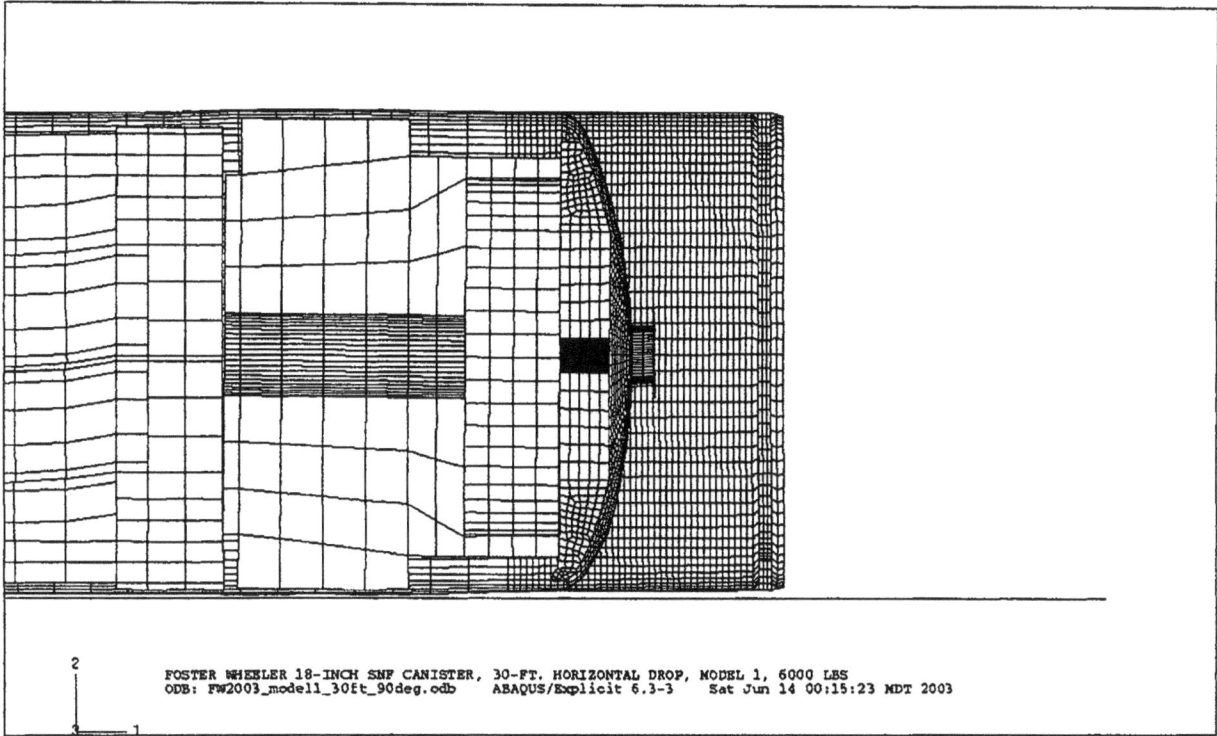

Fig. 13. Model #1 canister top end deformations, 90 degrees off-vertical, 30-foot (9-m) drop, 6000 pounds (2718 kg) (Snow and Morton 2003).

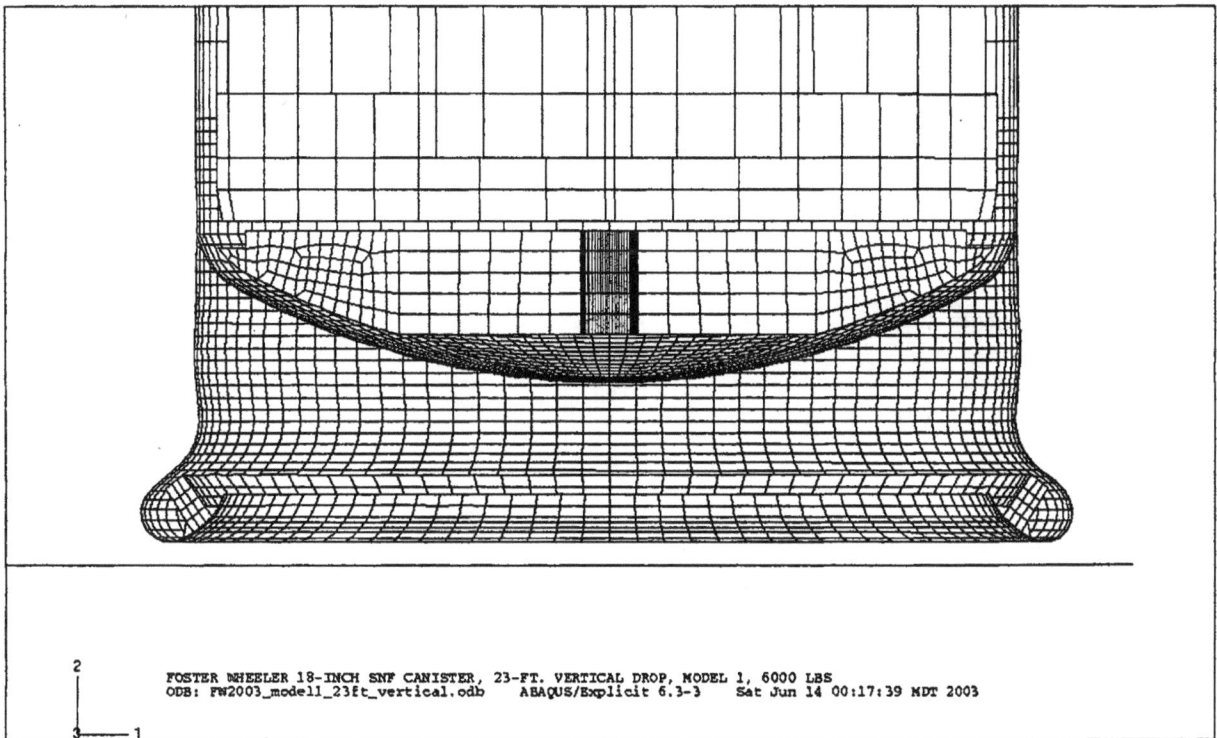

Fig. 14. Model #1 canister deformations, vertical, 23-foot (6.9-m) drop, 6000 pounds (2718 kg) (Snow and Morton 2003).

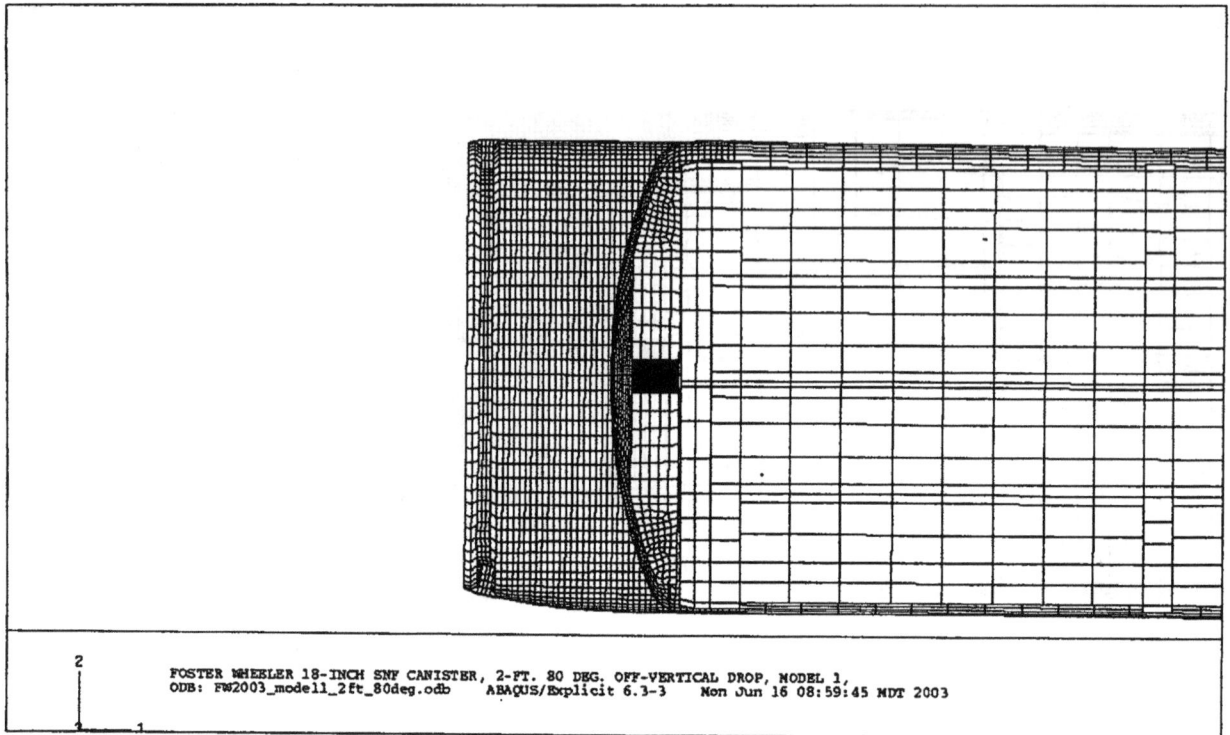

Fig. 15. Model #1 canister bottom end deformations, 80 degrees off-vertical, 2-foot (0.6-m) drop, 6000 pounds (2718 kg) (Snow and Morton 2003).

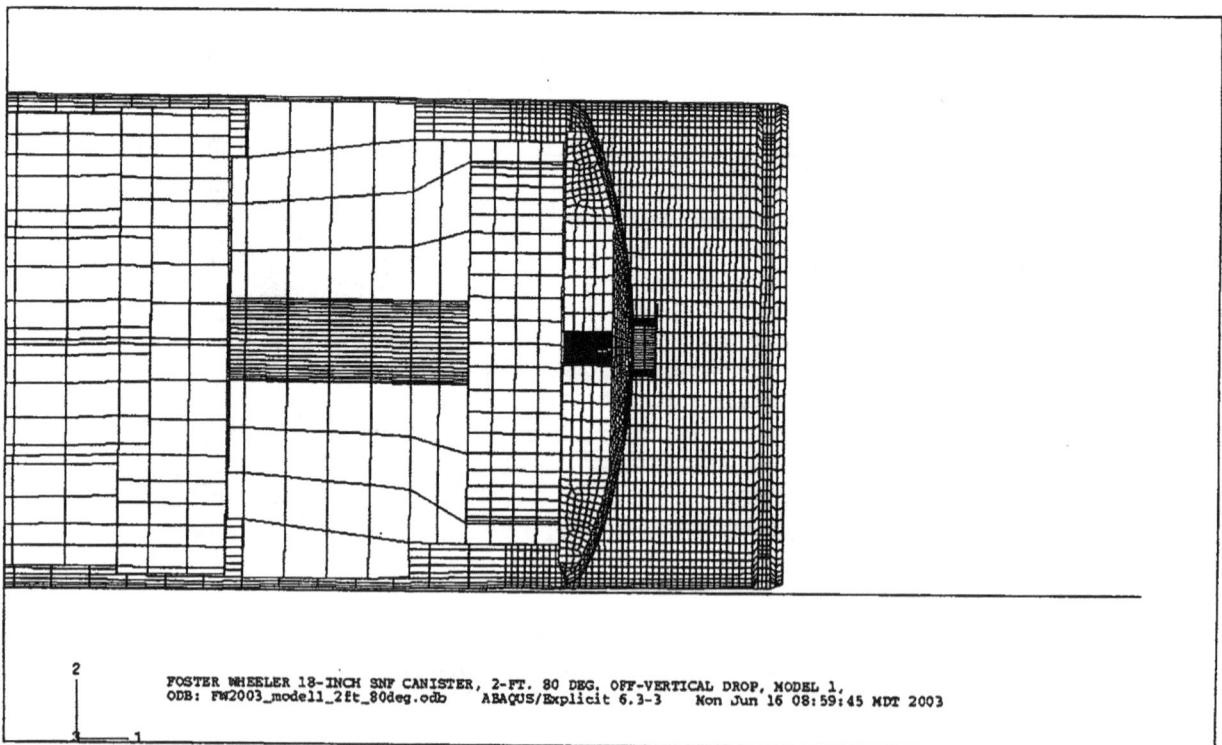

Fig. 16. Model #1 canister top end deformations, 80 degrees off-vertical, 2-foot (0.6-m) drop, 6000 pounds (2718 kg) (Snow and Morton 2003).

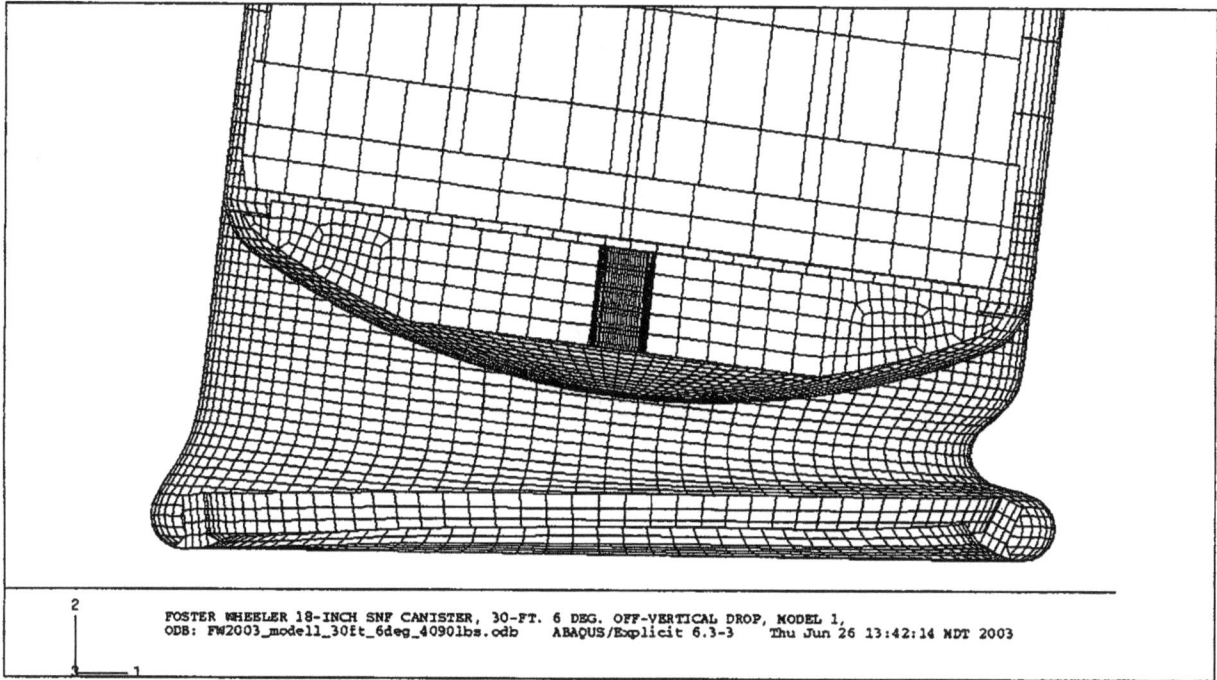

Fig. 17. Model #1 canister deformations, 6 degrees off-vertical, 30-foot (9-m) drop, 4090 pounds (1853 kg) (Snow and Morton 2003).

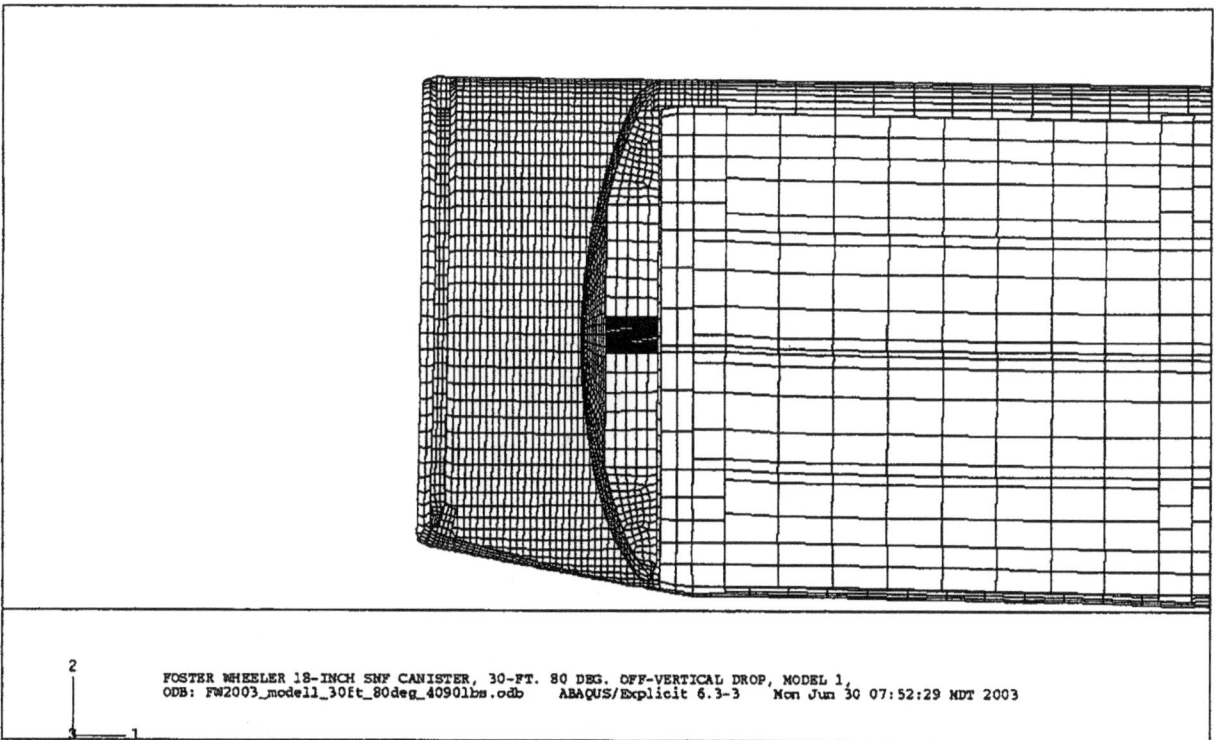

Fig. 18. Model #1 canister bottom end deformations, 80 degrees off-vertical, 30-foot (9-m) drop, 4090 pounds (1853 kg) (Snow and Morton 2003).

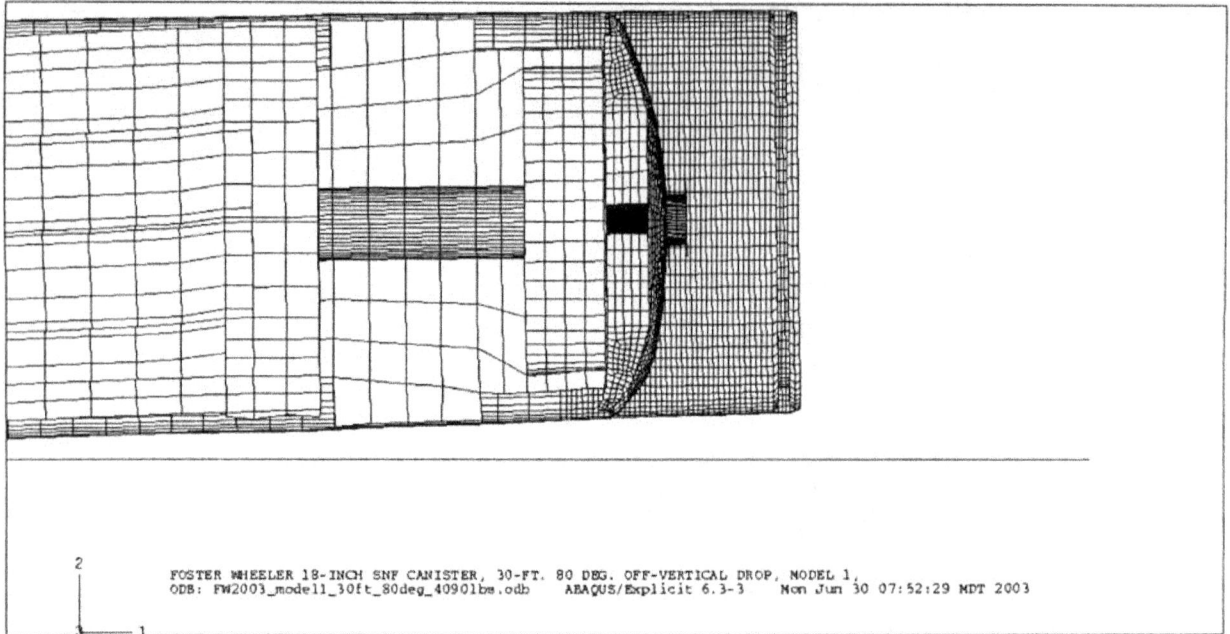

Fig. 19. Model #1 canister top end deformations, 80 degrees off-vertical, 30-foot (9-m) drop, 4090 pounds (1853 kg) (Snow and Morton 2003).

Figures 20 through 28 show plots of the energy history for each drop scenario. The plots show model kinetic energy history (ALLKE); plastic dissipation history (ALLPD); frictional dissipation history (ALLFD); elastic energy history (ALLSE); and artificial energy history (ALLAE). Artificial energy is the amount of drop energy used (taken away from the total model energy) to prevent finite element numerical instabilities. Normally, the artificial energy should remain below 5% in order for the results to be considered valid. When the artificial energy increases above 5%, a rerun of the model with a higher drop value is recommended to compensate for that artificial energy.

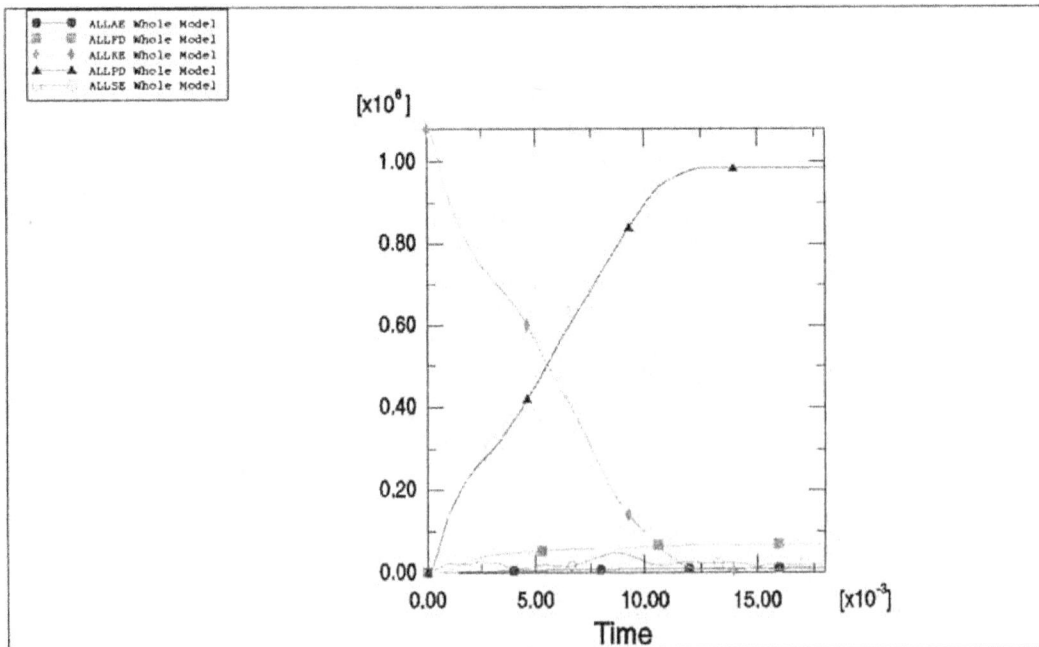

Fig. 20. Model #1, energy history, ½ degrees off-vertical, 30-foot (9-m) drop, 6000 pounds (2718 kg) (Snow and Morton 2003).

78

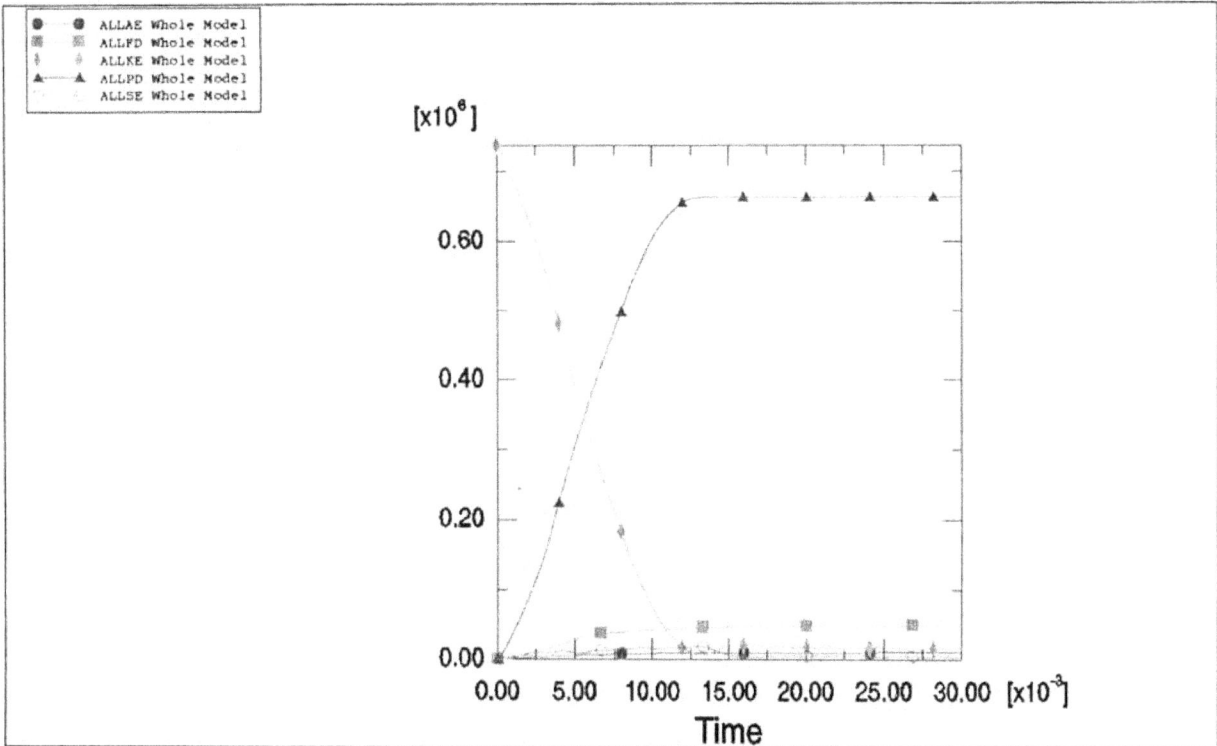

Fig. 21. Model #1, energy history, 6 degrees off-vertical, 30-foot (9-m) drop, 6000 pounds (2718 kg) (Snow and Morton 2003).

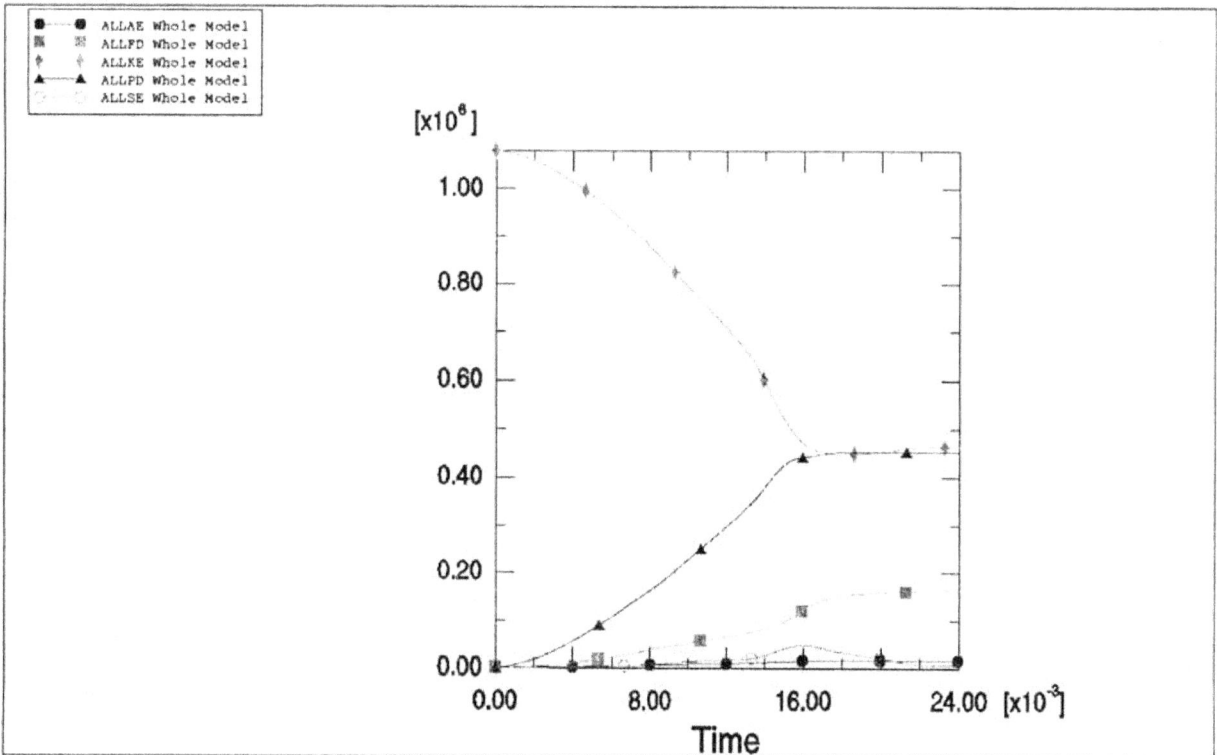

Fig. 22. Model #1, energy history, 45 degrees off-vertical, 30-foot (9-m) drop, 6000 pounds (2718 kg) (Snow and Morton 2003).

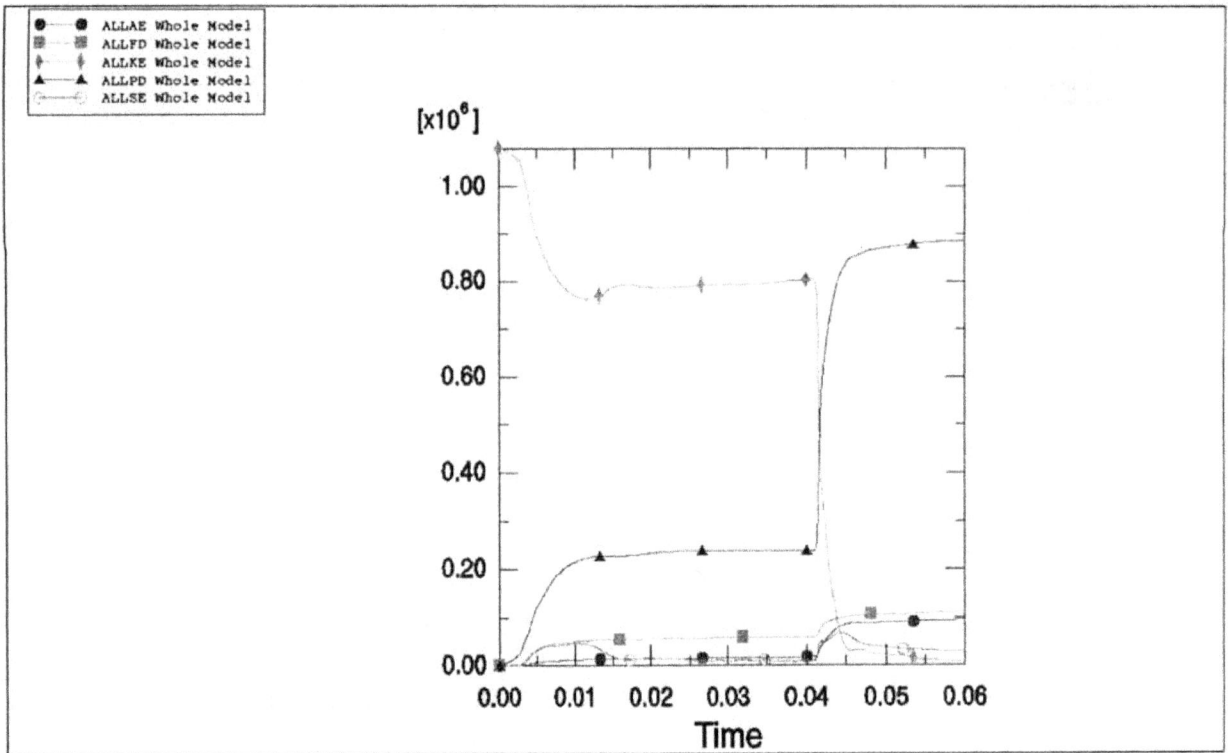

Fig. 23. Model #1, energy history, 80 degrees off-vertical, 30-foot (9-m) drop, 6000 pounds (2718 kg) (Snow and Morton 2003).

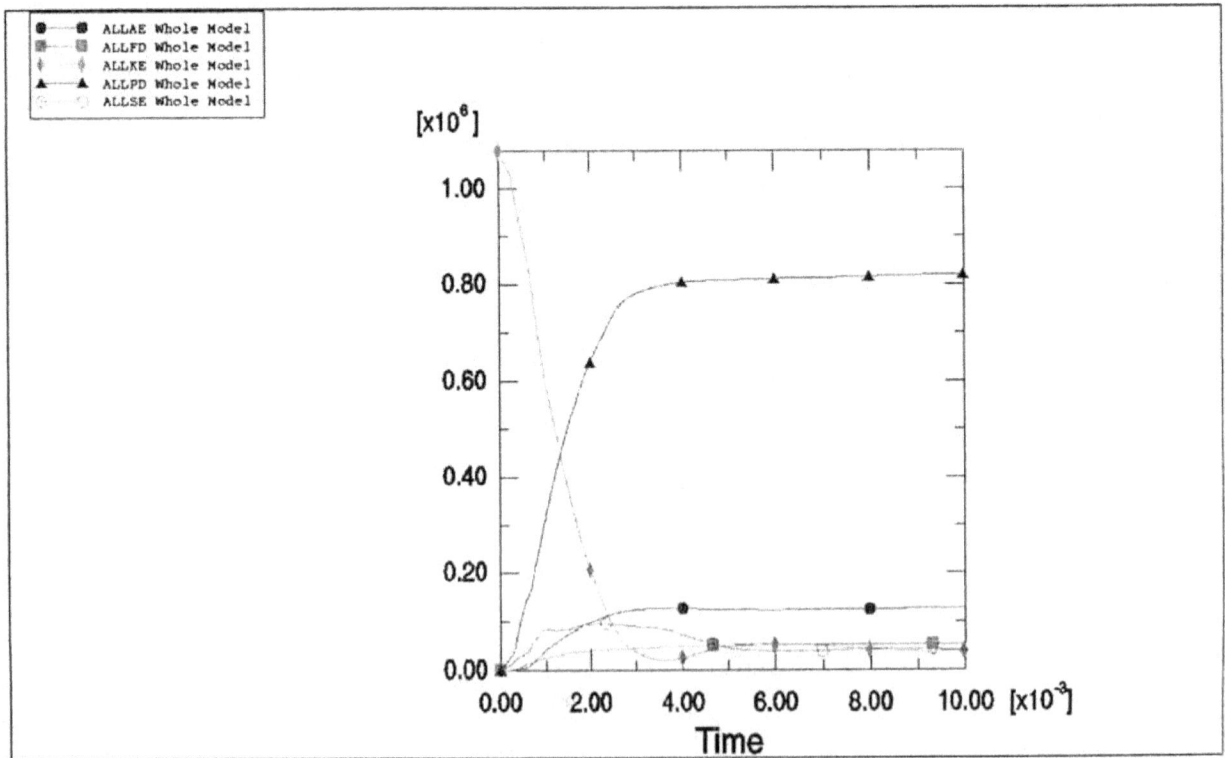

Fig. 24. Model #1, energy history, 90 degrees off-vertical, 30-foot (9-m) drop, 6000 pounds (2718 kg) (Snow and Morton 2003).

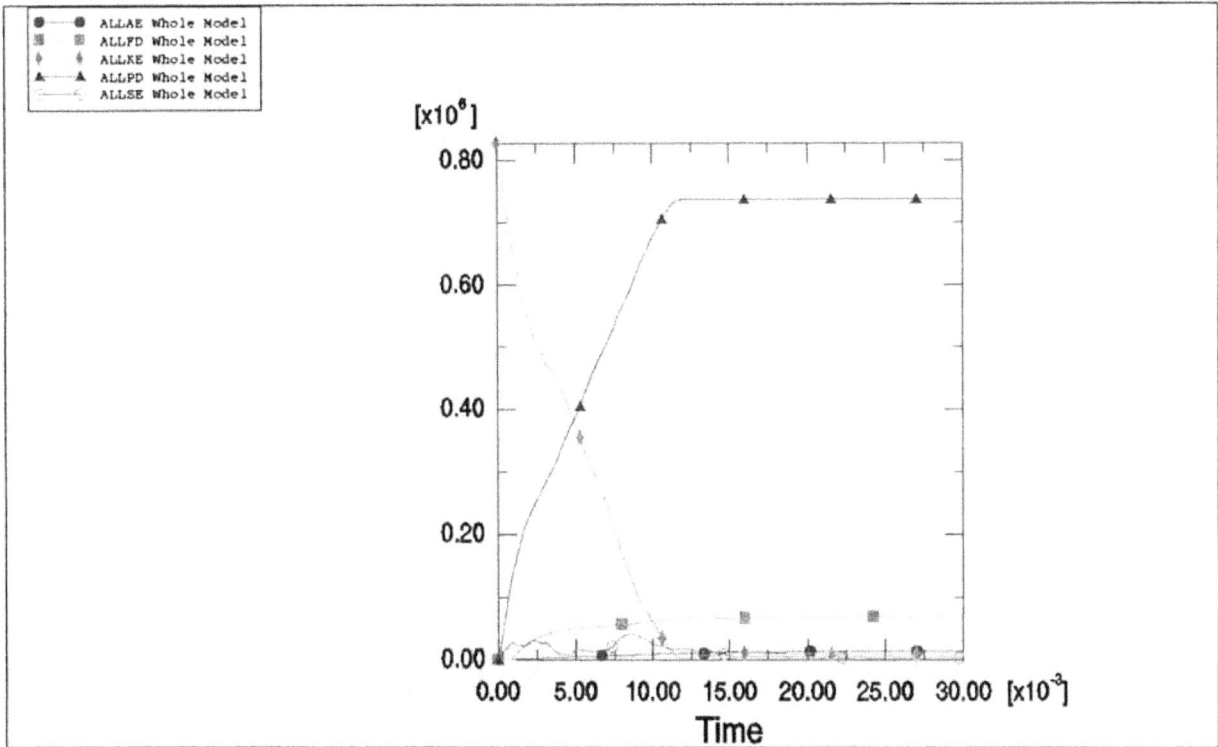

Fig. 25. Model #1, energy history, vertical, 23-foot (6.9-m) drop, 6000 pounds (2718 kg) (Snow and Morton 2003).

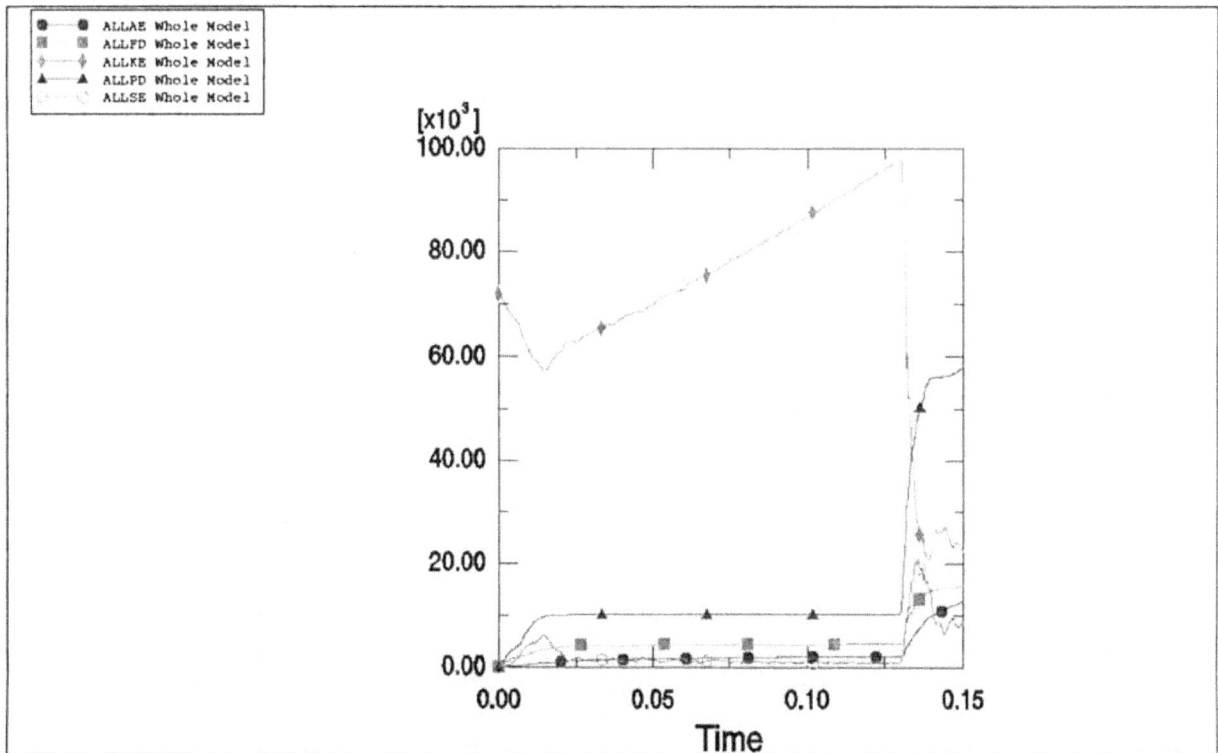

Fig. 26. Model #1, energy history, 80 degrees off-vertical, 2-foot (0.6-m) drop, 6000 pounds (2718 kg) (Snow and Morton 2003).

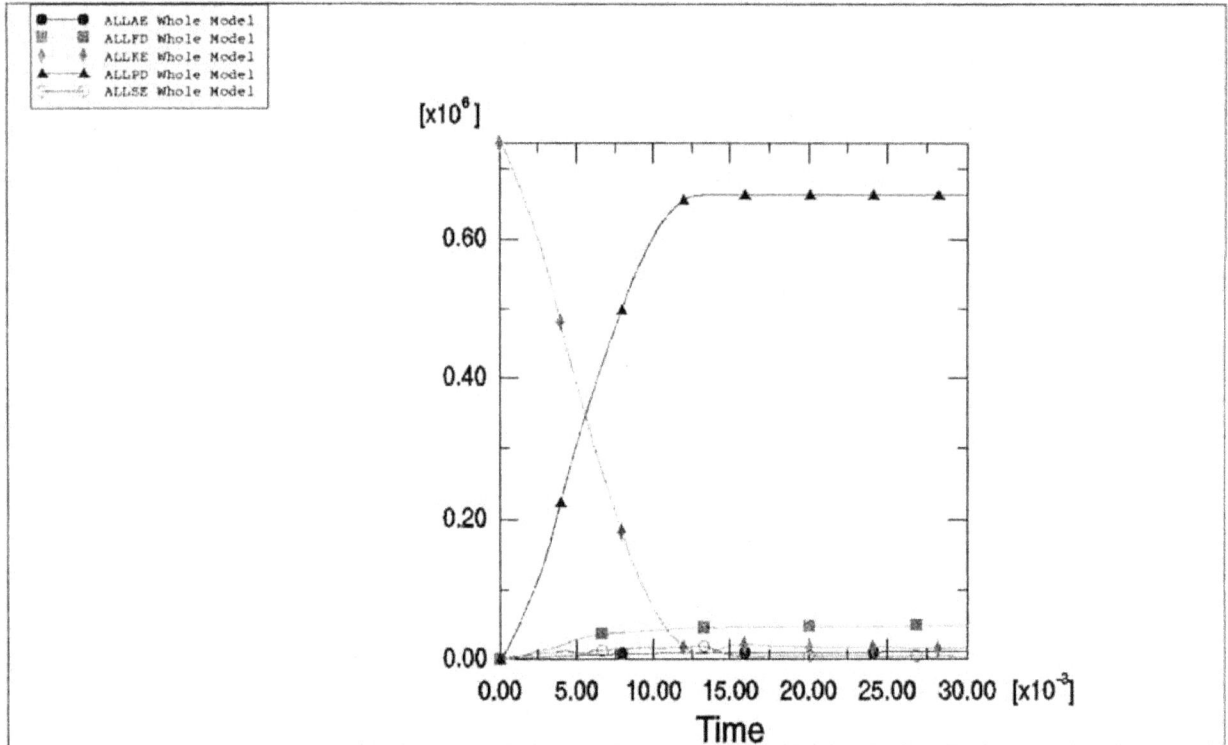

Fig. 27. Model #1, energy history, 6 degrees off-vertical, 30-foot (9-m) drop, 4090 pounds (1853 kg) (Snow and Morton 2003).

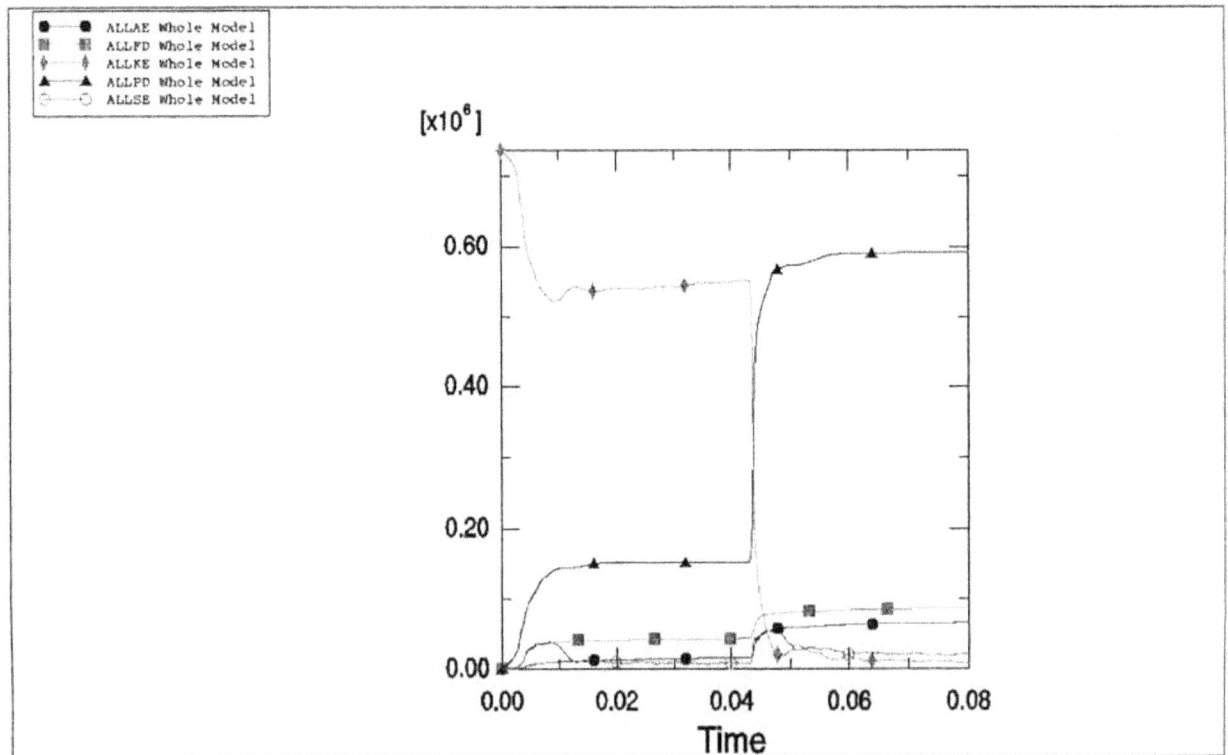

Fig. 28. Model #1, energy history, 80 degrees off-vertical, 30-foot (9-m) drop, 4090 pounds (1853 kg) (Snow and Morton 2003).

Two of the model #1 drops experienced a 13% artificial energy value for 6,000-pound (2,718 kg) canister dropped from 30 ft (9 m) at 90 degrees off-vertical (Fig. 24) and the same canister dropped 2 ft at 80 degrees off-vertical (Fig. 26). Two other models showed a 9% artificial energy [6,000-pound (2,718 kg) canister dropped 30 ft (9 m) at 80 degrees off-vertical (Fig. 23) and the same drop but with a 4,090-pound (1,853 kg) canister (Fig. 28)]. All four of these models were rerun with an increase in drop height to account for that artificial energy. These results appeared to be unchanged or small in deformations, while peak material strains were only slightly higher.

During these drop events, the bulk of the kinetic energy at impact is transformed into plastic work. The best measure of that plastic work is the equivalent plastic strain, which is a cumulative strain measure that takes into account the entire deformation history. Table 9 lists the peak equivalent plastic strains occurring in the various containment boundary and skirt components of the ISFP canisters during the specified drop events for a 6,000-pound (2,718 kg) canister (maximum design weight). Table 10 lists the same for a 4,090-pound (1,853 kg) canister (maximum actual weight). Peak strains are listed for the outside surface of a component, mid-plane, and inside surface. However, these peak strains did not necessarily occur at the same location (element) through the thickness.

Table 9. Peak plastic strains in Foster Wheeler 18-inch (457-mm) diameter, 15-foot (4.5-m) long canisters, 6,000 pounds (2,718 kg) maximum design weight (Snow and Morton 2003).

Drop Height		Angle From Vertical (Degrees)	Peak Equivalent Plastic Strain (PEEQ) %														
			Lower Head			Lower Skirt			Upper Head			Upper Skirt			Body		
(ft)	(m)		out*	mid*	in*	out*	mid*	in*	out*	mid*	in*	out*	mid*	in*	out*	mid*	in*
30	9	½ t = 0.018	3 [7]	1 [3]	2 [6]	61 [91]	21 [17]	53 [75]	5	0	5	-	-	-	-	-	-
30	9	6 t = 0.024	0.7 [9]	0.1 [3]	0.3 [10]	88 [107]	23 [21]	58 [94]	0.5	0	0.5	-	-	-	-	-	-
30	9	45 t = 0.024	22 [33]	10 [9]	16 [36]	59 [52]	31 [33]	106 [84]	-	-	-	-	-	-	2	1	1
30** (33)	9 (9.9)	80 t = 0.060	33 (34)	17 (17)	29 (30)	31 (33)	18 (19)	25 (26)	48 (48) [57]	24 (24) [19]	25 (25) [42]	15 (15) [24]	9 (9) [20]	9 (9) [19]	6 (7)	2 (2)	5 (6)
30*** (34)	9 (10.2)	90 t = 0.010	31 (32)	18 (19)	18 (19)	11 (12)	6 (6)	7 (7)	30 (32) [40]	17 (18) [15]	21 (23) [26]	7 (8) [10]	4 (4) [10]	5 (5) [10]	5 (6)	2 (2)	4 (4)
23 repository	6.9	0 t = 0.030	1 {10}	0.2 {3}	1 {6}	35 {40}	18 {16}	52 {48}	5	0	5	-	-	-	-	-	-
2*** (2.25) repository	0.6 (0.7)	80 t = 0.150	2 (3) {12}	1 (1) {7}	2 (3) {7}	9 (9) {6}	5 (5) {5}	4 (4) {9}	20 (20) {24}	11 (12) {11}	8 (8) {13}	5 (6) {5}	2 (3) {2}	3 (3) {6}	4 (4)	1 (1)	4 (4)

- Denotes that strains in these components are either zero or insignificant/not of interest.
* out = outside surface of component, mid = middle surface of component, in = inside surface of component.
** This model experienced about 9% artificial energy (see below).
*** These models experienced about 13% artificial energy. Artificial energy is drop energy used to prevent finite element numerical instabilities. The results shown in (parentheses) are for a higher drop that accounts for this lost drop energy. All other models experienced an artificial energy that was below 5% that is considered acceptable as is.
[Numbers reported in square brackets are those that were listed in Morton et al. (2000).]
{Numbers reported in these brackets are those that were listed in the Blandford (2003) standardized canister analysis.}

Table 10. Peak plastic strains in Foster Wheeler 18-inch (457-mm) diameter, 15-foot (4.5-m) long canisters, 4,090 pounds (1,853 kg) maximum actual weight (Snow and Morton 2003).

Drop Height (ft)	(m)	Angle From Vertical (Degrees)	Peak Equivalent Plastic Strain (PEEQ) %														
			Lower Head			Lower Skirt			Upper Head			Upper Skirt			Body		
			out*	mid*	in*	out*	mid*	in*	out*	mid*	in*	out*	mid*	in*	out*	mid*	in*
30	9	6 t = 0.030	0.4	0	0.2	57	19	51	0.8	0	0.8	-	-	-	-	-	-
30 (33)	9 (9.9)	80 t = 0.080	33 (35)	17 (17)	27 (28)	29 (31)	17 (18)	24 (24)	51 (50)	25 (24)	25 (24)	17 (15)	11 (9)	10 (9)	8 (8)	2 (2)	7 (6)

- Denotes that strains in these components are either zero or insignificant/not of interest.
* out = outside surface of component, mid = middle surface of component, in = inside surface of component.
** This model experienced a 9% artificial energy.
Artificial energy is drop energy used to prevent finite element numerical instabilities. The results shown in (parentheses) are for a higher drop that accounts for this lost drop energy. All other models experienced an artificial energy that was below 5%, which is considered acceptable as is.

All of the 18-in (457 mm) diameter, 15-ft (4.5 m) long, standardized canisters tested in 1999 were loaded to about 6,000 pounds (2,718 kg) total weight. This makes a comparison between those canister drops and the Table 9 ISFP (6,000 pound = 2,718 kg) canister drops valid. Table 9 shows the strain values reported in the 1999 evaluation of the standardized canister for the 30-ft (9 m) drop events. This table shows that the peak strains in many of the containment boundary components (lower head, body, upper head only) of the ISFP canister are lower than those for the 18-in (457 mm) diameter 6,000-pound (2,718 kg) standardized canister. The deformation and strain responses of a canister component during a drop event are dependent on component stiffness; adjacent component stiffness; interaction between adjacent components; material properties; canister loaded weight; drop height; drop angle; friction; and buckling. Varying one or more of these parameters results in a nonlinear (sometimes highly nonlinear) change in the canister component response. In the present case, the ISFP canister heads are stiffer than those on the standardized canister due to their increased thickness. As expected, the strain response is nonlinear in that peak strains sometimes decreased and sometimes increased in a given canister component relative to the standardized canister response. The following ISFP canister peak strains under the 30-ft (9 m) drop event were higher:

1. The ½ degrees and 6 degrees off-vertical drop produced strains in the upper head in the 18-in (457 mm) diameter ISFP canister due to the impact plate bending the retaining ring (which is welded to the upper head). Such strains were absent in the standardized canister because it has no impact plate retaining ring. These upper head strains are not considered a containment concern because they are much lower than the maximum 1999 strains calculated for the 80-degree drop.
2. The 45 degrees off-vertical drop produced a 1% strain increase on the mid-surface of the lower head.
3. The 80 degrees off-vertical drop produced a 5% strain increase on the mid-surface of the upper head. This is significant because it exceeds the maximum mid-surface strain calculated in the 1999 efforts.
4. The 90 degrees off-vertical (horizontal) drop produced a 3% strain increase on the mid-surface of the upper head. However, the 80 degrees off-vertical drop results were controlling.

The strain increase values indicated above (and in similar circumstances below) are actual strain increases (e.g., if the peak strain was 3% and is now 4%, then it would be reported above as a "1% strain increase").

Peak strains in the skirts, which were designed to absorb drop energy and are not containment retaining, increased or decreased from the 1999 results depending on the drop angle and location. Note that the strains in the lower skirt for the 45 degrees off-vertical drop did reflect a 22% strain increase on the inside surface over those reported for the 1999 evaluation.

Table 9 also shows, in {brackets}, the strain values reported in the 2003 evaluation of the 18-in (457 mm) diameter standardized canister for the repository drop events. It can be seen that the peak strains in many of the containment boundary components (lower and upper heads) of the ISFP canister are at or below those for the 18-in (457 mm) diameter 6,000-pound (2,718 kg) standardized canister under repository drop conditions, except for the following:

1. The vertical drop produced strains in the upper head in the 18-in (457 mm) diameter ISFP canister due to the impact plate bending the retaining ring as discussed for the 30-ft (9 m) drop. Such strains were absent in the 18-in (457 mm) diameter standardized canister because it has no impact plate retaining ring. These upper head strains are not considered a containment concern because they are much lower than the maximum strains calculated for the 80-degree drop.
2. The 80 degrees off-vertical drop of this ISFP canister produced a 1% strain increase on the mid-surface of the upper head over the 18-in (457 mm) diameter standardized canister.

Peak strains in the skirts, which were designed to absorb drop energy and are not containment retaining, increased or decreased from the 1999 results depending on the drop angle and location.

Table 10 lists the peak equivalent plastic strains occurring in the various containment boundary and skirt components of the ISFP canisters during the two 30-ft (9 m) drop events for a 4,090-pound (1,853 kg) canister (maximum actual ISFP canister weight). The results show that the peak strains in the lower head for the 6 degrees and 80 degrees off-vertical drop events were the same or lower than those listed in Table 9 for the 6,000-pound (2,718 kg) ISFP canister. However, the peak strains in the upper head were slightly higher than for the 6,000-pound (2,718 kg) canister, even though intuition based on linear analyses would suggest that the strains should be reduced with a lower canister weight (same drop height). However, the drop energy for one drop scenario (drop height, canister weight, impact angle, etc.) may be absorbed by more canister material than in another drop scenario. This results in different peak strains. However, the increases for the upper head were not more than a 3% strain increase.

Several features of the 18-in (457 mm) diameter ISFP canisters that are not included in the standardized canisters have the following impact on the strains resulting from the specified drop events:

1. The ISFP canister impact plate retaining ring on the upper head causes straining in that head during the vertical and near-vertical drop events due to the load from the upper impact plate bearing on it. These strains are the largest containment boundary strains for those drop events. However, because of the increased thickness of the ISFP heads over that used on the standardized canister, those peak strains are still below those reported in the 1999 evaluation for the worst-case drop (slap-down).
2. The ISFP canister impact plate retaining ring on the lower head causes straining in that head during the 80 degrees and 90 degrees off-vertical drop events. However, because of the increased thickness of the ISFP heads over that used on the standardized canister, those strains are not the highest on the head. The highest strains are still located on the head just below the skirt-to-head weld.

3. This ISFP canister uses a fuel basket that is composed of a number of tubes with thick intermediate plates, and a large shield plug. The standardized canister in the 1999 evaluations used a sleeved spoked-wheel basket with no shield plug. It is easy to see the importance of having an internal sleeve for a basket with sharp edges (as the spoked-wheel basket), to keep those edges from directly impacting the containment boundary. The ISFP canister does not use an internal sleeve. The usefulness of an internal sleeve for this ISFP canister is less obvious/questionable since the baskets and shield plugs do not have such sharp edges. However, significant straining did result in the ISFP canister body due to their internal components, but these strain levels were not the highest in the containment boundary.

In many of the drop events, the 18-in (457 mm) diameter 15-ft (4.5 m) long ISFP canister experienced lower material strains than the 18-in (457 mm) diameter standardized canister, due mainly to the thicker heads. However, the 80 degrees off-vertical (slap-down) 30-ft (9 m) drop of the ISFP canister both for the 6,000-pound (2,718 kg) and 4,090-pound (1,853 kg) loaded weights showed higher mid-surface strains (5% to 6% strain increase) in the upper head than the standardized canister. The same occurred for the 90 degrees off-vertical drops, though to a lesser extent. This is somewhat offset by lower strains on the inside and outside surfaces of the upper head. These analytical evaluations indicate that the 1999 standardized canister drop testing worst-case scenario (slap-down) does not entirely envelop the response of the 18-in (457 mm) diameter ISFP canister for this 80 degrees off-vertical drop event.

ISFP canister model #2: The ISFP canister model #2, at a maximum actual weight of 3,648 pounds (1,653 kg), was evaluated for a 30-ft (9 m) drop onto a flat, rigid surface at 80 degrees off-vertical only. Other drop angles were not evaluated because the 15-ft (4.5 m) long canister design (Model #1) was expected to produce enveloping results for those impact angles. The information reported from this drop evaluation is limited to: model deformations, model energy plot, and canister component strains. The condition of the internal components was not of interest in this evaluation. Therefore, the reported results include only those that relate to the containment boundary components and the canister skirts.

Because of the design of the internals (baskets, spacers, and shield plug), this canister appeared to have the greatest potential of having a center-of-gravity away from the geometric center (centroid) of the canister. Using the actual geometry of the canister and internals, with a uniformly-distributed fuel mass assumption, it was estimated that the loaded center-of-gravity was located about 5 in (126.9 mm) axially above the canister geometric center (64 in [1624.9 mm] above the canister bottom). This is within the 24-in (609 mm) allowance specified for the standardized canisters.

Figures 29 and 30 show the deformed shape of the canister as it was rebounding off of the rigid surface. This deformed shape is consistent with that obtained in the 1999 drop testing efforts.

Figure 31 shows a plot of the energy history for this 80 degrees off-vertical drop scenario. The plot shows model kinetic energy history (ALLKE); plastic dissipation history (ALLPD); frictional dissipation history (ALLFD); elastic energy history (ALLSE); and artificial energy history (ALLAE). When the artificial energy increases above 5%, a rerun of the model with a higher drop value is recommended to compensate for that artificial energy. Figure 31 shows a maximum of about 5% artificial energy in the model, indicating that the results are not invalidated by artificial energy concerns.

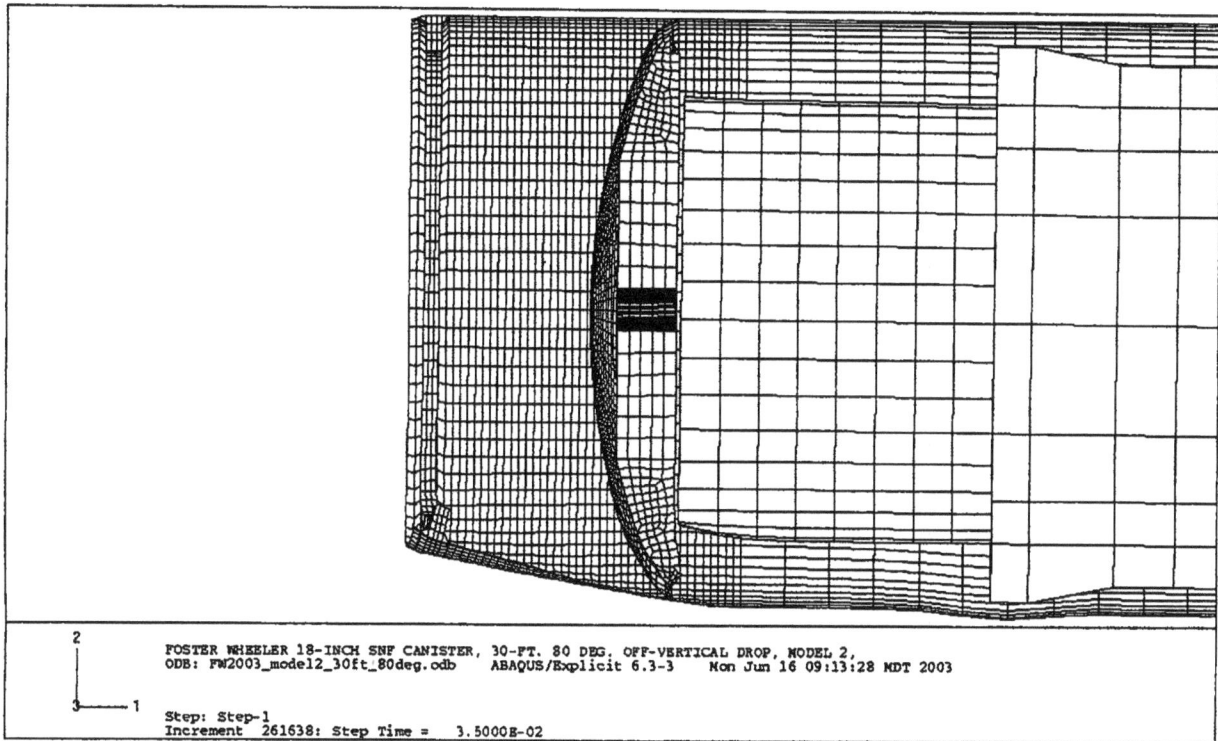

Fig. 29. Model #2 bottom end canister deformations, 80 degrees off-vertical, 30-foot (9-m) drop, 3648 pounds (1653 kg) (Snow and Morton 2003).

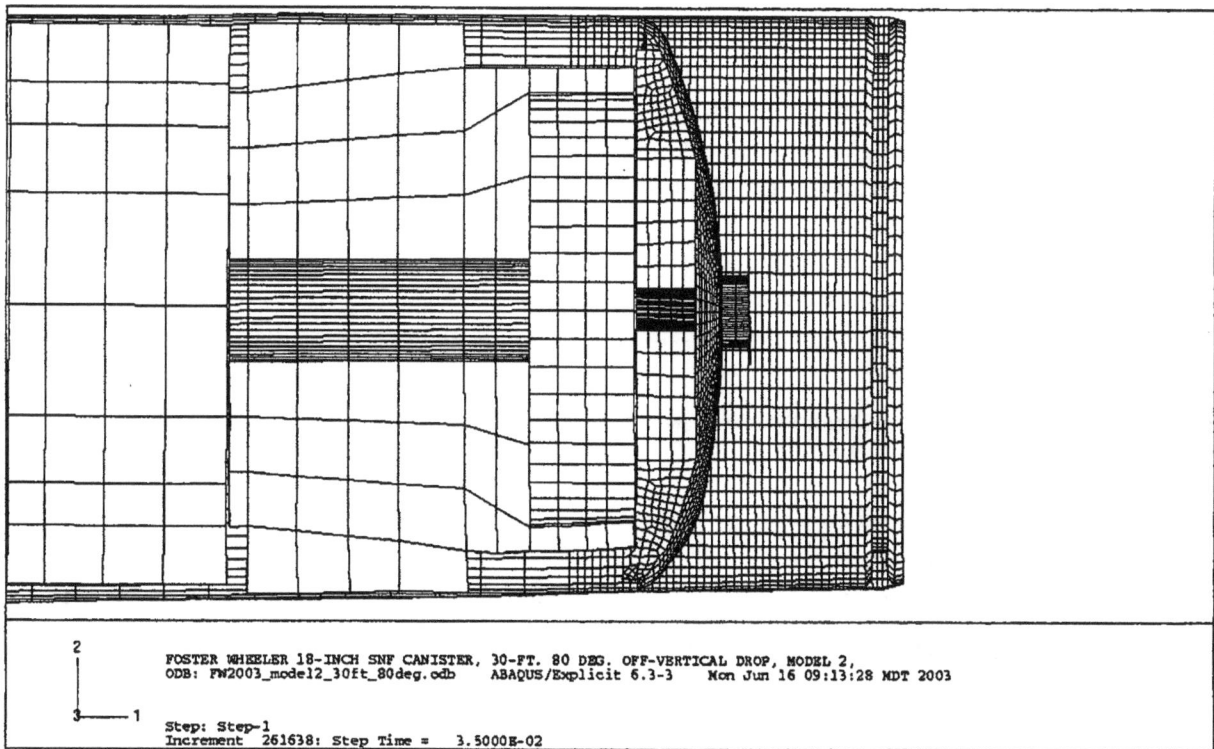

Fig. 30. Model #2 top end canister deformations, 80 degrees off-vertical, 30-foot (9-m) drop, 3648 pounds (1653 kg) (Snow and Morton 2003).

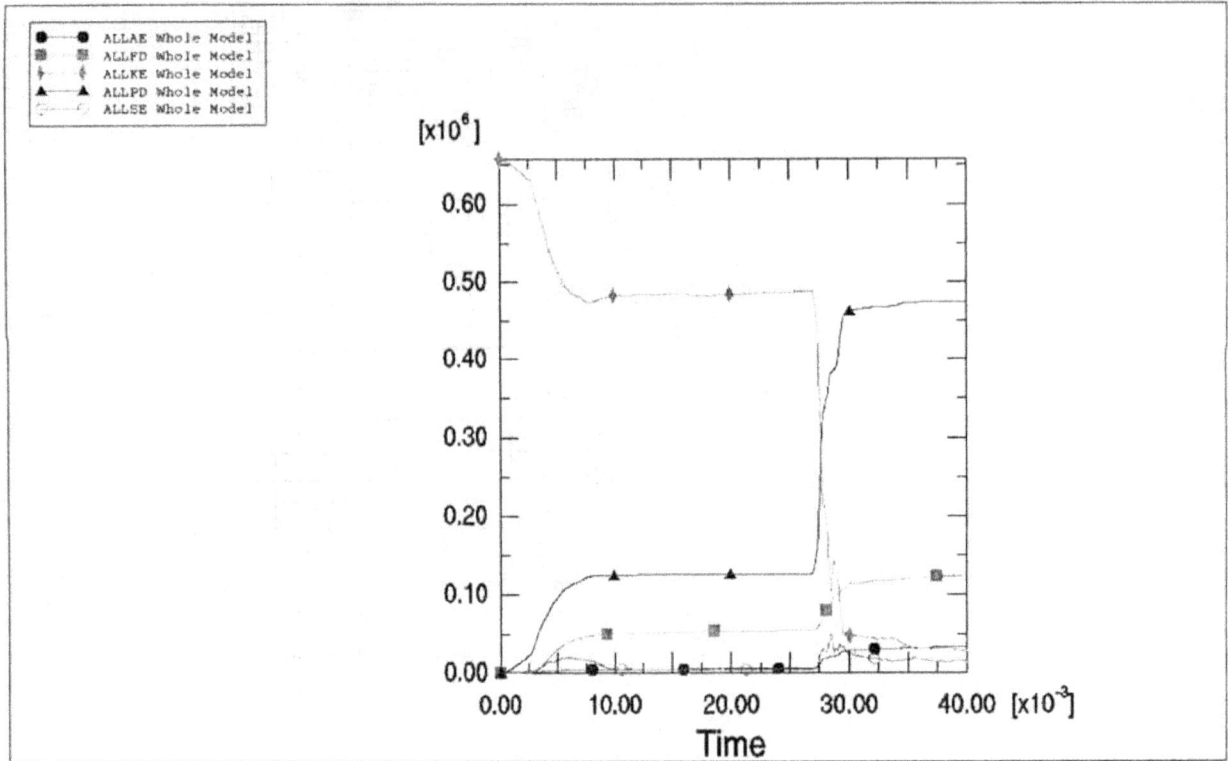

Fig. 31. Model #2 energy history, 80 degrees off-vertical, 30-foot (9-m) drop, 3648 pounds (1653 kg) (Snow and Morton 2003).

Table 11 lists the peak equivalent plastic strains occurring in the various containment boundary and skirt components of this ISFP canister during the specified drop event. Peak strains are listed for the outside surface of a component, mid-plane, and inside surface. However, these peak strains did not necessarily occur at the same location (element) through the thickness.

The maximum containment boundary strains occur in the upper head of this canister. Those peak strain levels are almost identical to those reported in Table 9 for the model #1 canister upper head under a 30-ft (9 m) 80-degree drop event. Strains in other components are also comparable between those two canister drops.

Table 11. Peak plastic strains in Foster Wheeler 18-inch (457-mm) diameter, 10-foot (3-m) long canister, 3,648 pounds (1,653 kg) maximum actual weight** (Snow and Morton 2003).

Drop Height		Angle From Vertical (Degrees)	Peak Equivalent Plastic Strain (PEEQ) %														
			Lower Head			Lower Skirt			Upper Head			Upper Skirt			Body		
(ft)	(m)		out*	mid*	in*	out*	mid*	in*	out*	mid	in*	out*	mid*	in*	out*	mid*	in*
30	9	80 t = 0.035	40	18	24	26	15	21	50	25	24	15	9	9	9	4	6

* out = outside surface of component, mid = middle surface of component, in = inside surface of component.
** Loaded center-of-gravity is estimated to be 5 inches (12.7 cm) above the geometric center of the canister, or 64 inches (1.62 m) above the bottom.

This analytical evaluation indicates that the 1999 standardized canister drop testing worst-case scenario (slap-down orientation) does not entirely envelop the response of the 18-in (457 mm) diameter ISFP 10-ft (3 m) long canister for this 80 degrees off-vertical drop event. This is the same conclusion reached for the 15-ft (4.5 m) long ISFP canister.

ISFP canister model #3: The ISFP canister model #3, at a maximum design weight of 10,000 pounds (4,530 kg), was evaluated for a 30-ft (9 m) drop onto a flat, rigid surface at five drop angles: 1) vertical; 2) 7 degrees off-vertical (center-of-gravity-over-corner); 3) 45 degrees off-vertical; 4) 70 degrees off-vertical; and 5) 90 degrees off-vertical (horizontal). This model was also evaluated for a 23-ft (6.9 m) repository vertical drop and a 2-ft (0.6 m) worst angle (70 degrees off-vertical) repository drop. Additionally, the model was evaluated at a minimum actual loaded weight of 8,280 pounds (3,751 kg) for the 30-ft (9 m) 7 degrees off-vertical and 30-ft (9 m) 70 degrees off-vertical drops as well. This is shown in Table 12.

Table 12. ISFP canister model #3 drop conditions (Snow and Morton 2003).

Model File Name	Drop Height		Drop Angle from Vertical (Degrees)
	(ft)	(m)	
10,000-Pound (4,530-kg) Canister Weight			
model3_30ft_vertical	30	9	0
model3_30ft_7deg	30	9	7
model3_30ft_45deg	30	9	45
model3_30ft_70deg	30	9	70
model3_30ft_90deg	30	9	90
model3_23ft_vertical	23	6.9	0
model3_2ft_70deg	2	0.6	70
8,280-Pound (3,751-kg) Canister Weight			
model3_30ft_7deg_8280lbs	30	9	7
model3_30ft_70deg_8280lbs	30	9	70

The information reported from these drop evaluations is limited to: model deformations; model energy plots; and canister component strains. The condition of the internal components was not of interest in this evaluation. Therefore, the reported results will include only those that relate to the containment boundary components and the canister skirts.

Figures 32 through 44 show the deformed shape of the canisters as they were rebounding off of the rigid surface. These deformed shapes are consistent with those obtained in the 1999 drop testing efforts (Morton et al. 2000) and the 2003 24-in (610 mm) diameter standardized canister evaluation (Blandford 2003).

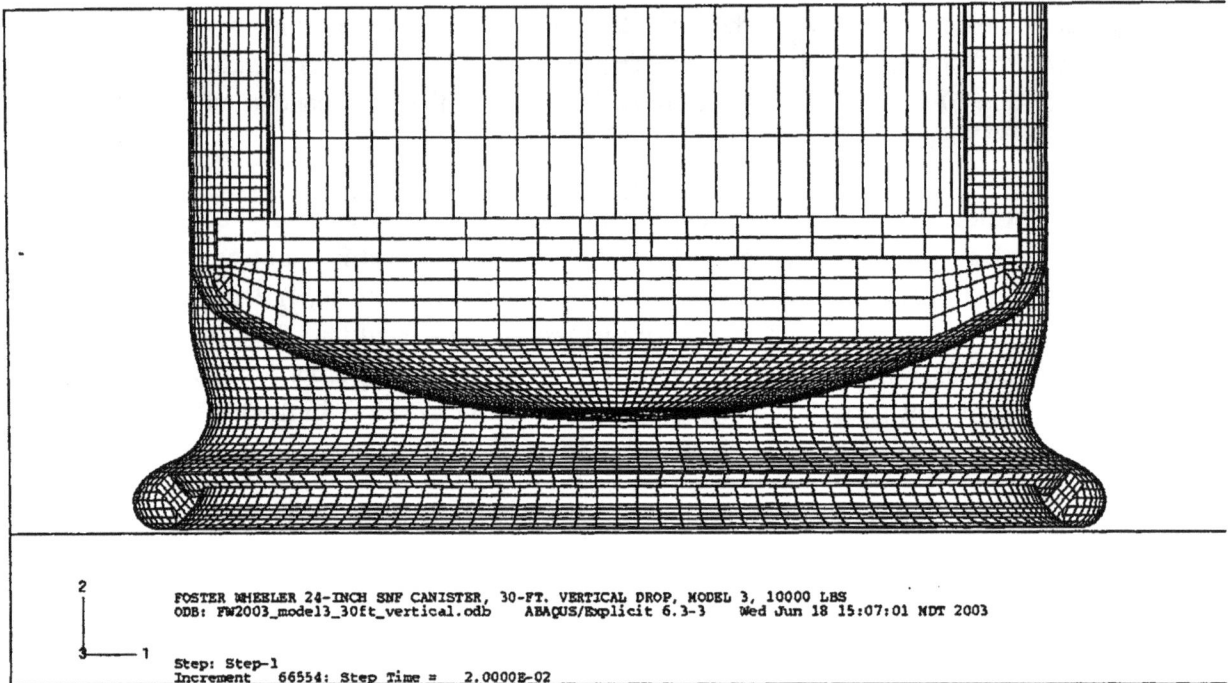

Fig. 32. Model #3 canister deformations, vertical, 30-foot (9-m) drop, 10000 pounds (4530 kg) (Snow and Morton 2003).

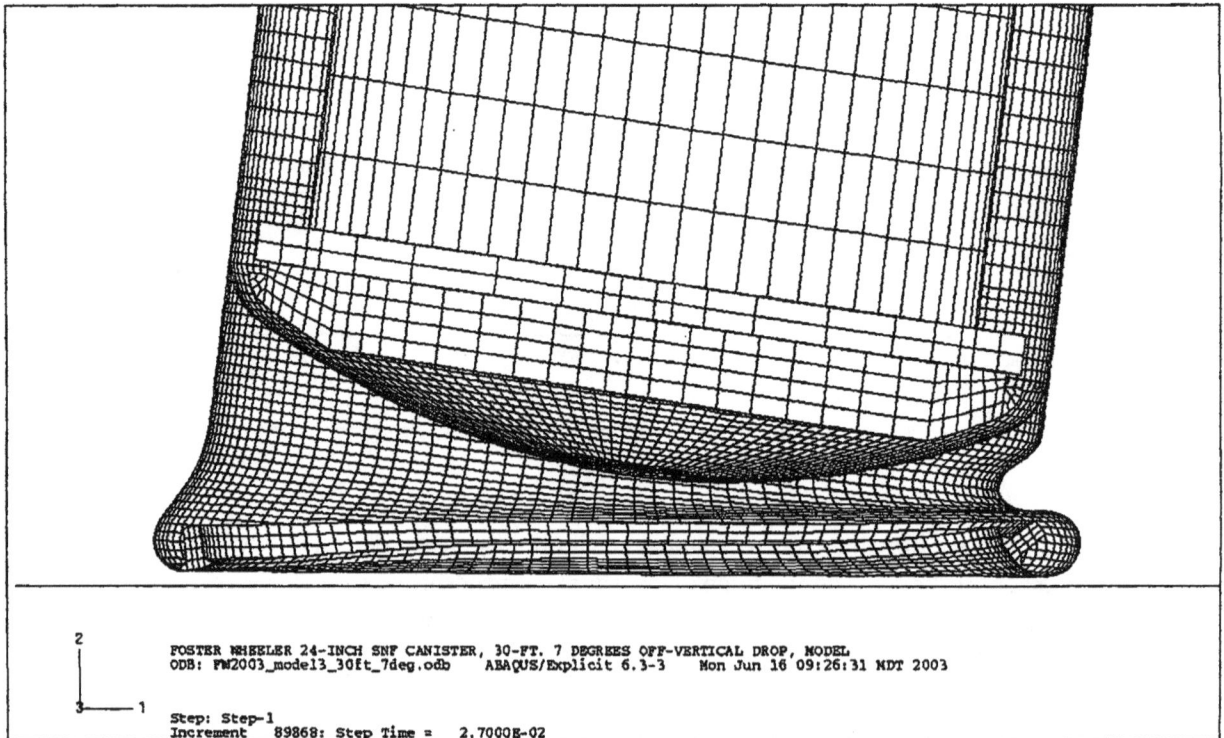

Fig. 33. Model #3 canister deformations, 7 degrees off-vertical, 30-foot (9-m) drop, 10000 pounds (4530 kg) (Snow and Morton 2003).

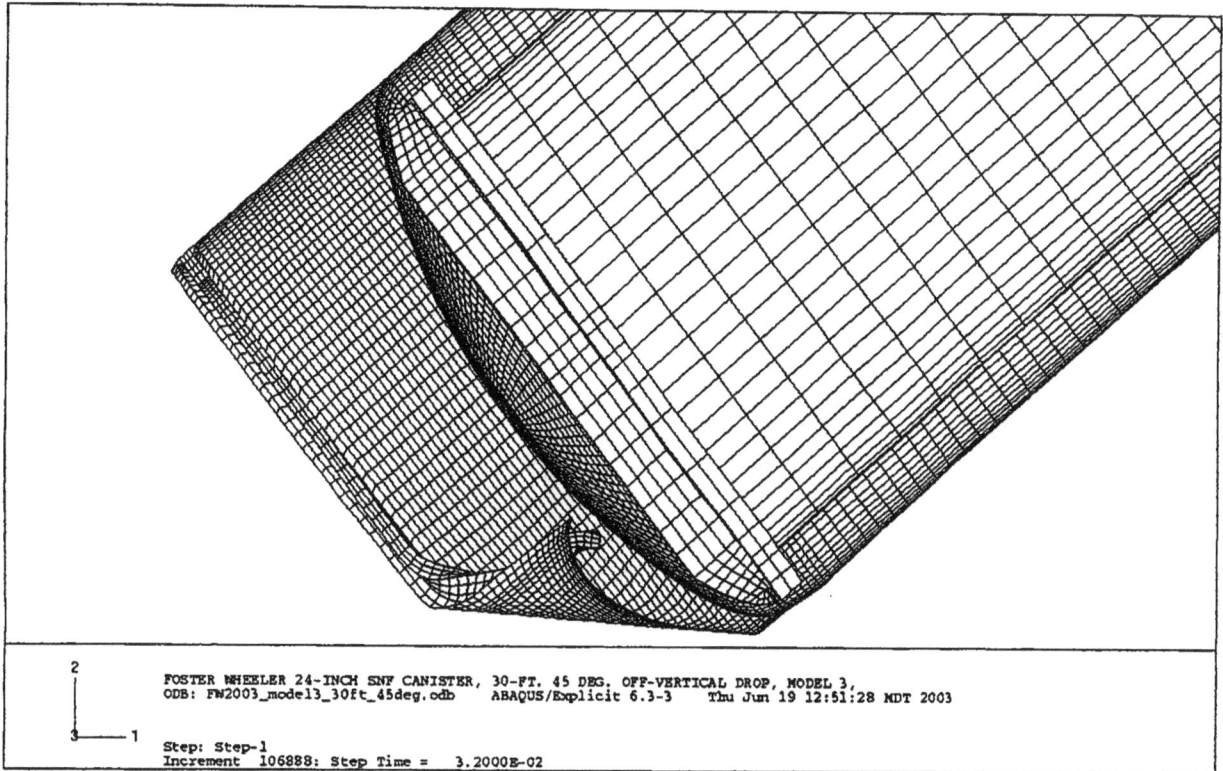

Fig. 34. Model #3 canister deformations, 45 degrees off-vertical, 30-foot (9-m) drop, 10000 pounds (4530 kg) (Snow and Morton 2003).

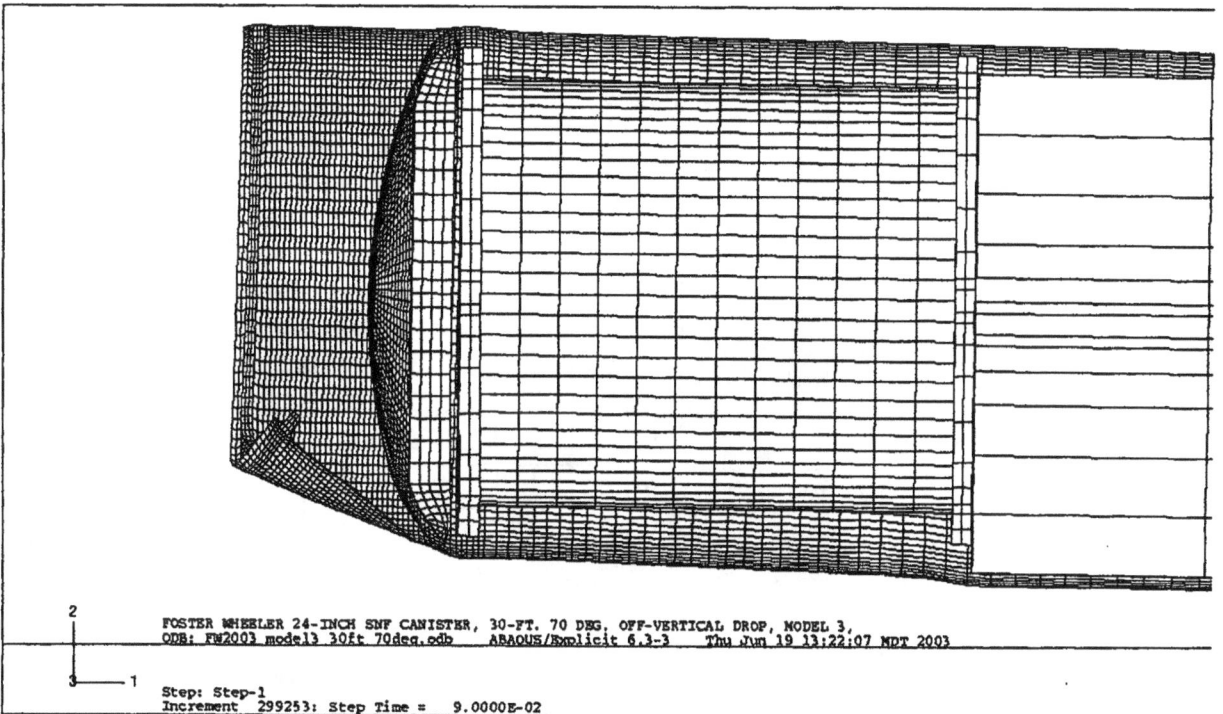

Fig. 35. Model #3 canister bottom end deformations, 70 degrees off-vertical, 30-foot (9-m) drop, 10000 pounds (4530 kg) (Snow and Morton 2003).

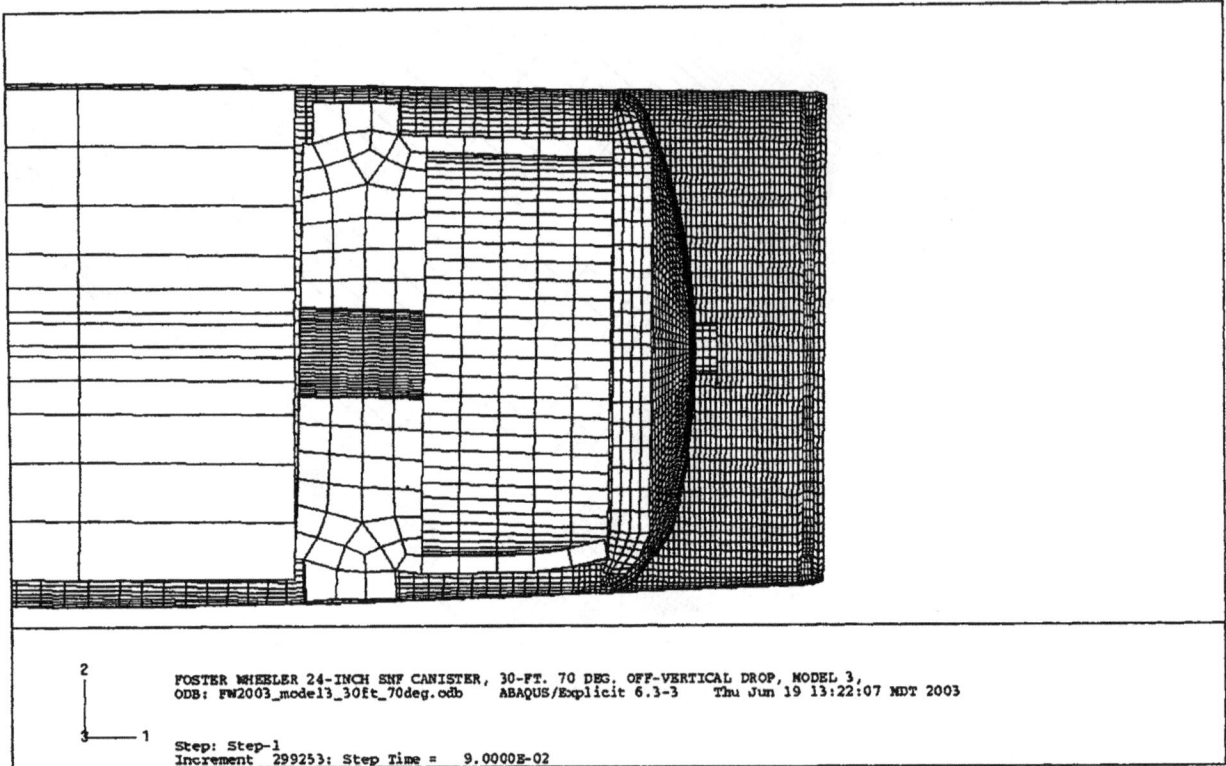

Fig. 36. Model #3 canister top end deformations, 70 degrees off-vertical, 30-foot (9-m) drop, 10000 pounds (4530 kg) (Snow and Morton 2003).

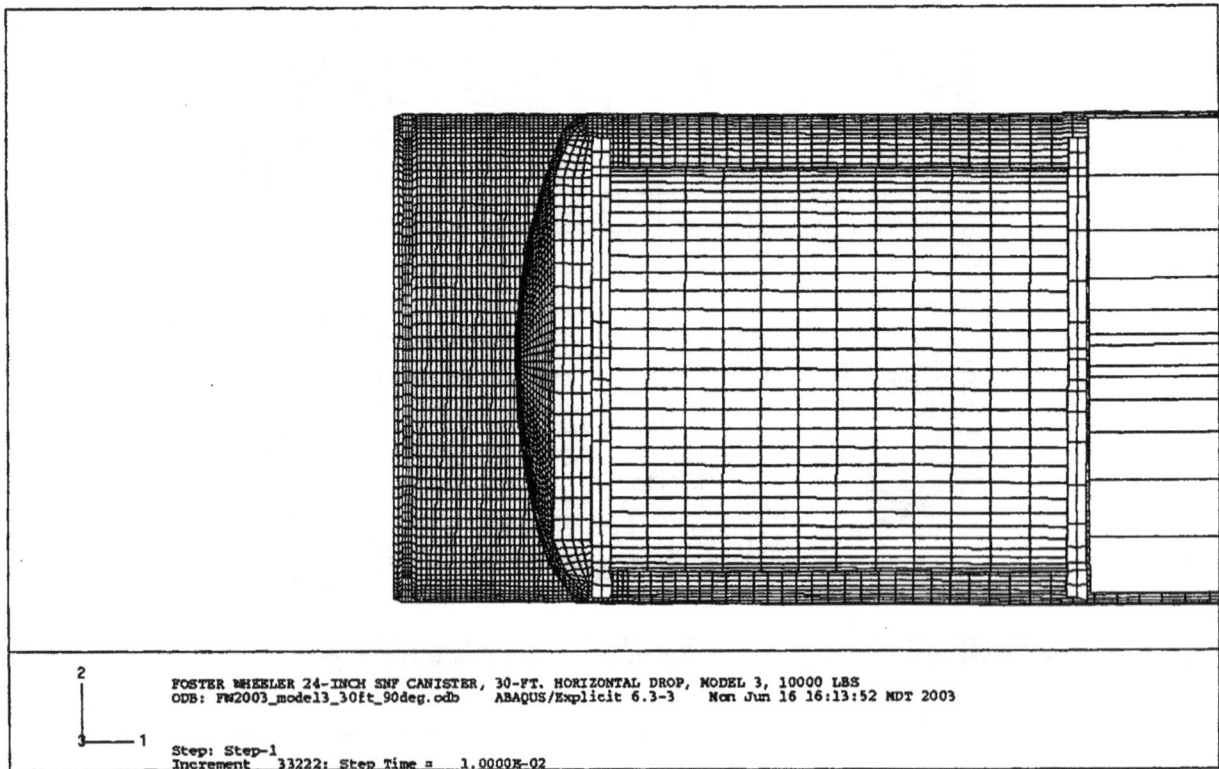

Fig. 37. Model #3 canister bottom end deformations, 90 degrees off-vertical, 30-foot (9-m) drop, 10000 pounds (4530 kg) (Snow and Morton 2003).

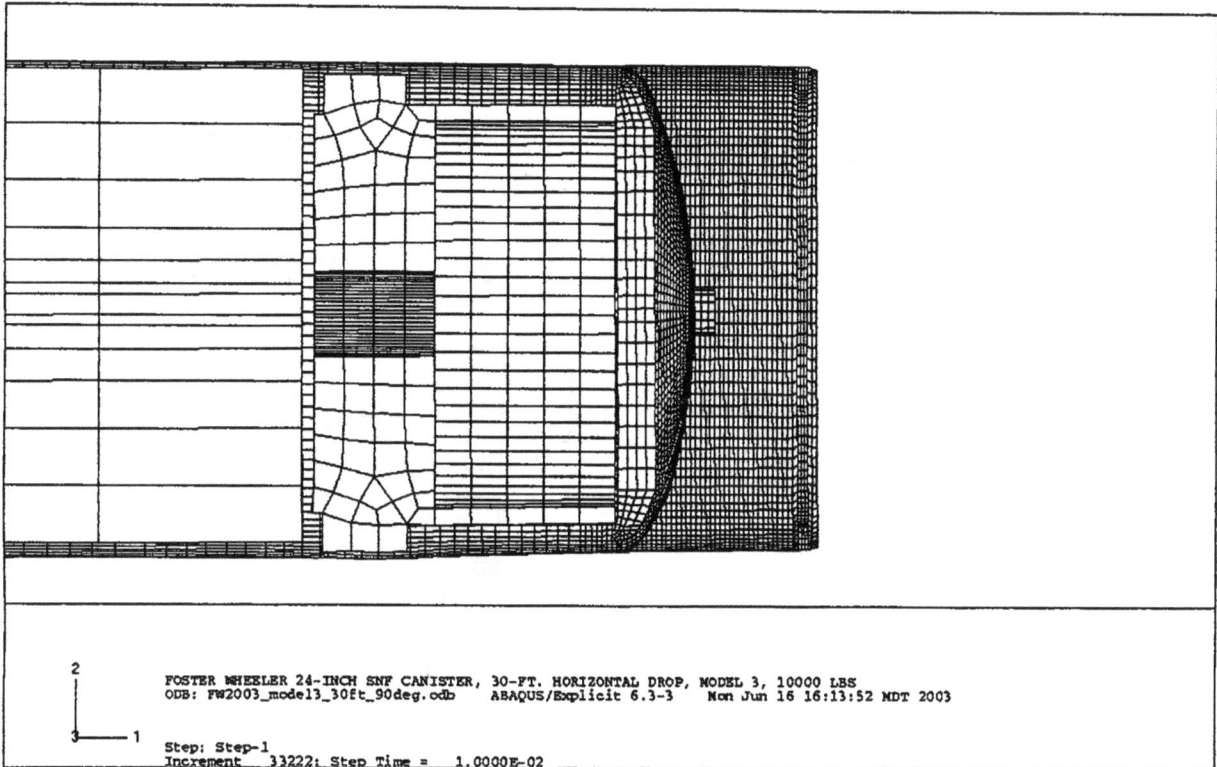

FOSTER WHEELER 24-INCH SNF CANISTER, 30-FT. HORIZONTAL DROP, MODEL 3, 10000 LBS
ODB: FW2003_model3_30ft_90deg.odb   ABAQUS/Explicit 6.3-3   Mon Jun 16 16:13:52 MDT 2003

Step: Step-1
Increment   33222: Step Time =   1.0000E-02

Fig. 38. Model #3 canister top end deformations, 90 degrees off-vertical, 30-foot (9-m) drop, 10000 pounds (4530 kg) (Snow and Morton 2003).

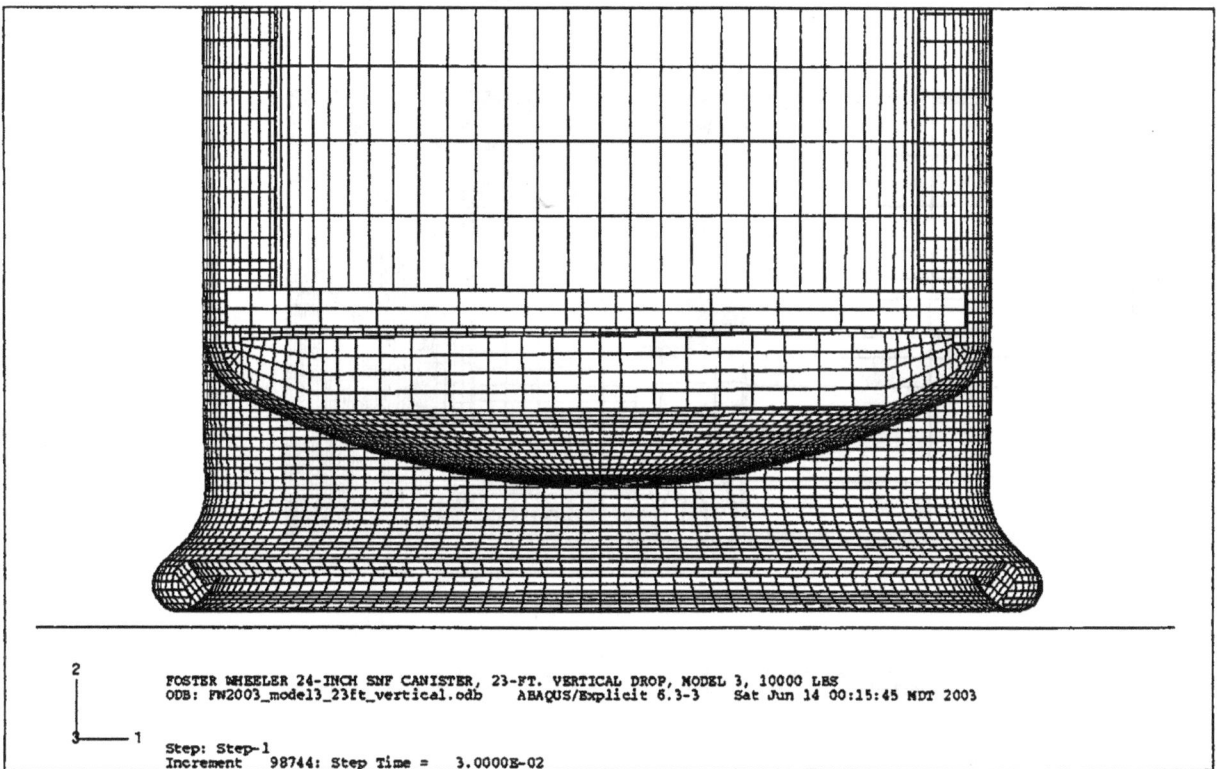

FOSTER WHEELER 24-INCH SNF CANISTER, 23-FT. VERTICAL DROP, MODEL 3, 10000 LBS
ODB: FW2003_model3_23ft_vertical.odb   ABAQUS/Explicit 6.3-3   Sat Jun 14 00:15:45 MDT 2003

Step: Step-1
Increment   98744: Step Time =   3.0000E-02

Fig. 39. Model #3 canister deformations, vertical, 23-foot (6.9-m) drop, 10000 pounds (4530 kg) (Snow and Morton 2003).

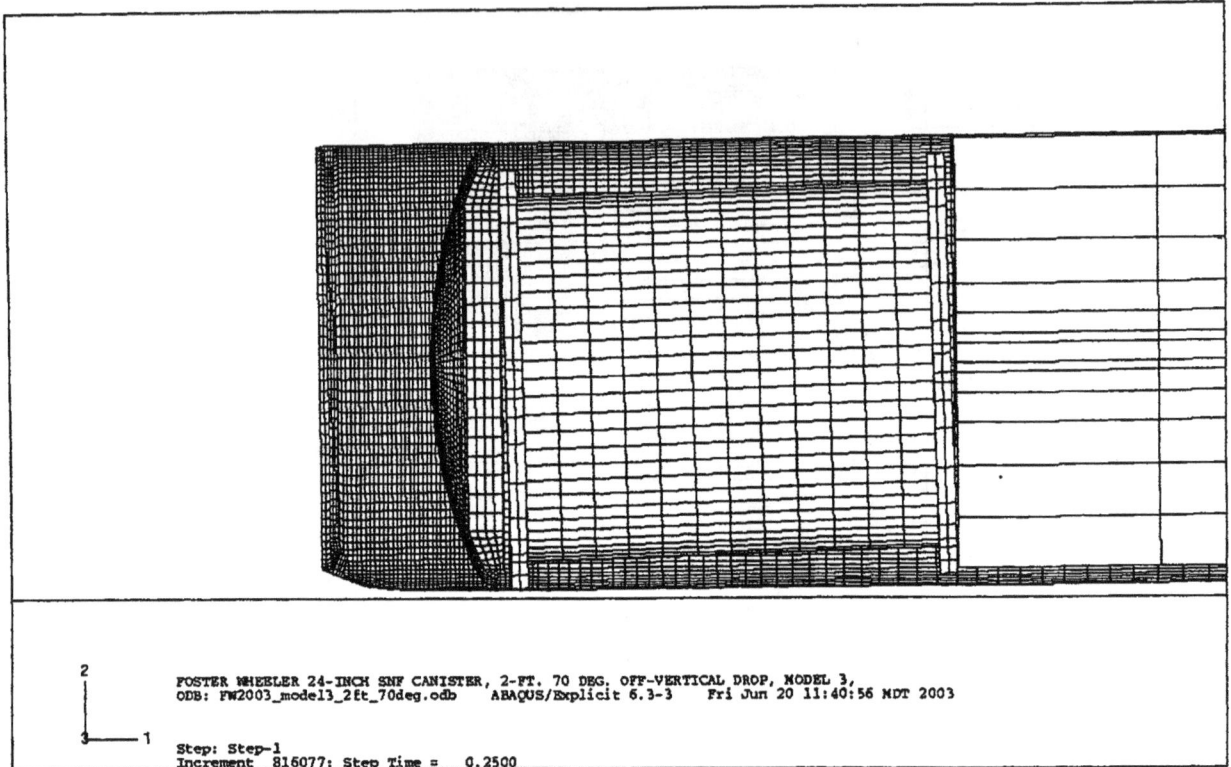

Fig. 40. Model #3 canister bottom end deformations, 70 degrees off-vertical, 2-foot (0.6-m) drop, 10000 pounds (4530 kg) (Snow and Morton 2003).

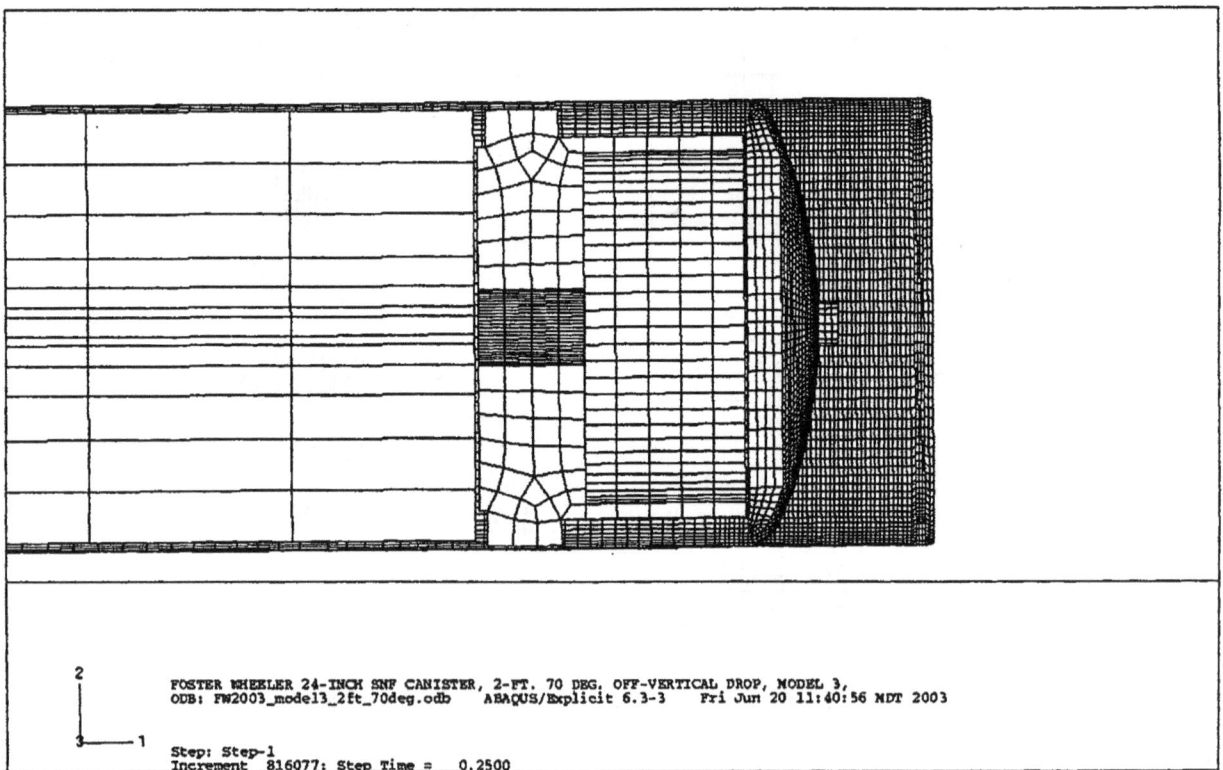

Fig. 41. Model #3 canister top end deformations, 70 degrees off-vertical, 2-foot (0.6-m) drop, 10000 pounds (4530 kg) (Snow and Morton 2003).

94

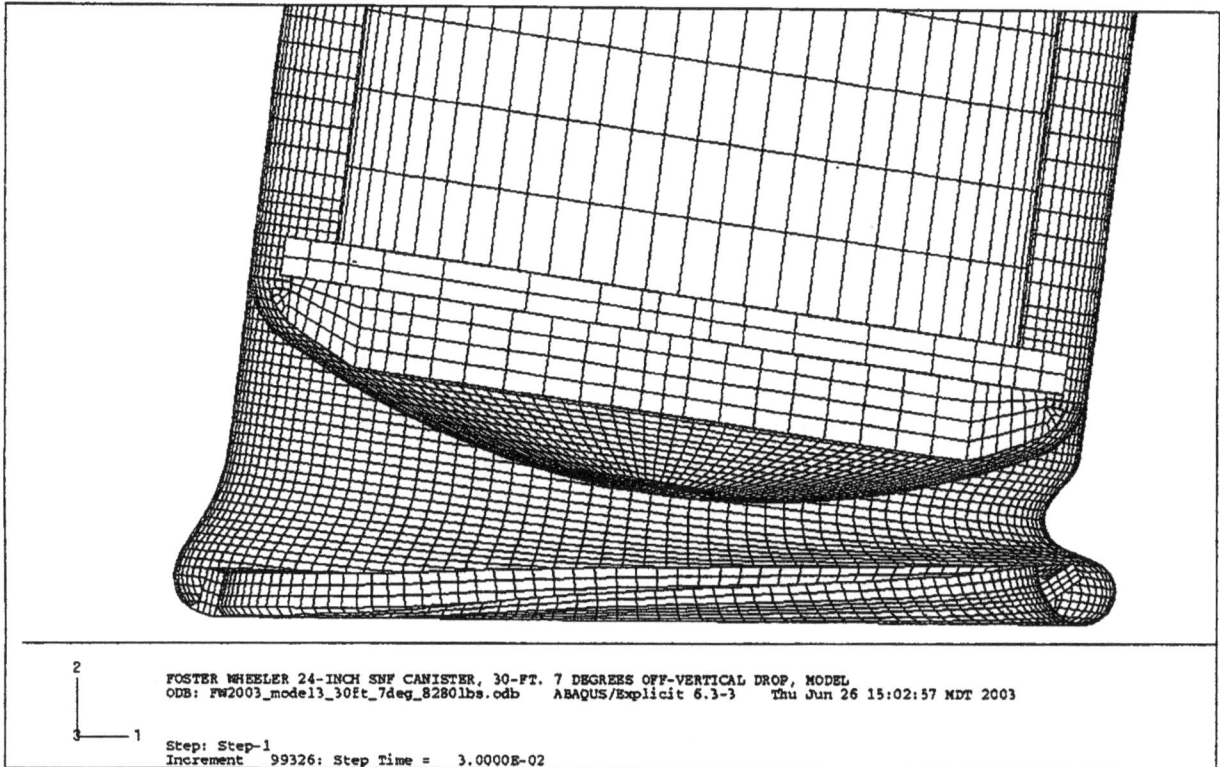

Fig. 42. Model #3 canister deformations, 7 degrees off-vertical, 30-foot (9-m) drop, 8280 pounds (3751 kg) (Snow and Morton 2003).

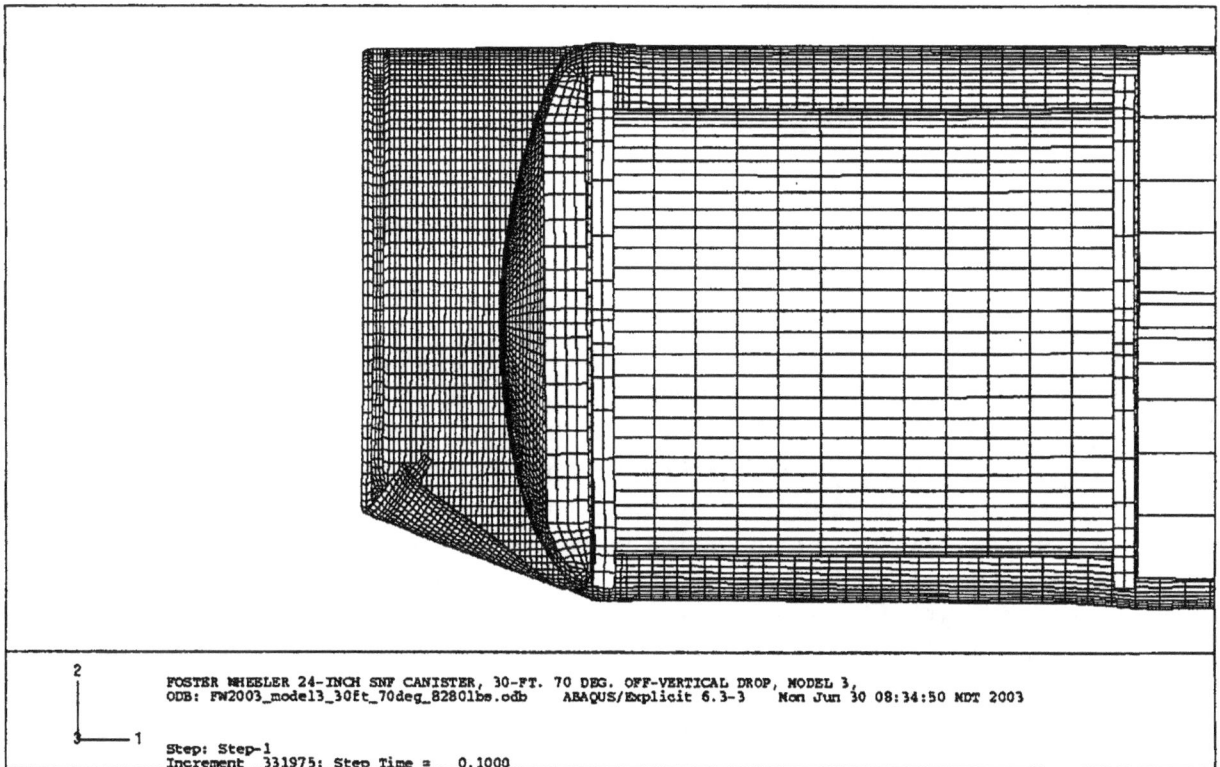

Fig. 43. Model #3 canister bottom end deformations, 70 degrees off-vertical, 30-foot (9-m) drop, 8280 pounds (3751 kg) (Snow and Morton 2003).

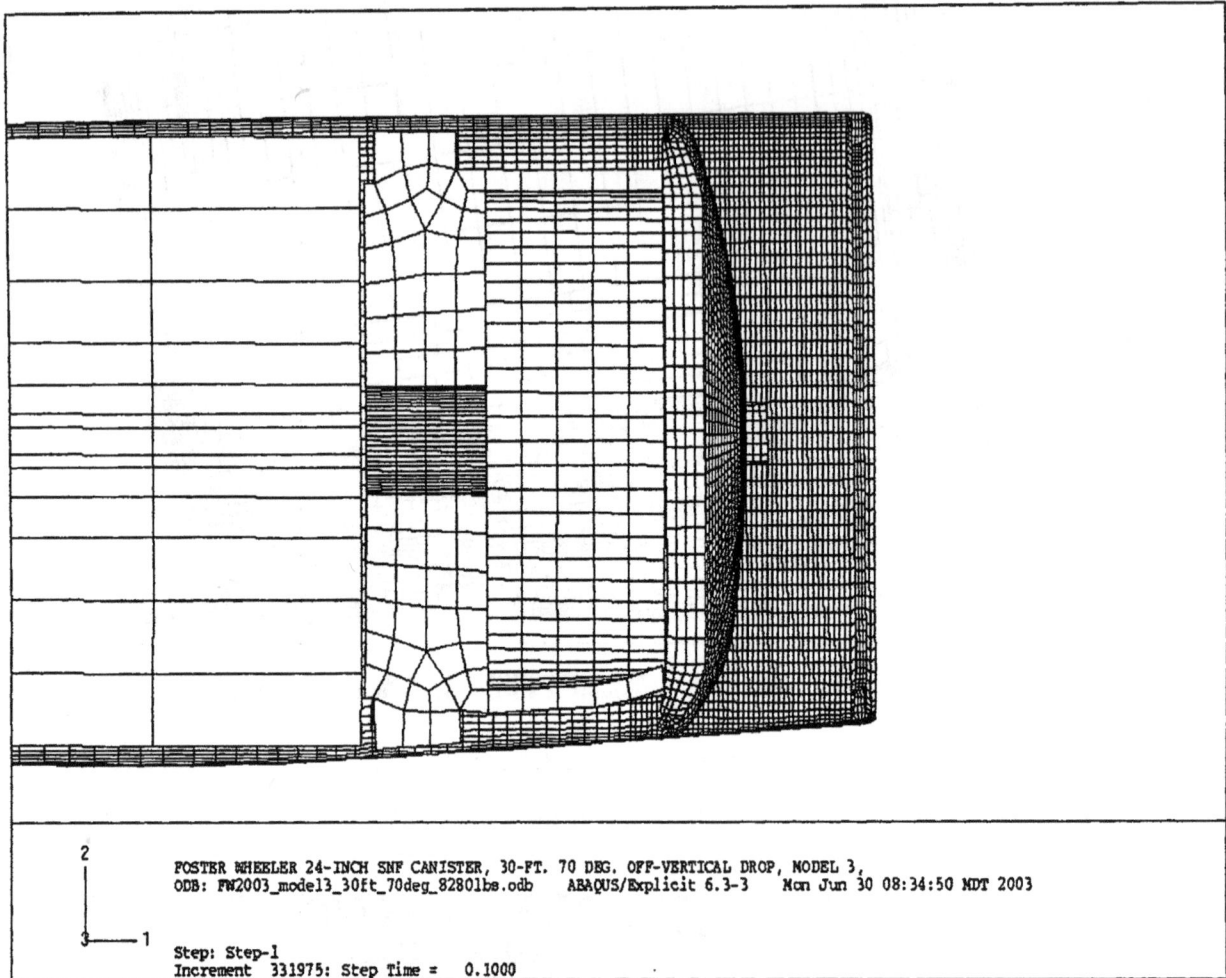

Fig. 44. Model #3 canister top end deformations, 70 degrees off-vertical, 30-foot (9-m) drop, 8280 pounds (3751 kg) (Snow and Morton 2003).

Figures 45 through 53 show plots of the energy history for each drop scenario. The plots show model kinetic energy history (ALLKE), plastic dissipation history (ALLPD), frictional dissipation history (ALLFD), elastic energy history (ALLSE), and artificial energy history (ALLAE). When the artificial energy increases above 5%, a rerun of the model with a higher drop value is recommended to compensate for that artificial energy. As shown in Figure 48, one of the canisters weighing 10,000-pound (4,530 kg) dropped from 30 ft (9 m) at 70 degrees off-vertical, experienced a 14% artificial energy. A second canister weighing 8,280-pound (3,751 kg) dropped 30 ft (9 m) at 70 degrees off-vertical, showed an 11% artificial energy (Fig. 53). The former canister model was rerun with an increase in drop height to account for that artificial energy. The results appeared to be unchanged or very small in deformations, while peak material strains (to be discussed next) were unchanged or slightly higher at all locations except on the outside surface of the upper head that saw an 8% strain increase. The 8,280-pound (3,751 kg) canister model was also rerun with a 10% higher drop. The results showed unchanged or slightly higher material peak strains.

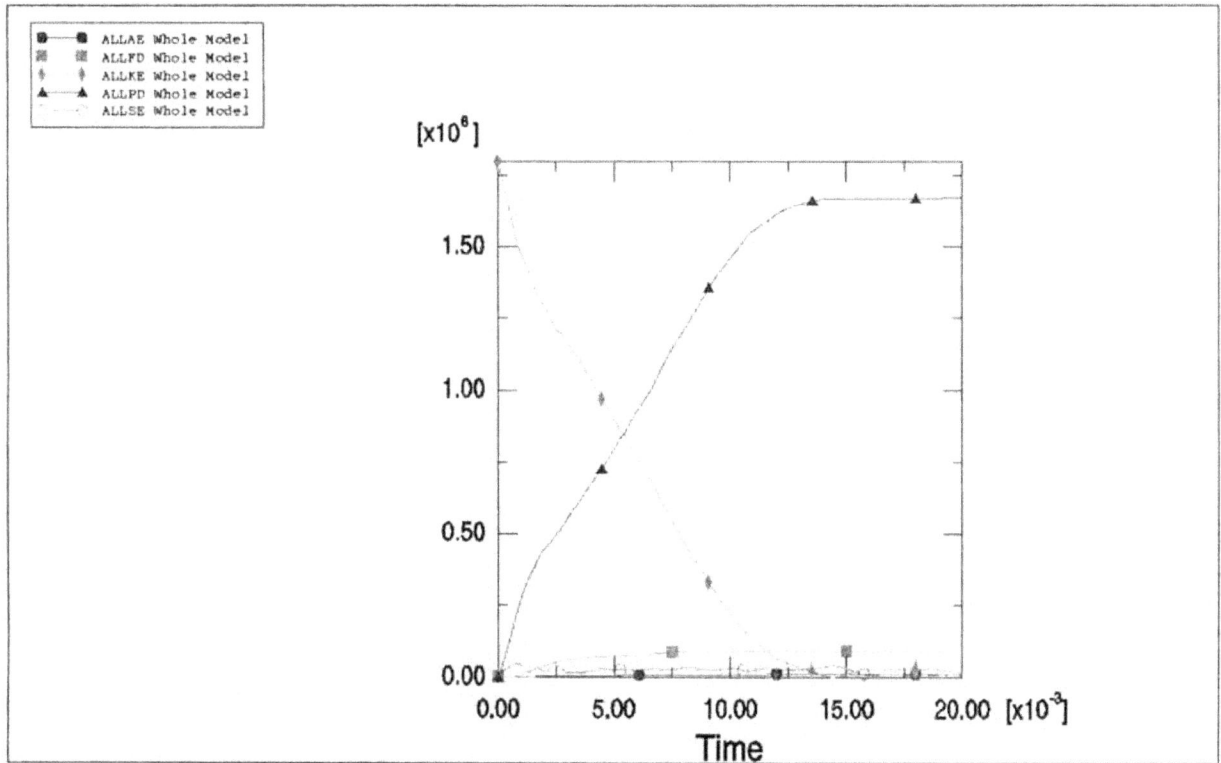

Fig. 45. Model #3 energy history, vertical, 30-foot (9-m) drop, 10000 pounds (4530 kg) (Snow and Morton 2003).

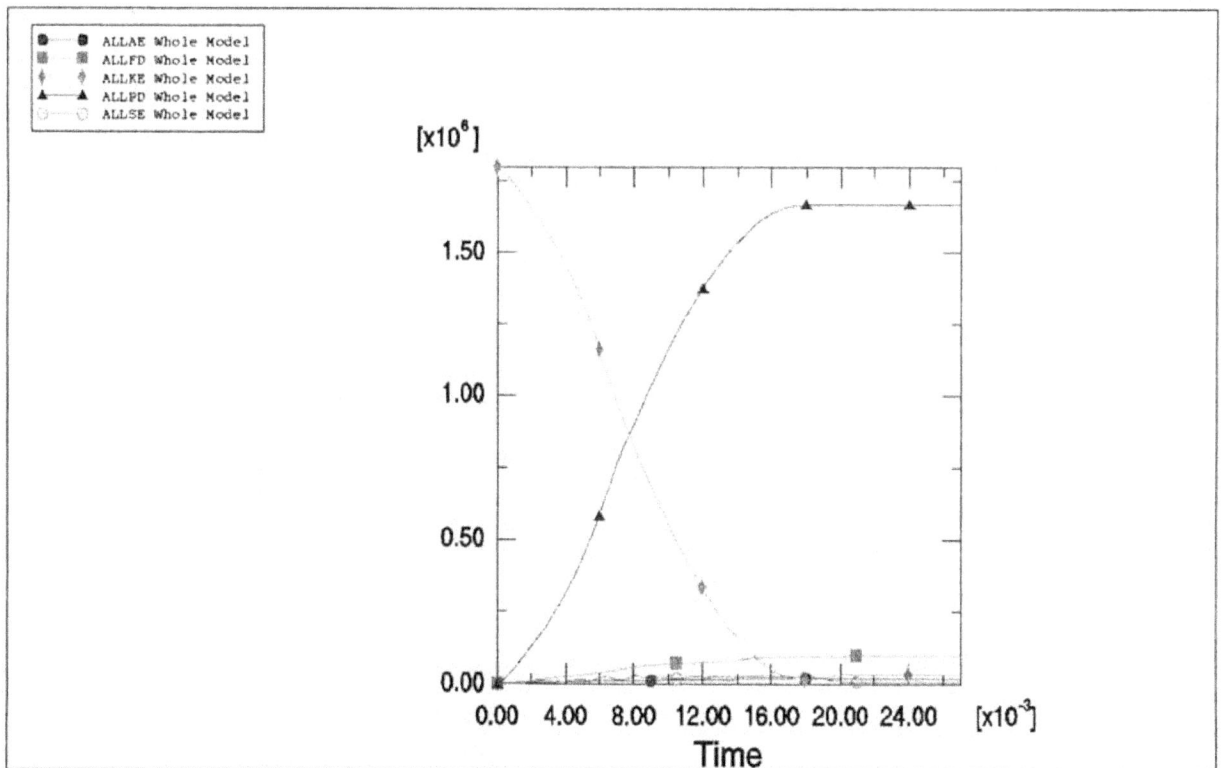

Fig. 46. Model #3 energy history, 7 degrees off-vertical, 30-foot (9-m) drop, 10000 pounds (4530 kg) (Snow and Morton 2003).

97

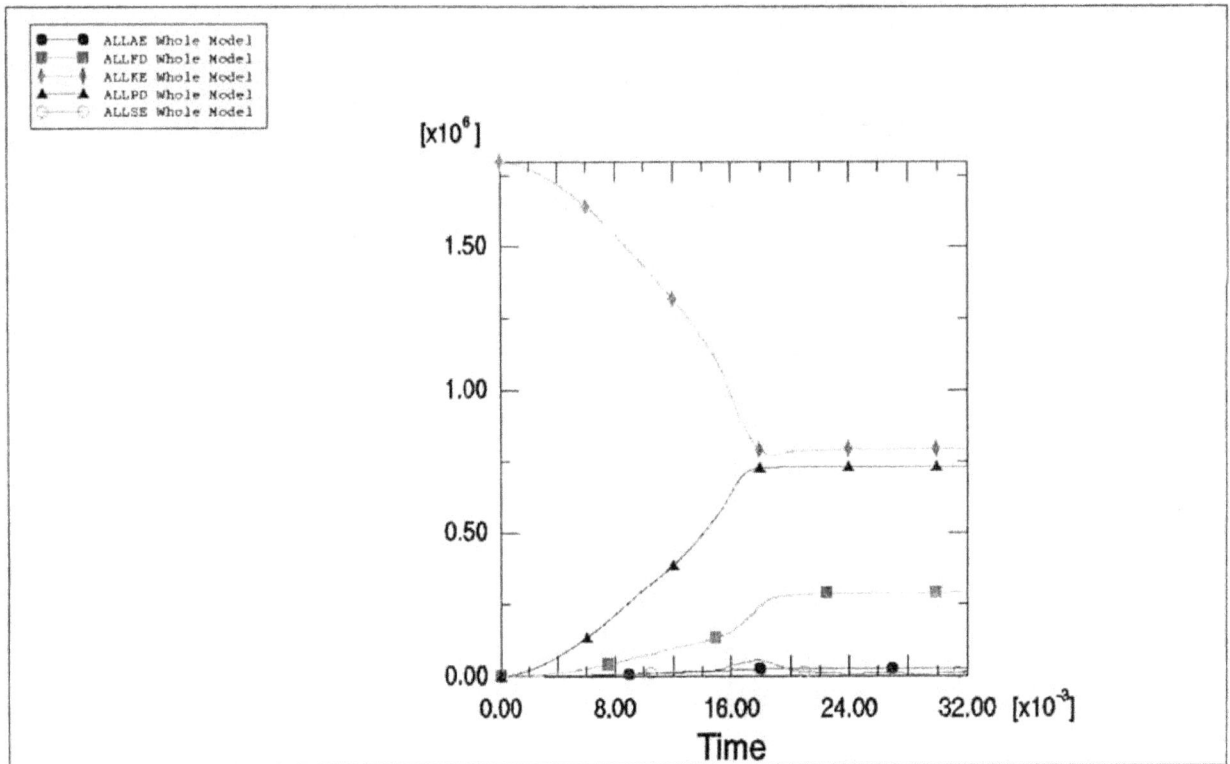

Fig. 47. Model #3 energy history, 45 degrees off-vertical, 30-foot (9-m) drop, 10000 pounds (4530 kg) (Snow and Morton 2003).

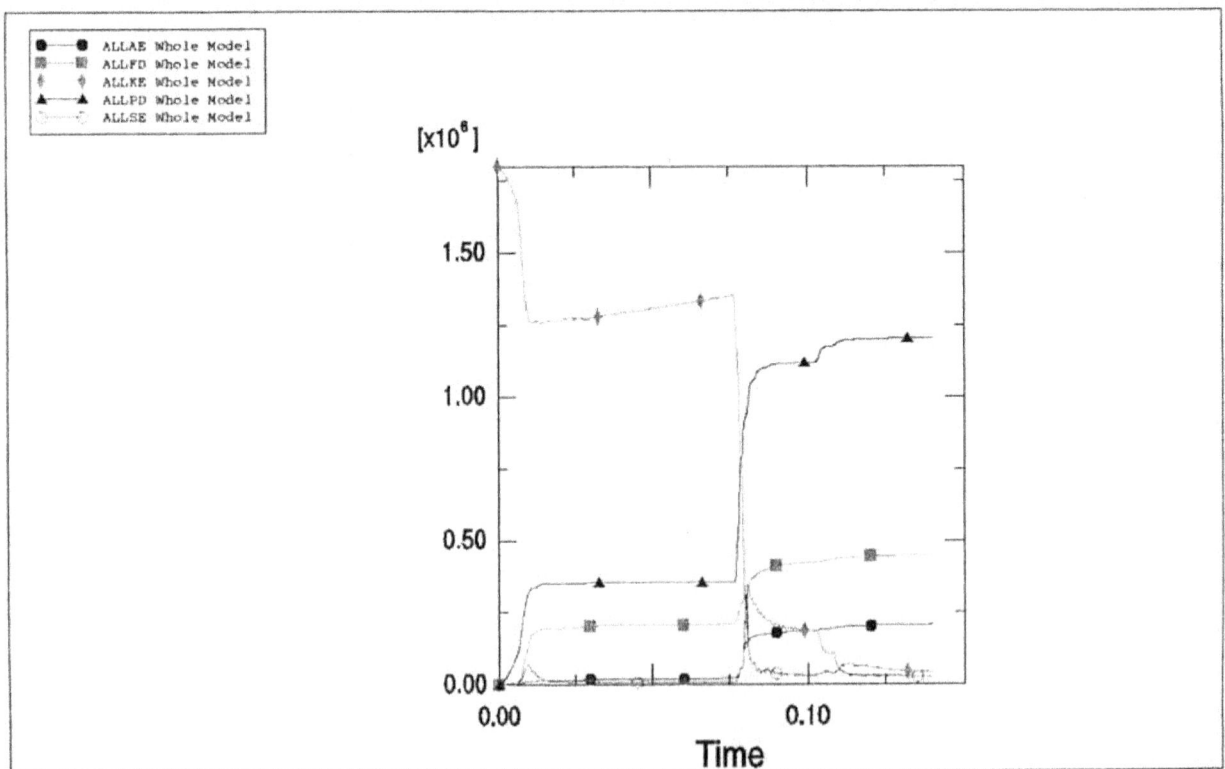

Fig. 48. Model #3 energy history, 70 degrees off-vertical, 30-foot (9-m) drop, 10000 pounds (4530 kg) (Snow and Morton 2003).

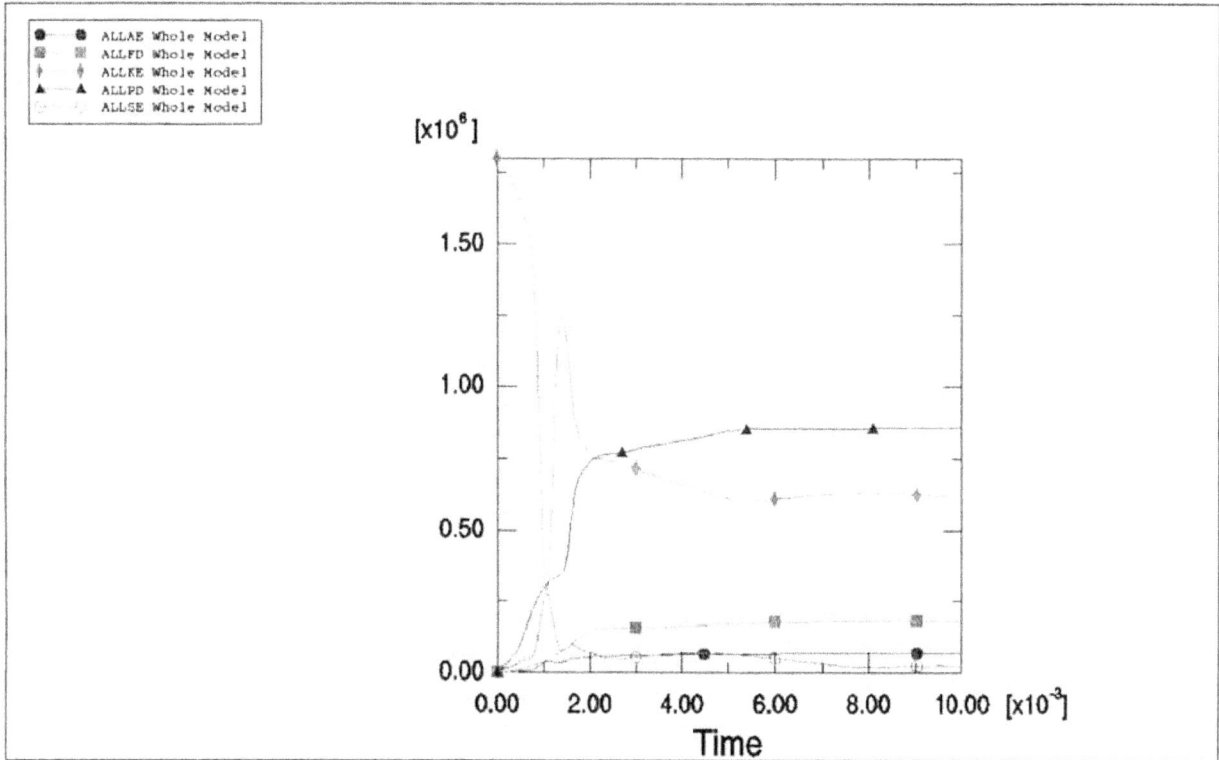

Fig. 49. Model #3 energy history, 90 degrees off-vertical, 30-foot (9-m) drop, 10000 pounds (4530 kg) (Snow and Morton 2003).

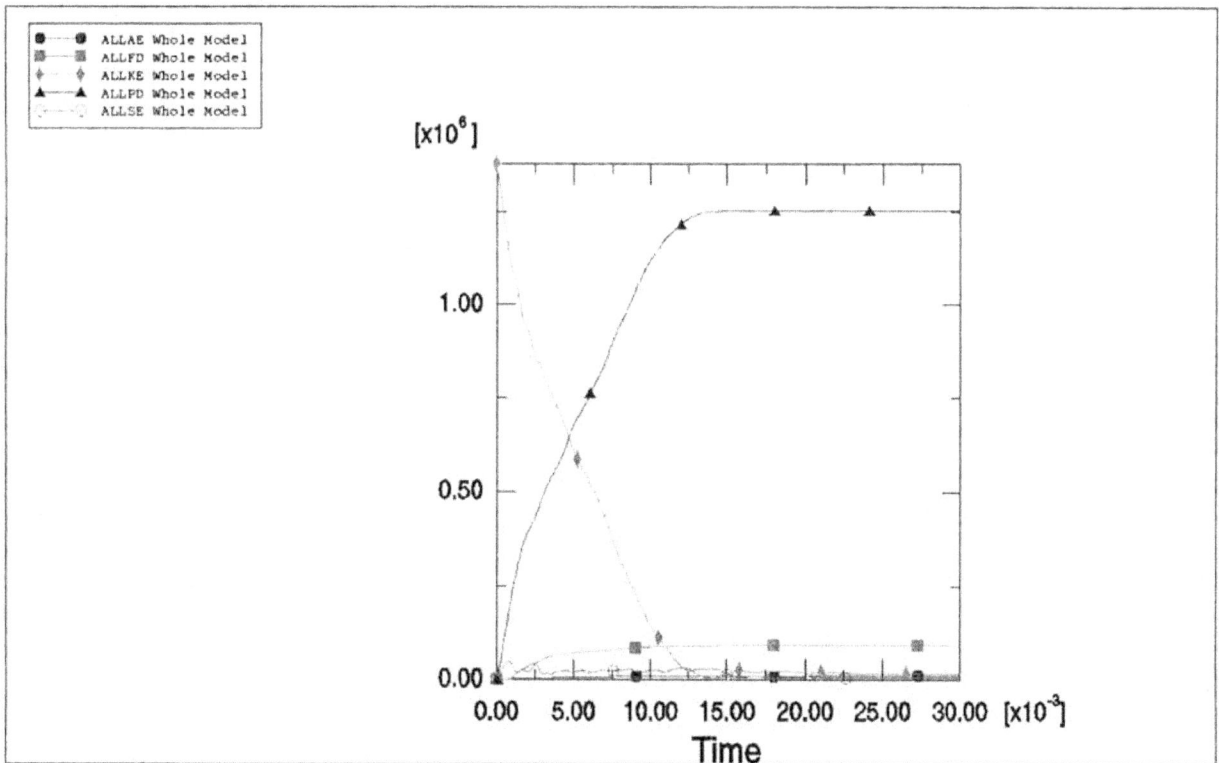

Fig. 50. Model #3 energy history, vertical, 23-foot (6.9-m) drop, 10000 pounds (4530 kg) (Snow and Morton 2003).

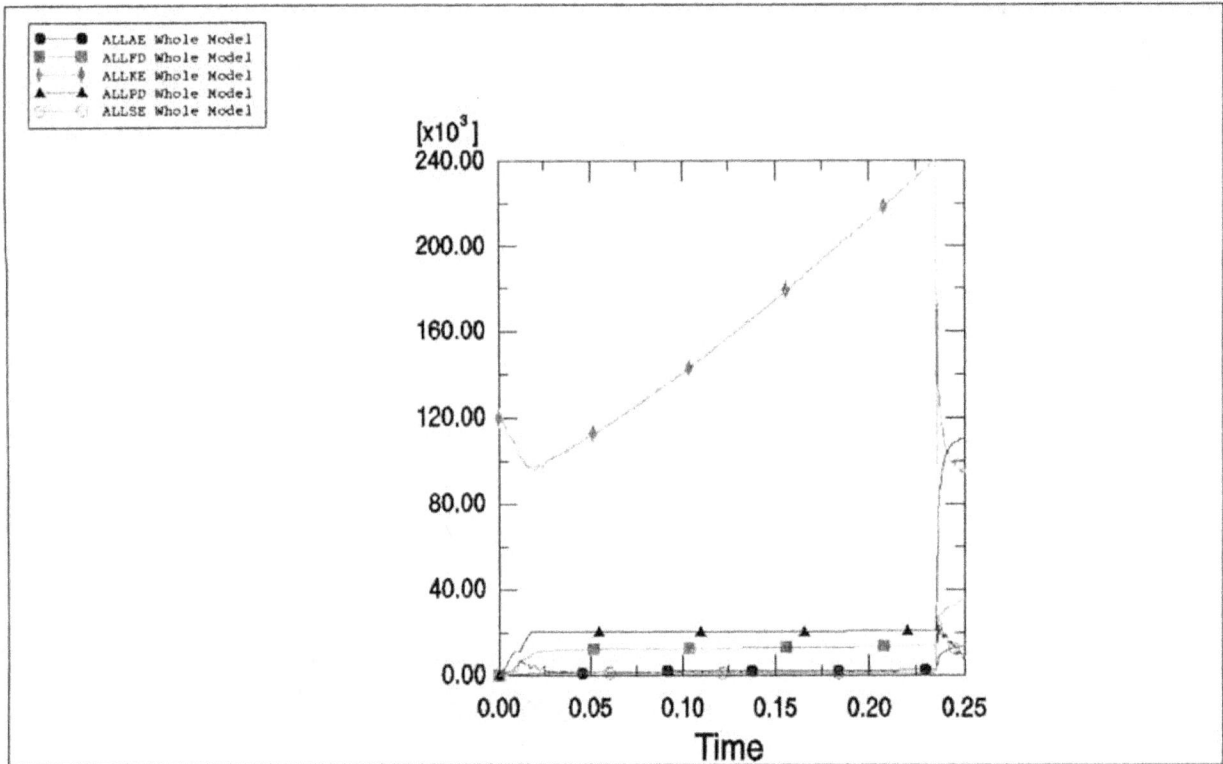

Fig. 51. Model #3 energy history, 70 degrees off-vertical, 2-foot (0.6-m) drop, 10000 pounds (4530 kg) (Snow and Morton 2003).

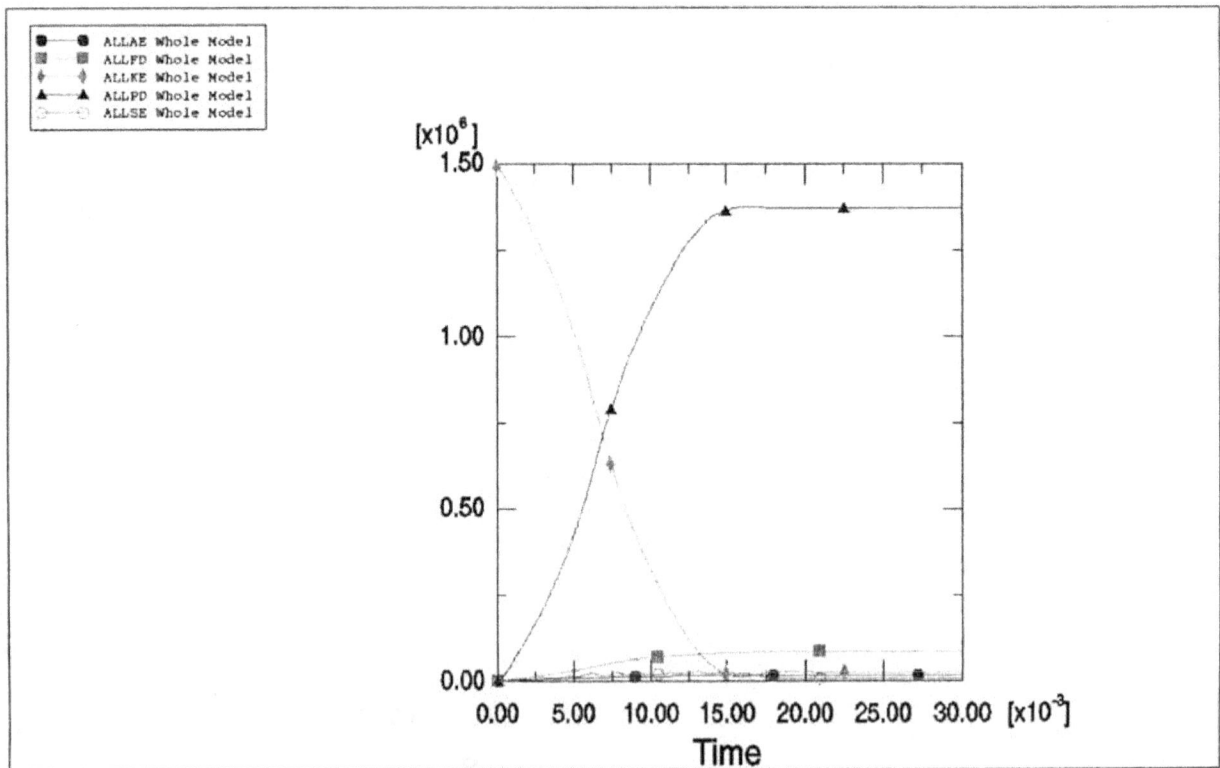

Fig. 52. Model #3 energy history, 7 degrees off-vertical, 30-foot (9-m) drop, 8280 pounds (3751 kg) (Snow and Morton 2003).

100

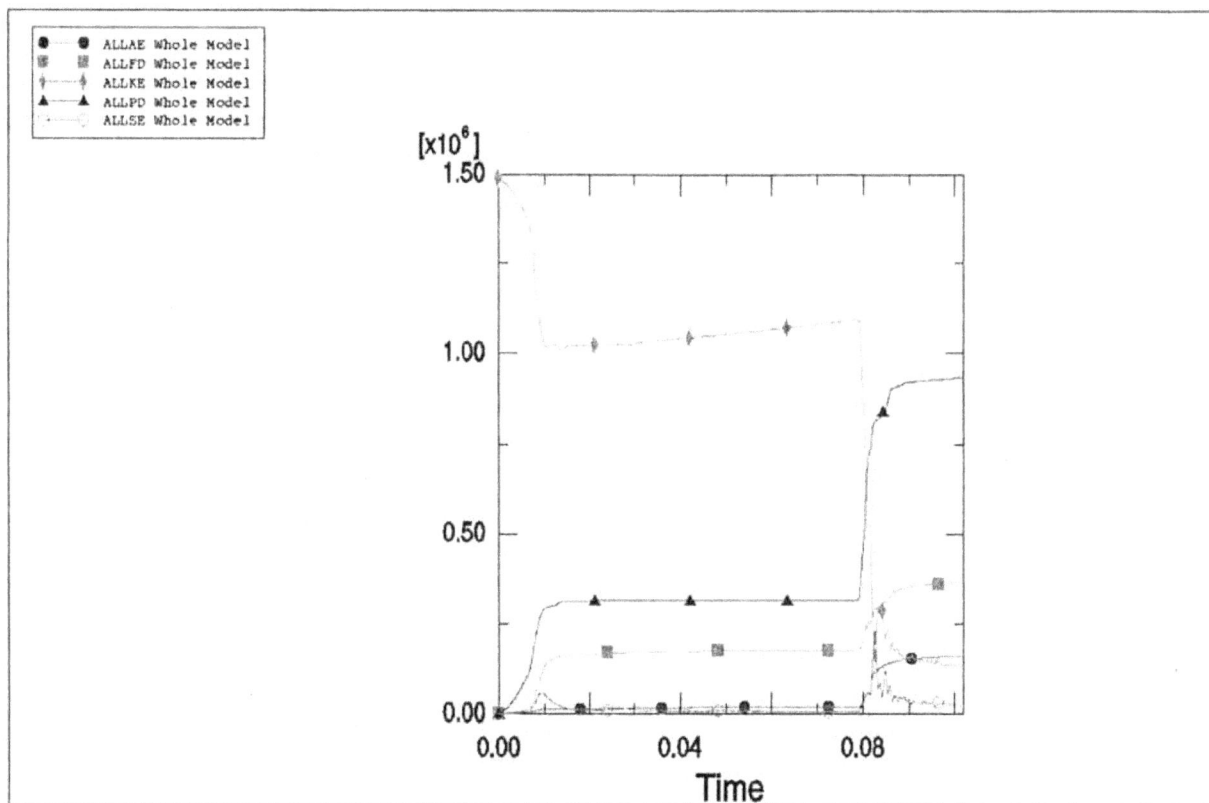

Fig. 53. Model #3 energy history, 70 degrees off-vertical, 30-foot (9-m) drop, 8280 pounds (3751 kg) (Snow and Morton 2003).

Table 13 lists the peak equivalent plastic strains occurring in the various containment boundary and skirt components of the ISFP canisters during the specified drop events for a 10,000-pound (4,530 kg) canister (maximum design weight). Table 14 lists the same for an 8,280-pound (3,751 kg) canister (minimum actual weight). Peak strains are listed for the outside surface of a component, mid-surface, and inside surface. However, these peak strains did not necessarily occur at the same location (element) through the thickness.

All of the 24-in (610 mm) diameter, 15-ft (4.5 m) long, standardized canisters evaluated by Blandford (2003) were loaded to about 10,000 pounds (4,530 kg) total weight. This makes a comparison between those canister drops, and the Table 13 ISFP with identical weights canister drops valid. Table 13 shows the strain values reported in the 2003 evaluation of the 24-in (610 mm) diameter standardized canister for the 30-ft (9 m) drop events. It is noted that the peak strains in many of the containment boundary components of the 24-in (610 mm) diameter ISFP canister were at or below those of the 24-in (610 mm) diameter standardized canister. The deformation and strain responses of a canister component during a drop event are dependent on component stiffness; adjacent component stiffness; interaction between adjacent components; material properties; canister loaded weight; drop height; drop angle; friction; and buckling. Varying one or more of these parameters results in a nonlinear (sometimes highly nonlinear) change in the canister component response.

Table 13. Peak plastic strains in Foster Wheeler 24-inch (610-mm) diameter, 15-foot (4.5-m) long canisters, 10,000 pounds (4,530 kg) maximum design weight (Snow and Morton 2003).

Drop Height (ft)	(m)	Angle From Vertical (Degrees)	Peak Equivalent Plastic Strain (PEEQ) %														
			Lower Head			Lower Skirt*			Upper Head			Upper Skirt			Body		
			out*	mid*	in*	out*	mid	in*	out*	mid*	in*	out*	mid*	in*	out*	mid*	in*
30	9	0 t = 0.020	3 {6}	0.3 {0.6}	2 {4}	41 {45}	18 {16}	56 {54}	5	0.4	2	-	-	-	-	-	-
30	9	7 t = 0.027	0.2 {0.7}	0 {0.1}	0.1 {0.6}	86 {76}	19 {31}	57 {60}	2	0	2	-	-	-	-	-	-
30	9	45 t = 0.032	18 {18}	10 {9}	22 {15}	57 {95}	30 {33}	106 {117}	1 {***}	0 {***}	1 {***}	- {***}	-	-	3	5	8
30** (34)	9 (10.2)	70 t = 0.090	47 (49) {36}	18 (18) {17}	41 (42) {35}	51 (52) {74}	31 (30) {41}	68 (67) {58}	58 (66) {57}	25 (27) {23}	27 (29) {48}	20 (21) {30}	11 (13) {15}	12 (13) {15}	10 (10)	4 (5)	9 (10)
30	9	90 t = 0.010	38 {34}	19 {16}	17 {22}	11 {7}	6 {3}	7 {5}	36	18	16	12	5	8	8	2	6
23 repository	6.9	0 t = 0.030	3 {6}	0.2 {0.5}	2 {4}	38 {44}	16 {14}	53 {53}	4	0.4	3	-	-	-	-	-	-
2 repository	0.6	70 t = 0.250	5 {16}	3 {8}	2 {12}	8 {24}	4 {23}	5 {23}	25 {23}	14 {15}	10 {16}	6 {22}	2 {21}	4 {21}	3	1	3

- Denotes that strains in these components are either zero or insignificant/not of interest.

* out = outside surface of component, mid = middle surface of component, in = inside surface of component.

** This model experienced a 14% artificial energy. Artificial energy is drop energy used to prevent finite element numerical instabilities. The results shown in (parentheses) are for a higher drop that accounts for this lost drop energy. All other models experienced an artificial energy that was below 5%, which is considered acceptable as is.

*** These results from the Blandford (2003) evaluation were for a top end impact. This current evaluation only looked at the bottom end impact. Top end impact results were enveloped by those from the 70 degrees off-vertical drop event.

(Numbers reported in these brackets are those that were listed in the Blandford (2003) standardized canister analysis).

Table 14. Peak plastic strains in Foster Wheeler 24-inch (610-mm) diameter, 15-foot (4.5-m) long canisters, 8,280 pounds (3,751 kg) actual weight (Snow and Morton 2003).

Drop Height (ft)	(m)	Angle From Vertical (Degrees)	Peak Equivalent Plastic Strain (PEEQ) %														
			Lower Head			Lower Skirt			Upper Head			Upper Skirt			Body		
			out*	mid*	in*	out*	mid*	in*	out*	mid*	in*	out*	mid*	in*	out*	mid*	in*
30	9	7 t = 0.030	0.1	0	0.1	59	18	57	1	0	1	-	-	-	-	-	-
30** (33)	9 (9.9)	70 t = 0.100	45 (46)	18 (18)	40 (41)	49 (51)	30 (30)	67 (67)	57 (59)	25 (25)	27 (28)	20 (19)	11 (12)	13 (13)	9 (10)	4 (5)	9 (9)

- Denotes that strains in these components are either zero or insignificant/not of interest.

* out = outside surface of component, mid = middle surface of component, in = inside surface of component.

** This model experienced an 11% artificial energy – drop energy used to prevent finite element numerical instabilities. The results shown in (parentheses) are for a higher drop that accounts for this lost drop energy.

In the present case, these ISFP canister heads are stiffer than those on the standardized canister due to their increased thickness. As expected, the strain response is nonlinear in that peak strains sometimes decreased and sometimes increased in a given canister component from the standardized canister response. The following ISFP canister peak strains under the 30-ft (9 m) drop event were higher:

1. The vertical and 7 degrees off-vertical drops produced strains in the upper head in the 24-in (610 mm) diameter ISFP canister due to the impact plate bending the retaining ring (which is welded to the upper head). These strains were absent in the standardized canister because it has no impact plate retaining ring. These upper head strains are not considered a containment concern because they are low in comparison to slap-down strains.
2. The 45 degrees off-vertical drop produced a 1% (mid-surface) and 7% (inside surface) strain increase on the lower head of the 24-in (610 mm) diameter ISFP canister than the 24-in (610 mm) diameter standardized canister.
3. The 70 degrees off-vertical drop produced a 13% (outside surface), 1% (mid-surface), and 7% (inside surface) strain increase on the ISFP canister lower head than occurred in the 24-in (610 mm) diameter standardized canister. The upper head of the ISFP canister experienced a 9% strain increase (outside surface) and a 4% mid-surface strain increase, than occurred on the 24-in (610 mm) diameter standardized canister. This is significant because these strain levels also exceeded those calculated for the 18-in (457 mm) diameter canister during the 1999 tests.
4. The 90 degrees off-vertical (horizontal) drop produced a 4% strain increase on the outside surface of the ISFP lower head, and a 3% strain increase in mid-surface of the lower head, than the 24-in (610 mm) diameter standardized canister results showed. Upper head strains are comparable to the lower head strains on this ISFP canister. However, the 70 degrees off-vertical drop results were controlling.

Peak strains in the skirts, which were designed to absorb drop energy and are not containment retaining, increased or decreased from the 2003 standardized canister results depending on the drop angle and location. Note that a 10% strain increase on the CGOC (7 degrees off-vertical) and 70 degrees off-vertical drops occurred in the lower skirt of the ISFP canister over that of the standardized canister.

Table 13 also shows the strain values reported in the 2003 evaluation of the 24-in (610 mm) diameter standardized canister for the repository drop events. It can be seen that the peak strains in the containment boundary components of the ISFP canister are generally at or lower than those for the standardized canister. There was one case in the 2 ft (0.6 m) worst orientation drop that the ISFP canister produced a 2% strain increase on the outer surface of the upper head over that of the standardized canister. The deformation and strain responses of a canister component during a drop event are dependent on component stiffness; adjacent component stiffness; interaction between adjacent components; material properties; canister loaded weight; drop height; drop angle; friction; and buckling. Varying one or more of these parameters results in a nonlinear (sometimes highly nonlinear) change in the canister component response. In the present case, the ISFP canister heads are stiffer than those on the standardized canister due to their increased thickness. This resulted in peak containment boundary strains that were typically at or lower than those for the standardized canister under the specified drop events. This might not be the case under other drop scenarios. Peak strains in the skirts, which were designed to absorb drop energy and are not containment retaining, increased or decreased from the 2003 results depending on the drop angle and location.

Table 14 lists the peak equivalent plastic strains occurring in the various containment boundary and skirt components of the 24-in (610 mm) diameter ISFP canisters during the specified 30-ft (9 m) drop events for an 8,280 pound (3,751 kg) canister (minimum actual weight). The results show that the 30-ft (9 m) 7 degrees and 70 degrees off-vertical drop strains in the containment boundary components of the 8,280 pound (3,751 kg) ISFP canister are at or below those reported in Table 13 for the 10,000 pound (4,530 kg) ISFP canister.

Several features of the 24-in (610 mm) diameter ISFP canisters that are not included in the standardized canisters have the following affects on the strains resulting from the specified drop events:

1. This ISFP canister impact plate retaining ring on the upper head causes straining in that head during the vertical and near-vertical drop events due to the load from the upper impact plate bearing on it. These strains are the largest containment boundary strains for those drop events. However, because of the increased thickness of the ISFP heads over that used on the 24-in (610 mm) diameter standardized canister, those peak strains are still low (<5%).
2. This ISFP canister impact plate retaining ring on the lower head causes straining in that head during the 70 degrees and 90 degrees off-vertical drop events. However, because of the increased thickness of the ISFP heads over that used on the 24-in (610 mm) diameter standardized canister, those strains are not the highest on the head. The highest strains are still located on the head just below the skirt-to-head weld.
3. This ISFP canister uses a fuel basket that either is composed of a number of tubes with thick intermediate plates, or is a Shippingport Reflector IV or V basket. All ISFP canisters include a large shield plug. The standardized canister in the 2003 evaluations used a sleeved spoked-wheel basket with no shield plug. It is easy to see the importance of having an internal sleeve for a basket with sharp edges (as the spoked-wheel basket), to keep those edges from directly impacting the containment boundary. Although an internal sleeve is determined to be useful for ISFP canisters, the usefulness of an internal sleeve for this ISFP canister is not obvious since the basket and shield plug do not appear to have similar sharp edges. Therefore, this ISFP canister does not use an internal sleeve. However, significant straining did result in this ISFP model #3 canister body due to the internal components, but these strain levels were not the highest in the containment boundary.
4. This ISFP canister with Shippingport Reflector IV and V baskets also included a shield plug support ring that was welded to the inside body of the canister. The model #3 canister also included this shield plug support ring. The 24-in (610 mm) diameter standardized canister has no such support ring. Some straining did occur in the canister body due to this ring under the vertical and near-vertical drop events. However, those strains were low (<1%).

In many of the drop events, the 24-in (610 mm) diameter 15-ft (4.5 m) long ISFP canister experienced lower material strains than the 18-in (457 mm) diameter standardized canister, due mainly to the thicker heads. However, the 45 degrees, 70 degrees, and 90 degrees off-vertical 30-ft (9 m) drops resulted in higher strains (up to 13% strain increase) for the 10,000 pound (4,530 kg) ISFP canister than for the 24-in (610 mm) diameter standardized canister. The 1999 worst-case scenario testing does appear to envelope all other 10,000-pound (4,530 kg) and 8,280-pound (3,751 kg) ISFP canister 30-ft (9 m) drops, and the 23-ft (6.9 m) vertical and 2-ft (0.6 m) 70 degrees off-vertical repository canister drops.

CONCLUSIONS

The design of the standardized canister was guided by the fact that not all of the uses or loading conditions for the standardized canisters were defined in 1997 through 1998. A robust canister design could accept a wider range of uses and loading conditions. Therefore, as much safety margin was incorporated into the design as was reasonably achievable. This was considered prudent, especially since the drop responses (plastic strains) of the standardized canisters could be high in the plastic regime. One prime component used to increase safety margins was the internal sleeve. Its dual-purpose uses were to reduce localized strains in the canister containment during accidental drop events and to help prevent or at least reduce containment material corrosion concerns if adequate dryness of the contents was not achieved. In order to maintain as much loading space as possible, welding support components (rings, ledges, etc.) to the inside surface of the canister were also deemed unacceptable. The original design goal was to have the internals locate and retain canister contents in acceptable positions, especially in the axial direction. Minimal welding to the containment material was also seen as a design goal since this would also minimize corrosion concerns; minimize the potential for localized strains; and minimize adverse interactions if undetected flaws were present.

Several features of the ISFP canisters that are not included in the standardized canisters have the following effects on the strains resulting from the specified drop events:

1. The ISFP canister impact plate retaining ring on the upper head causes straining in that head during the vertical and near-vertical drop events due to the load from the upper impact plate bearing on it. These strains are the largest containment boundary strains for those drop events. However, because of the increased thickness of the ISFP heads over that used on the standardized canister, those peak strains are still low.
2. The ISFP canister impact plate retaining ring on the lower head causes straining in that head during the 70 degrees through 90 degrees off-vertical drop events. However, because of the increased thickness of the ISFP heads over that used on the standardized canister, those strains are not the highest on the lower head. The highest strains are still located on the head just below the skirt-to-head weld.
3. The ISFP canisters use fuel baskets that are either composed of a number of tubes with thick intermediate plates, or is a Shippingport Reflector IV or V basket. All ISFP canisters include a large shield plug. In the 1999 evaluations, the standardized canister used a sleeved spoked-wheel basket with no shield plug. It is important having an internal sleeve for a basket with sharp edges (as the spoked-wheel basket) to keep those edges from directly impacting the containment boundary during a drop event. The ISFP canister does not use an internal sleeve. The usefulness of an internal sleeve for the ISFP canister (with the currently-defined internal configurations) is not obvious since the basket and shield plug do not appear to have equivalent sharp edges. However, significant straining did result in the ISFP canister body due to the internal components, but those strain levels were not the highest in the containment boundary.
4. The 24-in (610 mm) diameter ISFP canisters with Shippingport Reflector IV and V baskets also included a shield plug support ring that was welded to the inside body of the canister. The 24-in (610 mm) diameter ISFP canister model also included this shield plug support ring. The standardized canisters have no such support ring. Some straining did occur in the ISFP canister body due to this ring under the vertical and near-vertical drop events. However, those strains were low (<1%).

The analytical evaluations demonstrate that when design changes are made on a canister, significant changes may occur in the response. For the standardized and ISFP canisters, the design changes endorsed by Foster Wheeler Environmental Corportation (FWENC) have resulted in predicted plastic strains in the ISFP canister being higher for certain drop scenarios than those predicted for the standardized canisters. In addition, higher plastic strains resulted at more locations on the ISFP canisters.

The ISFP canister strain responses are enveloped by the worst-case 1999 standardized canister drop test for all drop events considered except for the 30-ft (9 m), slap-down drop events (using either design or actual weights). The standardized canister maintains containment as demonstrated via helium leak-testing for the 1999 drop testing effort. However, the ISFP canisters have not yet demonstrated a similar performance, but additional information can provide the necessary insights. The average total elongation determined from tensile testing of the 1999 test canister materials was about 48%. Total elongation is a very conservative estimate of the material rupture strain under uniaxial loading conditions. The 1999 drop-testing effort calculated external surface strains on a skirt in excess of 100%, with no indication of rupture or cracking on the actual component after the drop-test. It is expected that the ISFP canister material will have similar ductility. The largest containment boundary plastic strain on an exterior surface of these ISFP canisters is in the heads at 66%, while the largest middle surface plastic strain is only 27%. Additionally, these strains are due primarily to bending about one axis, though a triaxial load condition does exist. The peak strains in these ISFP canisters are below the expected material rupture strain level typically achieved with currently-available 316L material. Therefore, it is anticipated that the ISFP canister will maintain containment during all the drop events considered herein (including the defined repository drop events) provided acceptable materials are used for construction.

ANALYTICAL EVALUATION OF THE MULTI-CANISTER OVERPACK FOR REPOSITORY-DEFINED AND OTHER RELATED DROP EVENTS

INTRODUCTION

The Multi-Canister Overpack (MCO) is a container designed and fabricated for use at the U.S. Department of Energy (DOE) Hanford Site. The MCO is a stainless steel (SST) cylindrical vessel approximately 24 in (610 mm) in outer diameter and 166 in (4215 mm) long. Spent nuclear fuel (SNF) is placed in one of four types of baskets and then loaded into the MCO. A fully-loaded MCO holds five or six baskets (depending on type) and a shield plug fixed in place with a locking ring. A cover cap is welded on the top-end to complete the package. A fully-loaded MCO will weigh as much as 10 tons.

The MCO was intended to contain SNF from the Hanford K-Basins during interim storage at Hanford's Canister Storage Building for 40 years or more (Lorenz et al. 1999; Goldman et al. 2000a, 2000b; 2002). Analyses have been performed on the MCO to support its use at Hanford (Goldman et al. 2000a and 2000b; 2002). It is expected that the MCO will be shipped to the national repository for final disposal at some future time. However, the analyses performed to date on the MCO under accidental drop conditions do not envelope those required at the repository. Snow (2003) evaluated the MCO (with Mark 1A and Mark IV SNF fuel and scrap baskets) response under the accidental drop conditions defined for the repository (DOE 2003c). The accidental drop conditions at the repository, defined in DOE (2002), are as follows:

1. A 23-ft (6.9 m) drop of a container, with the container oriented vertically, onto a rigid (unyielding), flat surface
2. A 2-ft (0.6 m) drop of a container, with the container oriented so as to cause the most severe challenge to the containment boundary, onto a rigid, flat surface

The vertical drop event was evaluated for "slightly off-of-vertical" orientations as well. The vertical drop was also evaluated for an impact with a rigid edge at the collar region simulating an impact while falling into another cylinder such as a repository waste package. This was not a repository required drop event, but one that was suggested by NSNFP personnel. The effects of internal pressure and friction were also evaluated.

Room temperature (70°F = 21°C) material properties were applied for all MCO evaluations reported herein. This temperature was considered valid for this evaluation because the MCO was expected to experience very little internal heat generation by the time it arrives at the repository. Beginning of life material properties were applied for all MCO evaluations. Data showing significant material degradation on the MCOs during interim storage was not available. The MCO was assumed "dry" for all drop events. Drying processes prior to sealing confirms this assumption.

COMPONENTS OF THE MCO DESIGN

The MCO was designed to contain six Mark 1A or five Mark IV fuel baskets which hold intact fuel. As many as two scrap baskets, which hold fuel pieces or "scraps", may be placed one at each end of the MCO (each replacing a fuel basket).

MCO design

Figures 54 through 56 show the MCO design, with close-up views of the top and bottom ends. The main components of the MCO are as follows:

1. The main shell was made of 24-in (610 mm) nominal outer-diameter pipe with a 1/2-in (13 mm) nominal thickness (SA-312 TP304/304L SST).
2. The shell bottom was approximately 24 in (610 mm) in diameter and was about 2 in (51 mm) thick (SA-182 F304/304L SST).
3. The collar, which was about 15 in (381 mm) in height, was a continuation of the main shell and was threaded to accept the locking ring (SA-182 F304/304L SST).
4. The locking ring, which was about 6-1/2 in (165 mm) in height, threaded into the collar and held the shield plug in position in the collar (SA-182 F304N SST). The collar also included a ring for lifting the MCO.
5. The shield plug was about 16 in (406 mm) in height, and housed filters, rupture disks, and process valves (SA-182 F304L and SA-240 304L).
6. The process tube was made of 1-in (25 mm) schedule XXS pipe (146-1/2 in [3.72 m] in length), attached to the shield plug, and extended to the shell bottom (SA-312 TP304L SST).
7. Six basket support bars were welded to the shell bottom (SA-240 304L SST).
8. A guide cone was attached to the basket support bars to hold the bottom-end of the process tube (SA-479 304L SST).
9. The closure cover was about 9 in (229 mm) in height and attached to the collar to seal the container (SA-182 F304L). The cover also included a ring for lifting the sealed MCO.

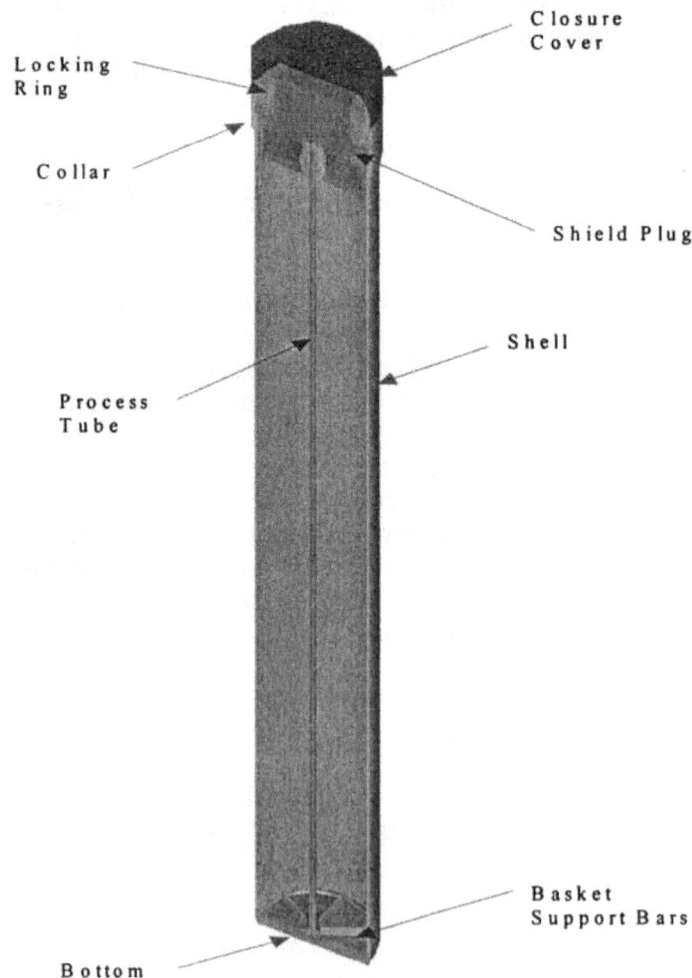

Fig. 54. MCO design (cross-section view) (Snow 2003).

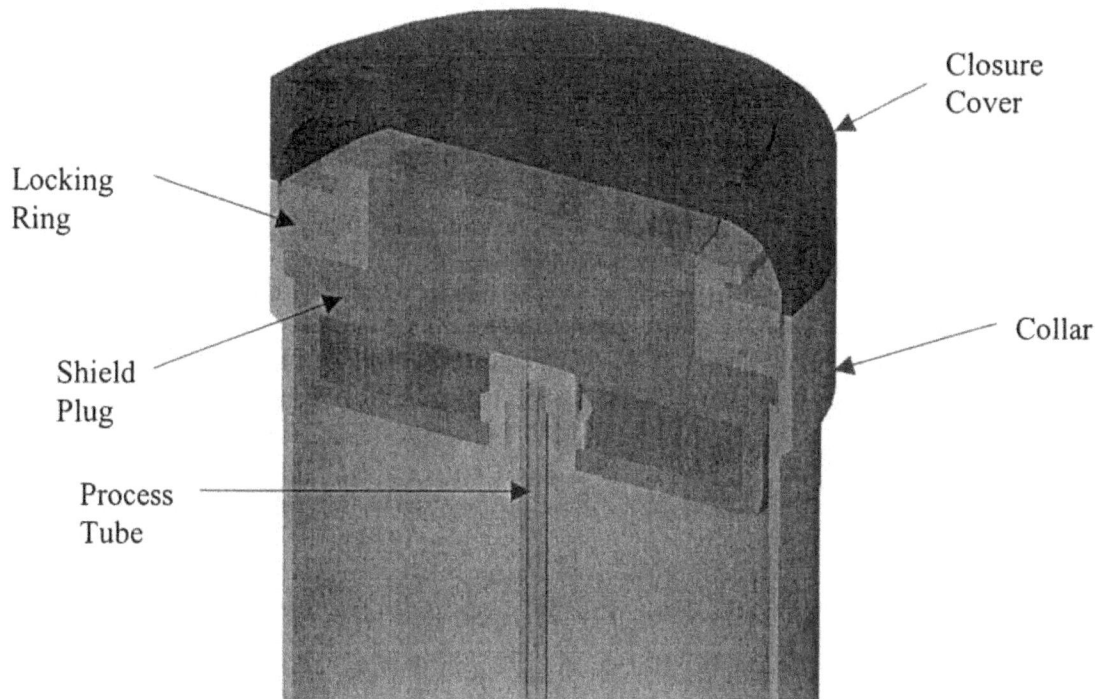

Fig. 55. Close-up of MCO top-end (cross-section view) (Snow 2003).

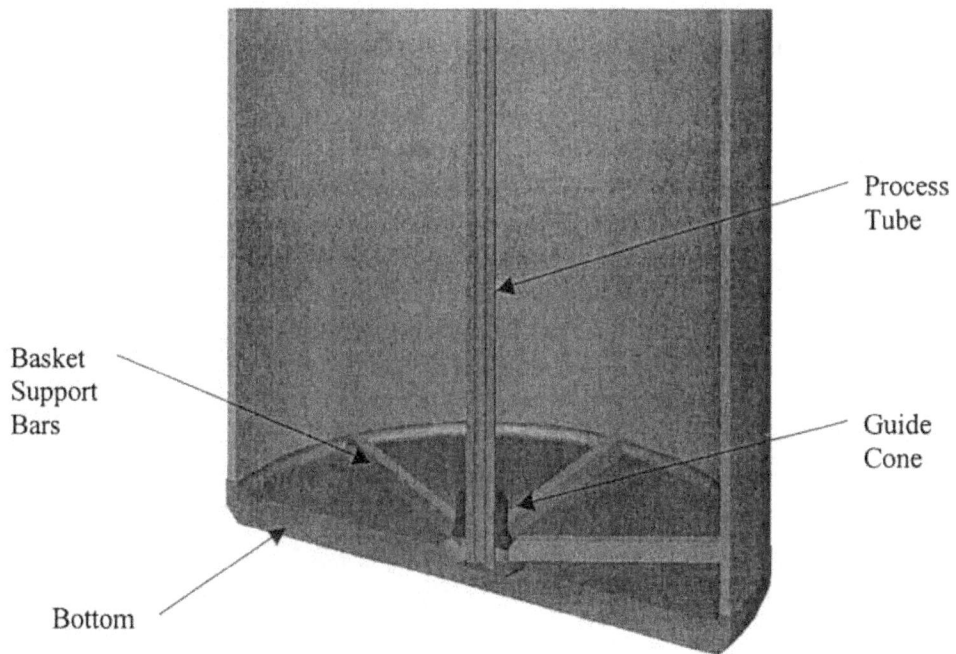

Fig. 56. Close-up of MCO bottom-end (cross-section view) (Snow 2003).

Mark 1A fuel basket design

The main components of the Mark 1A (Fig. 57) fuel basket are as follows:

1. The base plate was about 22-5/8 in (574 mm) in diameter and 1-1/4-in (32 mm) thick (SA-182 F304/ F304L SST).
2. The center post was 6-5/8 in (168 mm) in diameter and just under 25 in (635 mm) tall (SA-479 TP304/ 304L or 304 SST) which was threaded into the base plate.
3. Six trapezoidal-shaped perimeter bars/posts that were just under 22 in (559 mm) tall (SA-479 TP304/ 304L or 304 SST) which were bolted through the base plate.
4. Shroud sheet metal wall (0.048 in = 1.2 mm thick) that formed a wall 14 in (355 mm) tall around the basket perimeter (ASTM A240 TP304L SST) which was welded to the base plate.
5. An expanded metal spacer 0.300 in (8 mm) thick (AL 5005-H34 Ryerson) that rested immediately on top of the base plate.
6. A 2-1/2-in (64 mm) thick fuel plate rack with holes to accept each fuel element (ASTM B26 UNS A03560-T6 or A13560-T6).

The overall height, excluding the smaller diameter top portion of the center post, was just over 23 in (584 mm) for this Mark 1A fuel basket.

Mark 1A fuel scrap basket design

The Mark 1A scrap basket was similar to the fuel basket in that it had the base plate and metal spacer, perimeter bars, and center post. Instead of the simple cylinder-shaped shroud wall, it had a shroud that was fabricated of trapezoidal segments that separated the basket into six volumes. These segments were made of 1/8-in (3 mm) thick (ASTM F152 UNS C12200-060 sheet copper) and were welded together and the assembly was attached to the base plate with screws. The overall height of a Mark 1A scrap basket, excluding the smaller diameter top portion of the center post, matched that of the Mark 1A fuel basket at just over 23 in (584 mm).

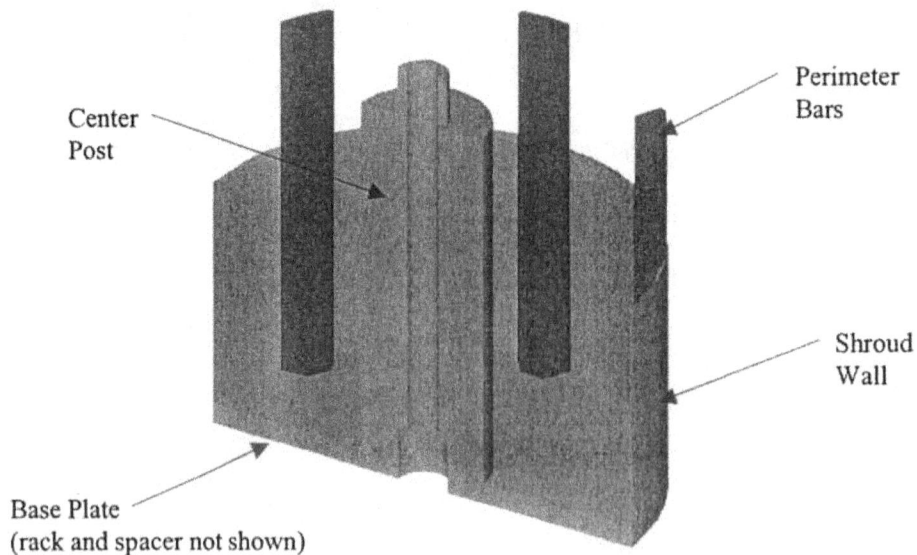

Fig. 57. MCO Mark 1A fuel basket (Snow 2003).

Mark IV SNF basket design

The main components of the Mark IV (Fig. 58) SNF basket are as follows:

1. The base plate was about 22-1/2 in (571 mm) in diameter and 1-1/4-in (32 mm) thick (ASTM A240 304/304L SST).
2. The center post was just under 2-7/8 in (73 mm) in diameter and just under 30-1/2 in (774.4 mm) tall (ASTM A511 304/304L or 304 SST) which was threaded into the base plate.
3. Six 1-5/16-in (33 mm) diameter perimeter bars/posts that were just under 22-1/2 in (571 mm) tall (ASTM A276 304/304L or 304 SST) which were bolted through the base plate.
4. Shroud sheet metal wall with a thickness of 0.048 in (1.2 mm) that formed a wall 14 in (355 mm) tall around the basket perimeter (ASTM A240 TP304L SST) which was welded to the base plate.
5. An expanded metal spacer 0.186 in (5 mm) thick (AL 5005-H34 Ryerson) that rested immediately on top of the base plate.
6. A 2-1/2-in (64 mm) thick fuel plate rack with holes to accept each fuel element (ASTM B26 UNS A03560-T6 or A13560-T6).

The overall height, excluding the smaller diameter top portion of the center post, was just less than 28 in (711 mm) for this Mark IV fuel basket.

Mark IV scrap basket design

The Mark IV scrap basket was similar to the fuel basket in that it had the base plate and metal spacer, perimeter bars, and center post. Instead of the simple cylinder-shaped shroud wall, it had a shroud that was fabricated of trapezoidal segments that separated the basket into six volumes. These segments were made of 1/8-in (3.2 mm) thick ASTM F152 UNS C12200-060 sheet copper and were welded together; and the assembly was attached to the base plate with screws. The overall height of a Mark IV scrap basket, excluding the smaller diameter top portion of the center post, matched that of the Mark IV fuel basket at just under 28 in (711 mm).

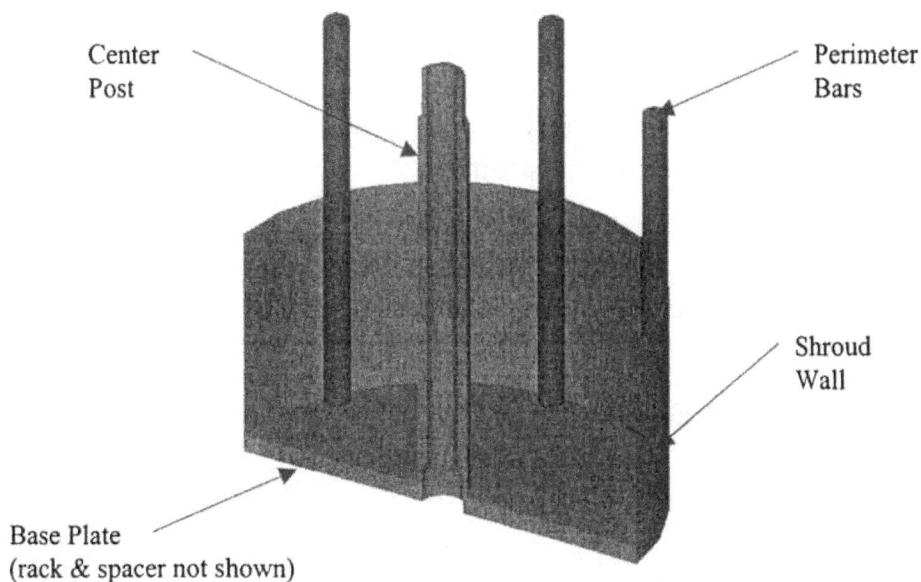

Fig. 58. MCO Mark IV fuel basket (Snow 2003).

MCO SNFs

The SNFs stored in the MCOs were Mark 1A and Mark IV fuel assemblies, elements, or pieces (Lorenz et al. 1999). The condition of the SNFs during and after an accidental drop event was not requested as part of this evaluation. Therefore, SNFs were modeled in a simplistic manner to represent their effect only on the MCO and basket structures during a drop event. This resulted in more drop energy being potentially absorbed by MCO containment components than would occur if the SNFs were modeled more accurately. This was conservative as far as the MCO containment was concerned.

MCO configurations evaluated

Lorenz et al. (1999) determined that the following two MCO configurations create the most challenging response to the MCO containment boundary:

1. **Configuration #1:** An MCO with cover cap, and six fully-loaded Mark 1A SNF baskets, at the maximum "dry" design weight of 18,260 pounds (8,272 kg).

2. **Configuration #2:** An MCO with cover cap, and five fully-loaded Mark IV SNF baskets, at the maximum "dry" design weight of 20,080 pounds (9,096 kg).

No scrap baskets were included in the evaluated configurations. The structure of the scrap baskets (thin trapezoidal volumes made of copper) provided a considerable amount of energy absorption capability —more than was available in the fuel baskets. Therefore, using SNF baskets only was conservative with respect to the MCO containment components.

ANALYTICAL MODELING OF THE MCO BASKETS

MCO

The MCO was modeled using solid linear brick elements (element type C3D8R); wedge elements (element type C3D6); and linear quadrilateral shell elements (element type S4R) as follows:

1. **Bottom:** The bottom used 2,944 solid (brick and wedge) elements, with four elements through the thickness of the base and four in the connection to the wall. This was done to ensure adequate modeling of bending responses.

2. **Main Shell:** The cylindrical shell employed 14,720 solid (brick only) elements, with four elements through the thickness. The connection between the shell and the bottom consisted of a full-penetration groove weld. This weld was modeled using nodes common to the shell and bottom elements.

3. **Collar:** The collar was modeled with 4,992 solid (brick only) elements, with a minimum of four elements through the thickness. The connection between the collar and the main shell, consisting of a full-penetration groove weld, was modeled using nodes common to the collar and main shell elements.

4. **Cover:** The cover used 2,144 solid (brick only) elements, with four elements through the thickness in the cylindrical portion and three elements through the flat top. The groove weld connection between the cover and the collar was also represented with common nodes.

5. **Shield Plug:** The shield plug utilized a total of 762 solid (brick only) elements. The mesh size in this component was quite coarse in order to simplify the model. The coarse mesh size was considered acceptable since the plug consisted of very thick members that were unlikely to deform significantly during any drop event—a coarse mesh would adequately simulate such a response. Valves, ports, filters, etc. that were part of the shield plug were not explicitly modeled because their influence on the adjacent components was considered negligible.

6. **Locking Ring:** The locking ring employed 432 solid (brick only) elements. This mesh was also coarse for the same reasons given for the shield plug. The threaded connection between the locking ring and the collar was represented by fixing the locking ring nodes (in the threaded portion) to the inside wall of the collar. This assumed that the threaded connection between the ring and collar would not fail during any drop event. This assumption was considered valid because more than 3 in (76 mm) of thread engagement length was far in excess of what is required to resist the worst-case loading during any drop event without failure. The setscrews on the locking ring were ignored in this evaluation because they had no significant effect on the MCO response during any drop event. Their purpose was to ensure a seal between the shield plug and the collar—which was not needed after the cover was welded onto the collar.

7. **Basket Support Bars:** The six basket support bars were each represented using 29 solid (brick only) elements. The fillet weld that attached each bar to the MCO bottom was represented by fixing the bar edge nodes to the top surface of the bottom. This was considered adequate because the exact condition of these welds was not of interest. This assumed that these welds would not fail during any drop event.

8. **Guide Cone:** The guide cone was modeled using 108 solid (brick only) elements. The welded connection between the guide cone and the six basket support bars was conservatively modeled using common nodes.

9. **Process Tube:** The process tube employed 462 quadrilateral shell elements.

Only half an MCO was explicitly modeled due to the symmetry in geometry, loading, and response during the drop events. Therefore, the number of elements listed above for the MCO components reflects a half-model only. Symmetry boundary conditions were applied to ensure that the half-model responded exactly like the complete MCO. This MCO model was used with six Mark 1A or five Mark IV basket models placed within to simulate the fully-loaded package.

Mark 1A basket

Each Mark 1A basket was modeled using solid linear brick elements (element type C3D8R); wedge elements (element type C3D6); and linear quadrilateral shell elements (element type S4R) as follows:

1. **Basket Base:** The basket base was represented with 252 solid (brick only) elements, with three elements through the thickness.

2. **Center Post:** The center post was modeled using 732 solid (brick only) elements, with five elements through the wall. The threaded connection between the center post and the basket base employed nodes common to both components. This assumed that the post would remain firmly attached to the base during all drop events. This assumption is valid due to small resulting deformations and material strains in these connections during all drop events.

3. **Perimeter Bars:** The trapezoidal perimeter bars were each represented using 308 solid (brick only) elements. The actual connection was made using one bolt through the base and into the end of the bar. The

model simulated this connection by fixing the bar nodes to the base. This made a connection that was more rigid than was provided by the bolt. This was acceptable because the objective of this evaluation was to determine the condition of the MCO containment boundary—not determine the exact condition of the baskets and fuels during a drop event. This method of modeling the perimeter bar connection was conservative as far as the MCO containment boundary was concerned.

4. **SNFs:** It was not the purpose of this evaluation to determine the condition of the fuels during and after a drop event. Therefore, the modeling of fuels was only sufficient to represent their effect on the basket and MCO structure. The fuels contained within the Mark 1A basket were represented in one of two manners, depending on the drop event being simulated:

4.1. For the vertical or near-vertical drop events (from 23 ft), the fuel was simply modeled as 49 mass elements attached to each basket base (element type MASS). This was quite conservative because it prevented the fuel from absorbing any drop energy—all energy was forced to the basket and MCO structures.

4.2. For the worst-case drops (2-ft = 0.6 m drop, with a secondary impact or slap-down of an end), the 48 fuel elements per basket were simply modeled as six 6.3-in (160 mm) diameter 20-in (508 mm) long solid bars (using 50 wedge elements). This approach adequately located the center-of-mass of the fuel elements in a basket; allowed the fuel movement within a basket during a drop event; and kept the fuel from transferring load from one basket to the next (modeled fuel was shorter than the actual fuel).

5. **Basket Walls:** The basket walls were simulated with 84 shell elements. The walls were connected to the basket base using common nodes to represent the attachment weld.

As with the MCO, only half a basket was explicitly modeled due to the symmetry in geometry, loading, and response during the drop events. Therefore, the number of elements listed above for the basket components reflects a half-model only. Symmetry boundary conditions were applied to ensure that the half-model responded exactly like the complete basket.

Mark IV basket

Each Mark IV basket was modeled using solid linear brick elements (element type C3D8R), wedge elements (element type C3D6), and linear quadrilateral shell elements (element type S4R) as follows:

1. **Basket Base:** The basket base was represented with 324 solid (brick only) elements, with three elements through the thickness.

2. **Center Post:** The center post was modeled using 1032 solid (brick only) elements, with three elements through the wall. The threaded connection between the center post and the basket base employed nodes common to both components. This assumed that the post would remain firmly attached to the base during all drop events. The design of this connection prevents the post from separating from the base during any of the specified drop events, though the vertical and near vertical drops do cause significant bending in the post just above this connection. Therefore, the modeling of this connection was considered valid.

3. **Perimeter Bars:** The round perimeter bars were each represented using 312 solid (brick only) elements. The actual connection was made using one bolt through the base and into the bar end. The model simulated this connection by fixing the bar nodes to the base. This made a connection that was more rigid than was provided by the bolt. This was acceptable because the objective of this evaluation was to determine the condition of the MCO containment boundary—not determine the exact condition of the baskets

114

and fuels during a drop event. This method of modeling the perimeter bar connection was conservative as far as the MCO containment boundary was concerned.

4. **SNFs:** As stated previously, it was not the purpose of this evaluation to determine the condition of the SNFs during and after a drop event. Therefore, the modeling of SNFs was only sufficient to represent their effect on the basket and MCO structure. The SNFs contained within the Mark IV basket were represented in one of two manners, depending on the drop event being simulated:

4.1. For the vertical or near vertical drop events, the SNF was simply modeled as 62 mass elements on each basket base (element type MASS). This prevented the fuel from absorbing any drop energy, thereby forcing all energy to the basket and MCO structures.

4.2. For the worst-case drops (2-ft = 0.6 m drop, with a secondary impact or slap-down of an end), the 54 fuel elements per basket were simply modeled as six 6.3-in (160 mm) diameter 26-in (660 mm) long solid bars. This approach adequately located the center-of-mass of the fuel elements in a basket; allowed the fuel movement within a basket during a drop event; and kept the fuel from transferring load from one basket to the next (modeled fuel was slightly shorter than the actual fuel).

5. **Basket Walls:** The basket walls were simulated with 84 shell elements. The walls were connected to the basket base using common nodes to represent the attachment weld.

As with the MCO, only half a basket was explicitly modeled due to the symmetry in geometry, loading, and response during the drop events. Therefore, the number of elements listed above for the basket components reflects a half-model only. Plane symmetry boundary conditions were applied.

Finite element mesh size

The element sizes for the MCO models were chosen based on the type of event being simulated and the expected response. Because large plastic deformations were expected, the element sizes could not be too small; large elements would distort excessively causing the calculation to terminate before the event was completed. Small element size would also require many elements, resulting in excessive solution times. At the other extreme, elements that were too large would not respond properly (e.g., a bulge in a component would be shown as a sharp edge instead of a smooth curve) and the results would be in question. This was particularly important in areas where significant deformations would occur. Additionally, large elements in areas of high deformation required excessive artificial energy (model energy required to maintain solution stability). Some iteration in preliminary modeling was required to arrive at elements sufficiently small to provide acceptable results. Resulting solution times were 1-2 days for vertical and near-vertical impact orientations, and 3-4 weeks for the worst-case orientation events. All components were modeled using nominal dimensions.

Material density

The basic density of 304/304L stainless steel used was 0.285 psi (7,889 kg/m³). The density of the fuel, simulated with solid round rods, was adjusted so that the correct maximum-loaded weight was achieved.

Contact modeling

Contact between components was simulated using the ABAQUS General Contact option supplemented by the Contact Pairs option in areas of interest (impact locations). This was one of the approved methods detailed in DOE (2003b). These contact options employed penalty contact stiffness. Preliminary evaluations increased the default stiffness calculated within ABAQUS/Explicit Version 6.3-3 by a factor of ten.

The results were the same as those obtained using the default stiffness values. This indicated that the default penalty stiffness calculated within ABAQUS was adequately stiff to simulate a "hard impact" for these MCO evaluations.

Flat, rigid impact surface

The flat, rigid impact surface was modeled using one large rigid quadrilateral element (element type R3D4) that was fixed in space. The surface used in the edge-contact vertical drop used several rigid elements in a flat ring shape with a small chamber, all fixed in space.

Friction

The coefficient of friction (COF) between two steel surfaces during an impact event can vary widely. Snow et al. (2001) reported that the COF could vary significantly and still predict similar deformations (and thus material strains) for a stainless steel canister drop that was oriented vertically (or near vertical) or from about 65 degrees off-vertical to horizontal, impacting a flat, rigid surface. The range of drops evaluated herein fall into that category. A COF of 0.3 should give valid deformations and strains for the drops specified herein (Snow et al. 2001). A COF of 0.3 was therefore used in all of the subsequent analyses, with one exception: one drop was evaluated with a COF of 0.1 and 0.5 to show consistent results, indicating that the results for the impact orientations considered were not highly dependent on the COF value used (i.e., that a COF of 0.3 was acceptable).

Initial conditions

The finite element (FE) models began the drop event by locating the MCO model just above the rigid surface and applying a gravitational acceleration and an initial velocity. This allowed the elimination of calculations while the MCO was freely falling through air. The initial velocity was calculated by equating the potential energy of the MCO at the beginning of the drop to the kinetic energy just before impact. For example, at a drop height of 23 ft (7 m) the velocity at impact of the MCO would be 462 in/s (11.7 m/s).

Model solution termination

Model solution was terminated when the MCO had progressed through the first impact for the vertical or near vertical drops. The worst-orientation drops were continued through the secondary impact of the MCO.

Material properties

The MCO and baskets were constructed of various forms of 304 and 304/304L stainless steel. Basic properties at 70°F (21°C) are as follows (ASME 2001):

1. Modulus of Elasticity (E) = 28.3 x 10^6 psi (195 x 10^6 kPa)

2. Poisson's Ratio (μ) = 0.29

The purpose of this evaluation was to determine the condition of the MCO during and after an accidental drop event. It was, therefore, desirable to use actual "average" material properties instead of minimum-specified properties. A survey of actual material test reports for the MCO and basket materials was performed to determine the average properties for use in the FE analysis. A summary of that survey is shown in Table 15 (Snow 2003). The FE software requires that material stress and strain data be given as true values, not engineering (or "nominal") values. Therefore, the Table 15 data will be used to form simple bi-linear stress-strain curves for each component. This is done as follows:

1. True Stress $(\sigma_{true}) = \sigma_{nominal} \times [1 + \varepsilon_{nominal}]$

2. True Strain $(\varepsilon_{true}) = \ln [1 + \varepsilon_{nominal}] - \sigma_{true} / E$

The yield strength from Table 15 is defined at 0.2% offset, which was a nominal plastic strain of 0.002. This meant that the actual true stress at the nominal yield strength, using the previous equation, was a factor of 1.002 higher than the nominal yield strength. The difference between the two was considered negligible. Therefore, for the sake of simplicity in creating bi-linear stress-strain curves, the nominal yield strength was used as the true stress at a plastic strain of zero. The strain at fracture was calculated as follows:

True Fracture Strain $(\varepsilon_{f\,true}) = \ln [1 / (1-R/100)]$

Table 15. MCO and baskets material properties (Snow 2003).

Component	Yield Strength (σ_y, psi)	Ultimate Strength (σ_u, psi)	Elongation (%)	Area Reduction (R)	Average Yield Strength $(\sigma_{y\,ave} psi)$	True Fracture Strain $(\varepsilon_{f\,true})$
			MCO Components			
Main Shell SA-240/	41388	88145	59.2	70.1	44052	1.207
SA-312 (304/304L)	46067	91401	60.9	69.4		1.184
	44702	89164	61.8	70.8		1.231
Collar SA-182	43200	86000	56.0	77.0	41366	1.470
(F304/F304L)	40700	82000	60.0	79.0		1.561
	40200	86000	54.0	76.0		1.427
Bottom SA-182	35800	80800	56.0	75.0	36650	1.386
(F304/F304L)	35100	80400	63.0	75.0		1.386
	35600	80500	58.0	75.0		1.386
	40100	82900	57.0	73.0		1.309
Cover SA-182 (F304L)	40500	84500	62.0	78.0	44000	1.514
	47500	89000	55.0	75.0		1.386
			Mark 1A Basket Components			
Base Plate SA-182	44800	86500	57.0	78.0	42467	1.514
(F304/F304L)	42800	86000	58.0	77.0		1.470
	39800	80500	59.0	78.0		1.514
Perimeter Bar SA-479	64940	95730	47.1	75.1	64190	1.390
(304/304L)	63440	96100	47.6	77.4		1.487
Center Post SA-479	41200	89500	57.0	75.0	43800	1.386
(304/304L)	44000	88500	55.0	78.0		1.514
	46200	88000	54.0	77.0		1.470
			Mark IV Basket Components			
Base Plate A240	33252	79969	62.5	74.9	34357	1.382
(304/304L)	35100	85500	60.0	78.0		1.514
	34719	83504	62.3	73.3		1.321
Perimeter Bar SA-479	39200	91200	54.1	74.6	46225	1.370
(304/304L)	48500	95200	55.3	77.2		1.478
	55000	94000	48.0	74.0		1.347
	42200	89500	48.8	72.0		1.273
Center Post A511	33649	80239	54.8	Not available	39465	Insufficient data
(304/304L)	45282	84788	49.6			

117

Table 15 also lists the calculated true fracture strain for each material as calculated using the above equation. The matching true fracture stress must be determined or calculated next. In order to calculate the true fracture stress, the nominal stress (or force) at fracture must be known. For materials where the engineering stress-strain curve is always increasing (positive slope) to fracture (or at least not decreasing), the ultimate strength is also the fracture strength. However, with 304 and 304L stainless steels, the engineering stress-strain curve reaches an ultimate strength (highest strength on the curve) and then the curve decreases (negative slope)—meaning that the load decreases to fracture. In this case, using the engineering ultimate strength as the fracture strength would give a higher-than-actual true fracture strength ranging from 300,000-400,000 psi (2×10^6 to 2.7×10^6 kPa). Recourse to another source is therefore required. Holt (1995) shows a typical true stress-strain curve for 304 stainless steel with a true fracture stress of about 240,000 psi (1.6×10^6 kPa). This value is used in the FE models, simulating the MCO material property for all 304 and 304/304L stainless steels for the considered drop events.

The material stress-strain data discussed thus far has been based on a quasi-static strain rate. During an MCO drop event, the material strain rate will not be quasi-static—but comparatively quite high. Many materials, including stainless steels, are sensitive to strain rate and experience a significant dynamic strengthening due to high strain rates. Morton et al. (2000) documented the actual drop testing of nine representative standardized DOE SNF canisters and the accompanying FE analyses. A dynamic increase in strength of 20% was included in those analyses in order to match analytical to actual results. Morton et al. (2000) present the documentation and justification for the 20% strength increase. This MCO evaluation also includes a 20% increase in strength to account for the dynamic strengthening of the 304 and 304/304L stainless steels during the specified drop events.

Table 16 shows the actual material properties employed in the MCO analyses. The true stresses were increased by 20% to account for the dynamic strengthening. Several components were not included in the material survey. Those components used the material properties of adjacent components in the FE evaluation. Specifically:

1. The properties of the MCO bottom were applied to the guide cone and basket support bars
2. The properties of the cover were applied to the lockring, the shield plug, and the process tube
3. The properties of the basket base plate were applied to the shroud wall

Due to the similarity in properties, the Mark 1A base plate properties were applied to the center post as well.

The fuels were modeled as mass elements attached to the base plate of each basket in the vertical and near-vertical drops, and as six solid rods per basket for the remaining drops. The six solid rods per basket were representing the 48-54 fuel elements. As an approximation, these solid rods were defined to yield at 36,000 psi (0.25×10^6 kPa) and subsequently respond in a perfectly plastic manner (see Table 16).

Modeling software

The I-DEAS Master Series Version 9 m2 computer program was used to create the finite element models of the MCO, baskets, and fuels. A solid model was created and then used to generate the finite element model. The models were checked in the calculation software.

Table 16. True Stress-Strain curves employed in FE models (Snow 2003).

Component	True Stress/Matching Strain Points for Bi-Linear Curve	
	Point 1* (psi, MPa, in./in.)	Point** 2 (psi, GPa, in./in.)
MCO Components		
Bottom, guide cone, basket support bars	43800, 302, 0	288000, 1.98, 1.37
Main shell	52800, 364, 0	288000, 1.98, 1.21
Collar	49800, 343, 0	288000, 1.98, 1.49
Cover, locking ring, process tube, shield plug	52800, 364, 0	288000, 1.98, 1.45
Mark 1A Basket Components		
Base plates, center posts, shroud walls	51600, 356, 0	288000, 1.98, 1.48
Perimeter bars	76820, 530, 0	288000, 1.98, 1.44
Mark IV Basket Components		
Base plates, shroud walls	41400, 285, 0	288000, 1.98, 1.41
Perimeter bars	55200, 381, 0	288000, 1.98, 1.37
Center posts	47400, 327, 0	288000, 1.98, 1.40
Fuels		
Fuel	36000, 248, 0	36000, 0.248, 0

* This true stress was the Table 15 average yield strength multiplied by 1.20.
** This true stress was the selected true fracture stress multiplied by 1.20. The matching true fracture strain was the Table 15 average value.

Calculation software

The computer program ABAQUS/Explicit Version 6.3-3, a nonlinear FE analysis software package that is widely used in many industries, was employed to calculate the response of the MCOs to the specified drop events. Extensive validation and verification (DOE 2003b) has been performed by the NSNFP on this software, approving it for such container drop evaluations. This rigorous checking process eliminated the need to control or validate the solid modeling software. All models were run in double precision.

EVALUATION DROP EVENTS

The MCO finite element model with six Mark 1A fuel baskets, and the MCO finite element model with five Mark IV fuel baskets were prepared for evaluations of the repository drop events. Though only a vertical orientation was required for the 23-ft (6.9 m) drop event, it seemed prudent to look at a slightly off-vertical orientation as well. Therefore, drops from 1 degree and 3 degrees off-vertical were also evaluated. The vertical drop was also evaluated for an impact with a rigid edge at the collar region (simulating an impact while falling into another cylinder, such as a waste package). Additionally, the 3 degrees off-vertical drop was evaluated with a coefficient of friction of 0.1 and 0.5 at the rigid surface. Finally, the vertical drop was evaluated with a 450 psig (3.10 MPa) internal pressure to determine the effect of internal pressure on the resulting deformations and strains in the MCO containment components.

In order to determine the 2-ft (0.6 m) drop orientation that would cause the most severe challenge to the containment boundary of the MCO, a simplified FE model of the MCO was created and evaluated. The results showed that an impact angle of about 60 degrees off-vertical (30 degrees from horizontal) caused the MCO to rotate with the greatest velocity for the secondary impact where the greatest challenge to the containment boundary would occur. This was comparable to the 65 degrees off-vertical angle determined to be the worst case for the 24-in (610 mm) diameter standardized DOE SNF canister (Blandford 2003). Therefore, a drop from 60 degrees off-vertical was performed on the MCO models (bottom-end hits, followed by a slap-down of the top end). Additionally, drops with the MCO in a horizontal orientation and oriented at 115 degrees off-vertical were performed since the MCO was not symmetric—the top end and bottom end were not of the same design. Table 17 summarizes the drop events evaluated herein.

Table 17. Evaluated drop conditions on the MCO models (Snow 2003).

Model File Name	Drop Height (ft)	Drop Height (m)	Drop Angle (deg.)	Brief Description
MCO with Mark 1A Baskets				
MCO_markIA_m11_vertical_mass	23	6.9	0	Vertical drop
MCO_markIA_m11_1deg_mass	23	6.9	1	near vertical drop
MCO_markIA_m11_3deg_mass	23	6.9	3	near vertical drop
MCO_markIA_m11_hole_edge_finer	23	6.9	0	collar hole-edge drop**
MCO_markIA_m11_vertical_mass_pressure	23	6.9	0	vertical drop w/450 psig (3.1 MPa) internal pressure**
MCO_markIA_m11_3deg_mass_friction1	23	6.9	3	near vertical drop w/COF=0.1**
MCO_markIA_m11_3deg_mass_friction5	23	6.9	3	near vertical drop w/COF=0.5**
MCO_markIA_m11_60deg_fuel	2	0.6	60	worst (slapdown) drop angle
MCO_markIA_m11_90deg_fuel	2	0.6	90	Horizontal drop
MCO with Mark IV Baskets				
MCO_markIV_m11_vertical_mass	23	6.9	0	Vertical drop
MCO_markIV_m11_1deg_mass	23	6.9	1	near vertical drop
MCO_markIV_m11_3deg_mass	23	6.9	3	near vertical drop
MCO_markIV_m11_60deg_fuel	2	0.6	60	worst (slapdown) drop angle
MCO_markIV_m11_90deg_fuel	2	0.6	90	Horizontal drop
MCO_markIV_m11_115deg_fuel	2	0.6	115	additional slapdown drop***

* Angle measured from vertical. For example, 0 degrees is vertical, and 90 degrees is horizontal.
** Only the MCO with Mark 1A baskets was evaluated for this drop.
*** Only the MCO with Mark IV baskets was evaluated for this drop.

RESULTS OF VERTICAL AND NEAR-VERTICAL 23 FT (6.9 M) DROP

Vertical and near-vertical drop model energies

Figures 59 through 64 show plots of the energy history for each vertical or near-vertical drop scenario. The plots show model kinetic energy history (ALLKE); plastic dissipation history (ALLPD); frictional dissipation history (ALLFD); elastic energy history (ALLSE), and artificial energy history (ALLAE). An artificial energy total of 3%-5% for a drop evaluation is typical. Figures 59 through 61 show that the kinetic energy had bottomed-out near zero and that the plastic energy dissipation had leveled-off, indicating that the MCO with Mark 1A fuel baskets impacted and then rebounded off the surface. Minor oscillations of the elastic energy occurred, as expected. Artificial energy reached the following levels: 3.7% for the vertical drops; 4.2% for 1 degree off-vertical drops; and 5.5% for 3 degrees off-vertical drops, respectively. From the standpoint of model energy, these FE model results were considered valid.

Figures 62 through 64 show that the kinetic energy had not quite bottomed-out and that the plastic energy dissipation had not completely leveled-off, indicating that the MCO with Mark IV fuel baskets had not yet completed the impact event. However, a review of the plastic straining occurring in the MCO containment boundary components showed that they had stopped absorbing drop energy in plastic deformation, and that plastic energy was increasing in the internal baskets only. Continuing the FE model evaluations would only increase the damage to the baskets, but would not change the results for the containment boundary components. Therefore, the drop evaluations were considered complete. Minor oscillations of the elastic energy occurred, as expected. Artificial energy reached the following levels: 4.8% for the vertical drops; 5.7% for 1 degree off-vertical drops; and 6% for 3 degrees off-vertical drops, repectively. From the standpoint of model energy, these FE model results were considered valid. Though slightly higher than desirable, these artificial energy levels were considered acceptable.

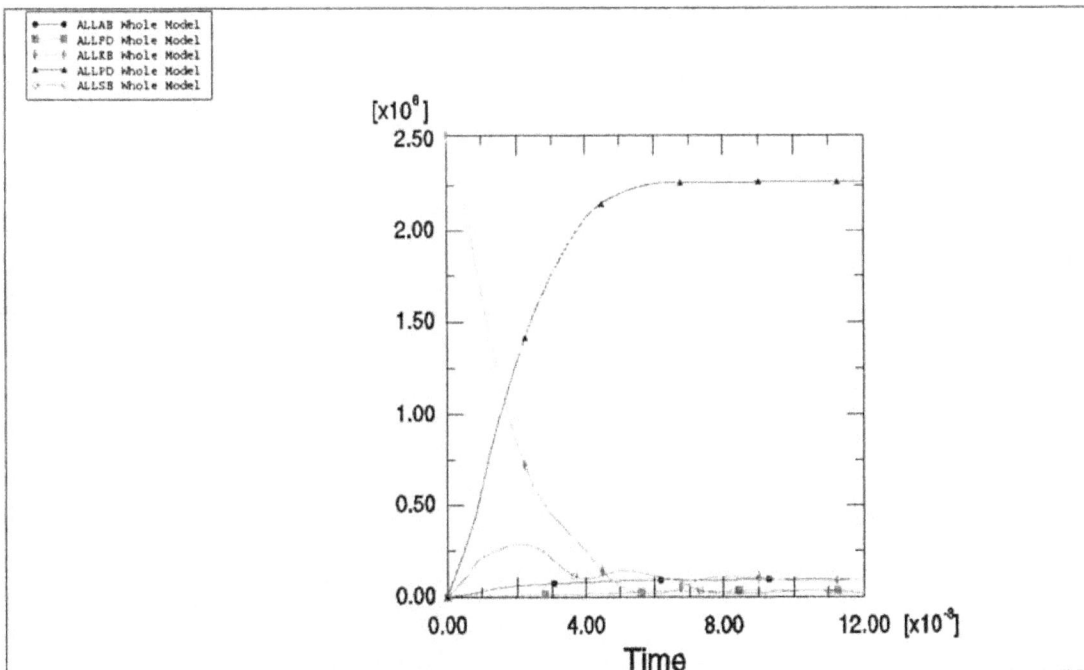

Fig. 59. MCO Mark 1A, 23-foot (6.9-m) vertical drop – model energy (Snow 2003).

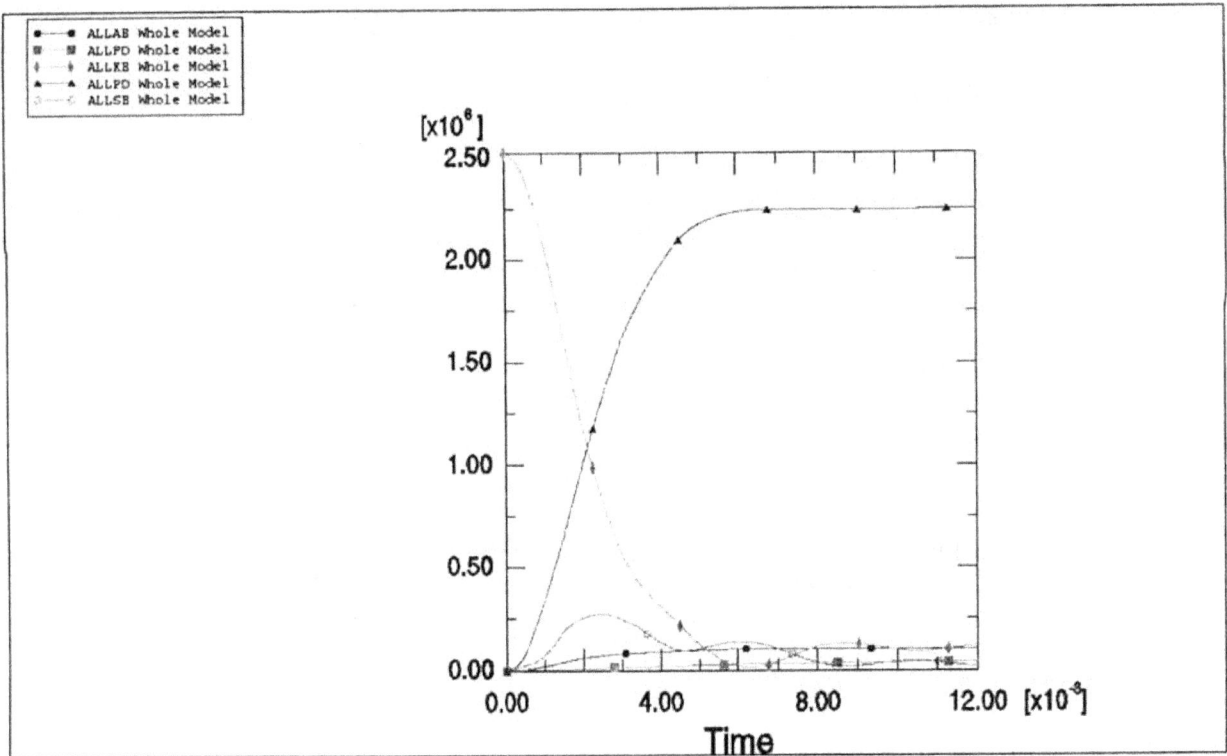

Fig. 60. MCO Mark 1A, 23-foot (6.9-m) 1 deg. off-vertical drop – model energy (Snow 2003).

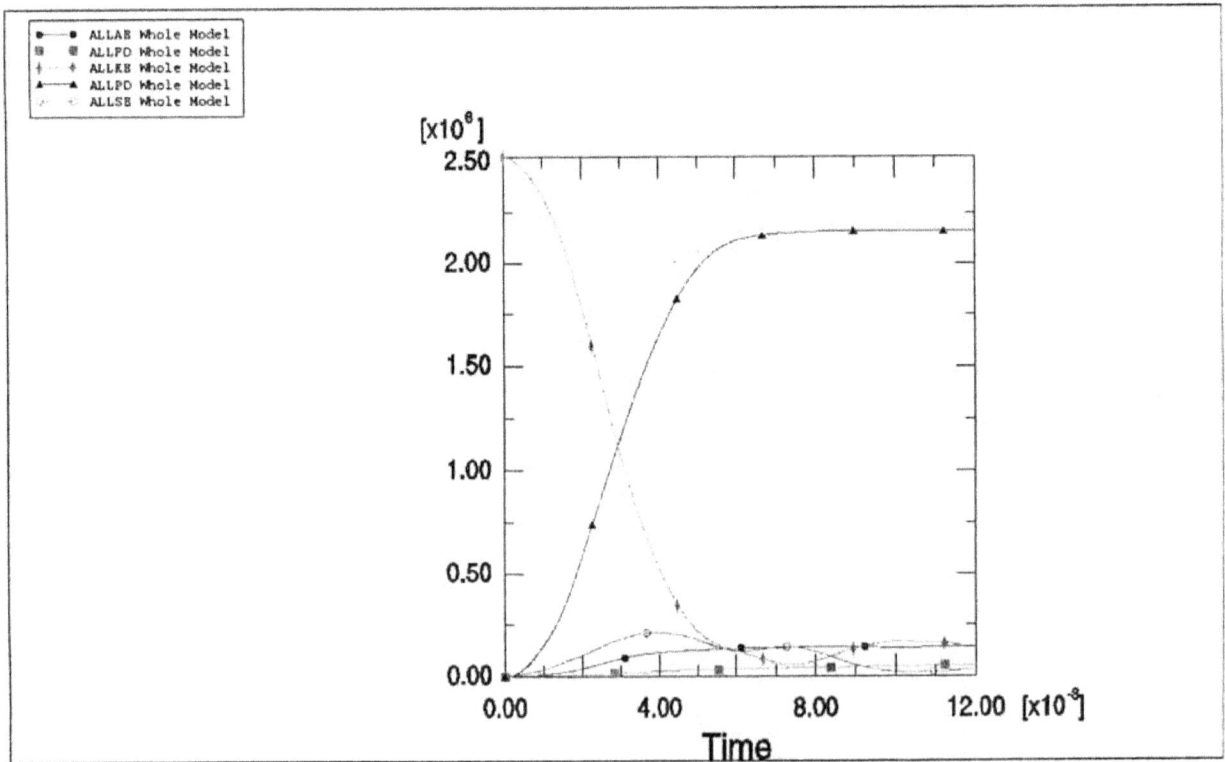

Fig. 61. MCO Mark 1A, 23-foot (6.9-m) 3 deg. off-vertical drop – model energy (Snow 2003).

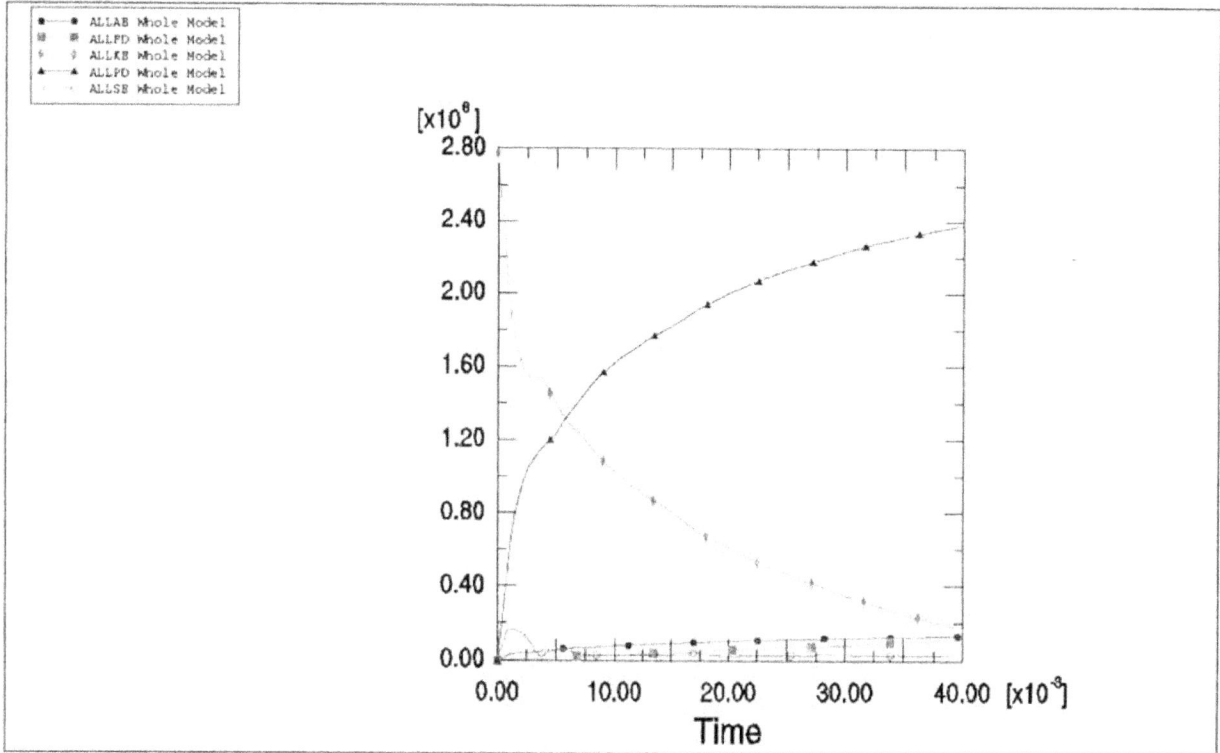

Fig. 62. MCO Mark IV, 23-foot (6.9-m) vertical drop – model energy (Snow 2003).

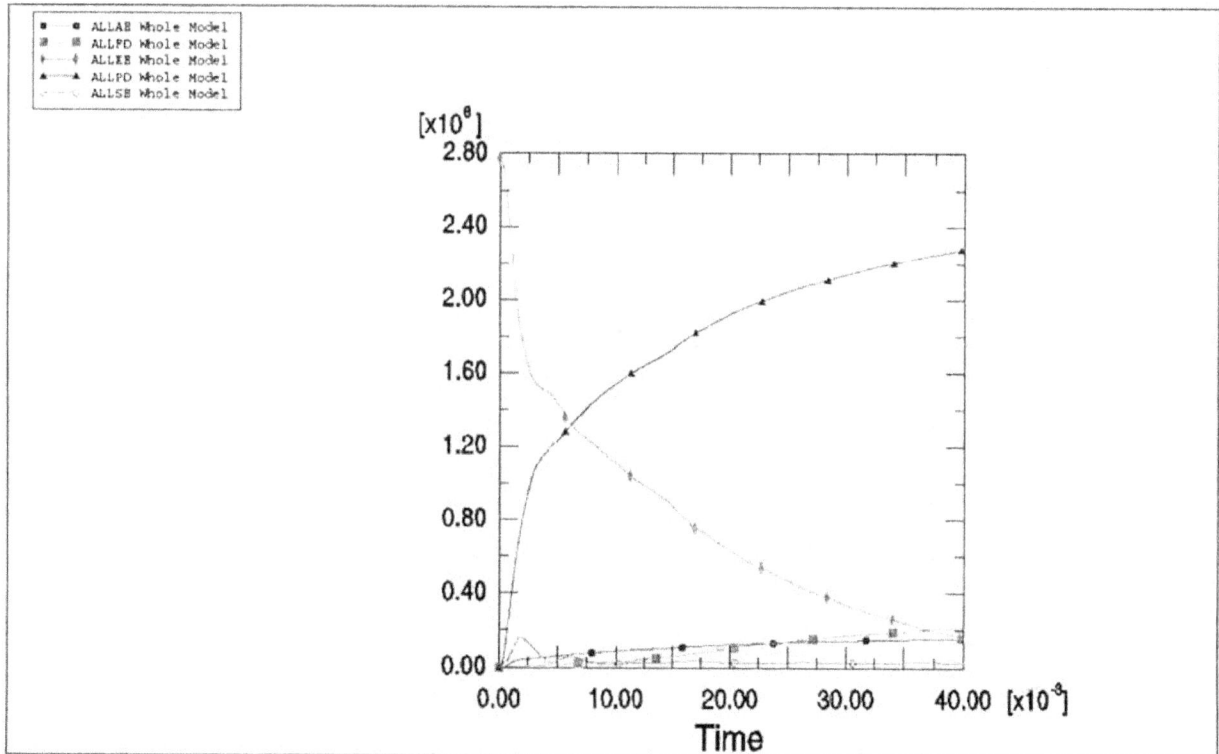

Fig. 63. MCO Mark IV, 23-foot (6.9-m) 1 deg. off-vertical drop – model energy (Snow 2003).

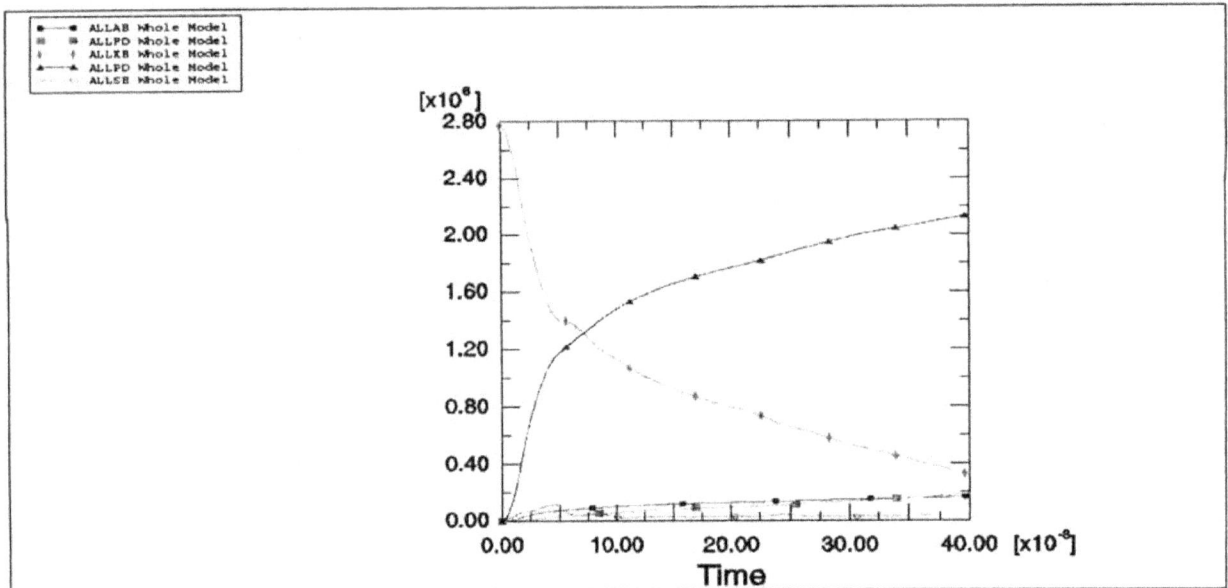

Fig. 64. MCO Mark IV, 23-foot (6.9-m) 3 deg. off-vertical drop – model energy (Snow 2003).

Vertical and near-vertical drop model deformations

Figures 65 through 70 show the deformed shape of the MCO models, for the vertical and near-vertical drops, at the end of the drop evaluation. Overall, the predicted deformations were in agreement with the expected results. Permanent deformation to the MCO bottom and main shell for the vertical drop orientation were minimal. The 1 and 3 degrees off-vertical drop orientations caused one bulge in the main shell near the bottom.

The bottom Mark 1A baskets showed some bending in the perimeter bars while the thick center post showed minimal deformation. This was in contrast to the bottom Mark IV baskets, which showed significant bending of the perimeter bars and the thinner center post. Clearly, the fuels within the bottom Mark IV basket would be more severely damaged than those in the Mark 1A basket during these drop events.

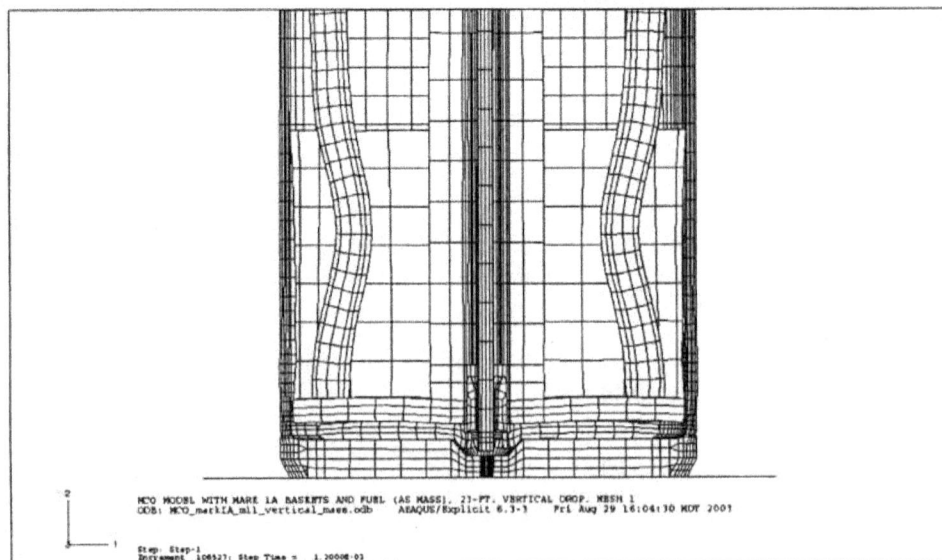

Fig. 65. MCO Mark 1A, 23-foot (6.9-m) vertical drop – deformed shape (bottom end) (Snow 2003).

124

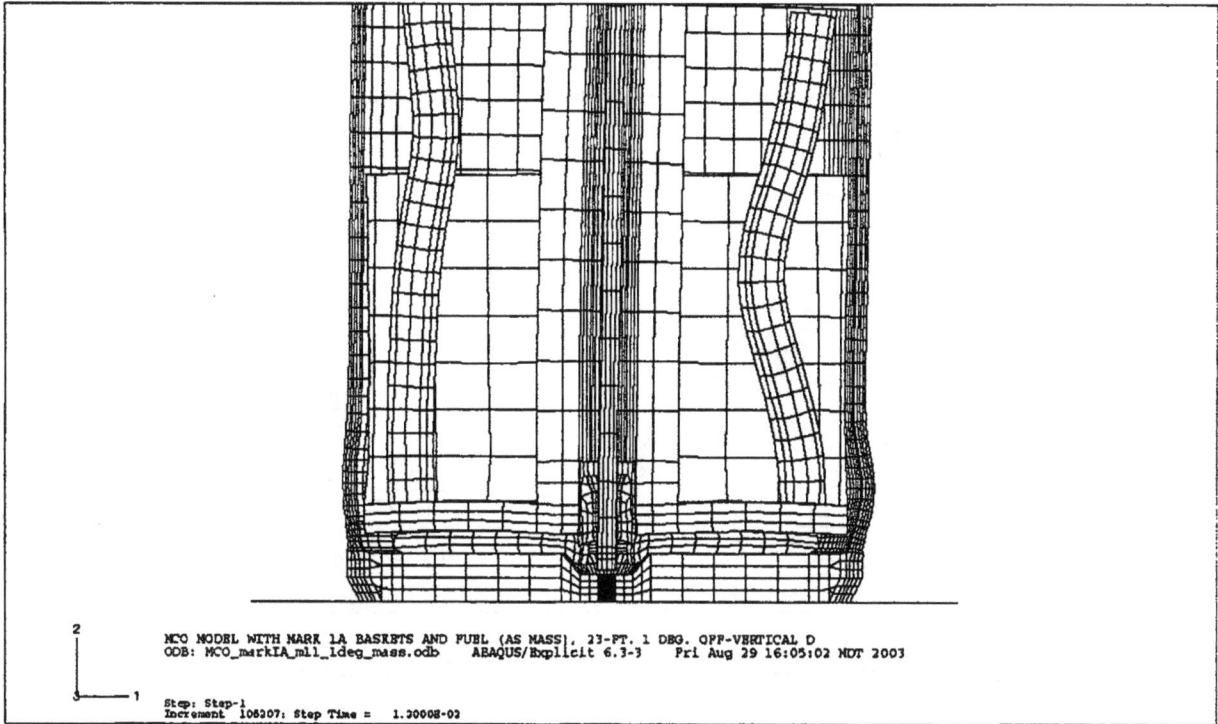

Fig. 66. MCO Mark 1A, 23-foot (6.9-m) 1 deg. off-vertical drop – deformed shape (bottom end) (Snow 2003).

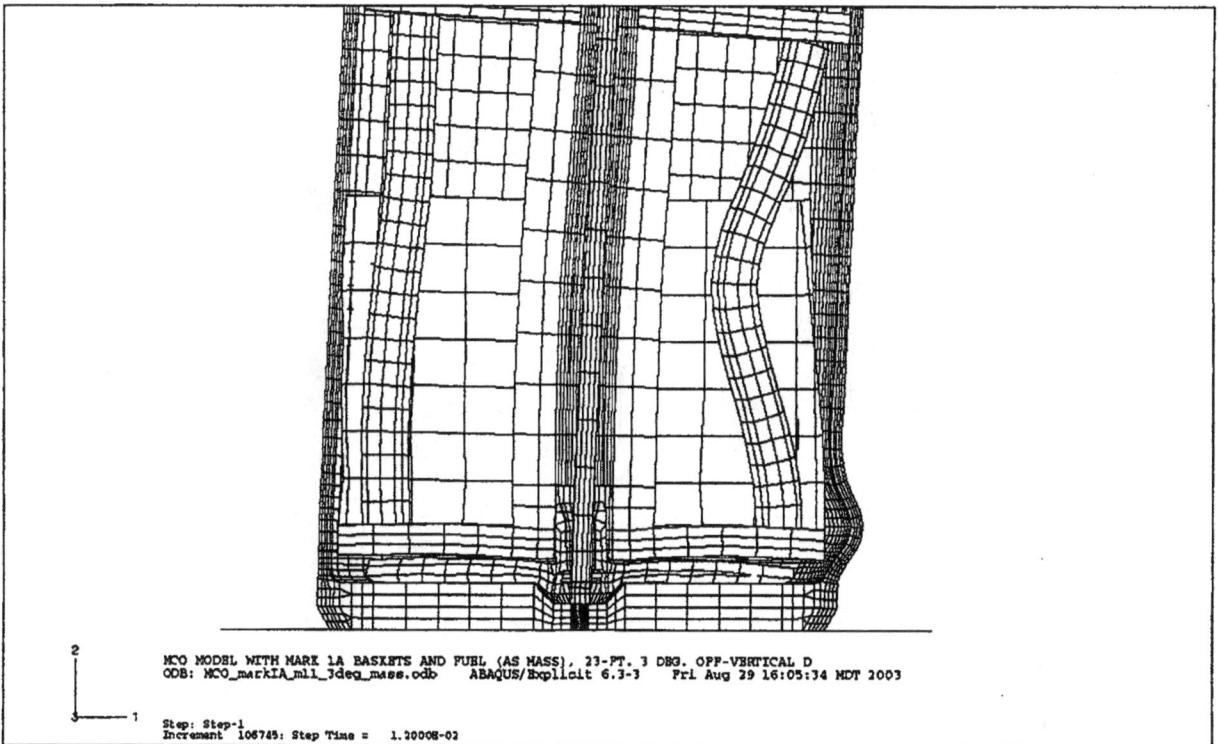

Fig. 67. MCO Mark 1A, 23-foot (6.9-m) 3 deg. off-vertical drop – deformed shape (bottom end) (Snow 2003).

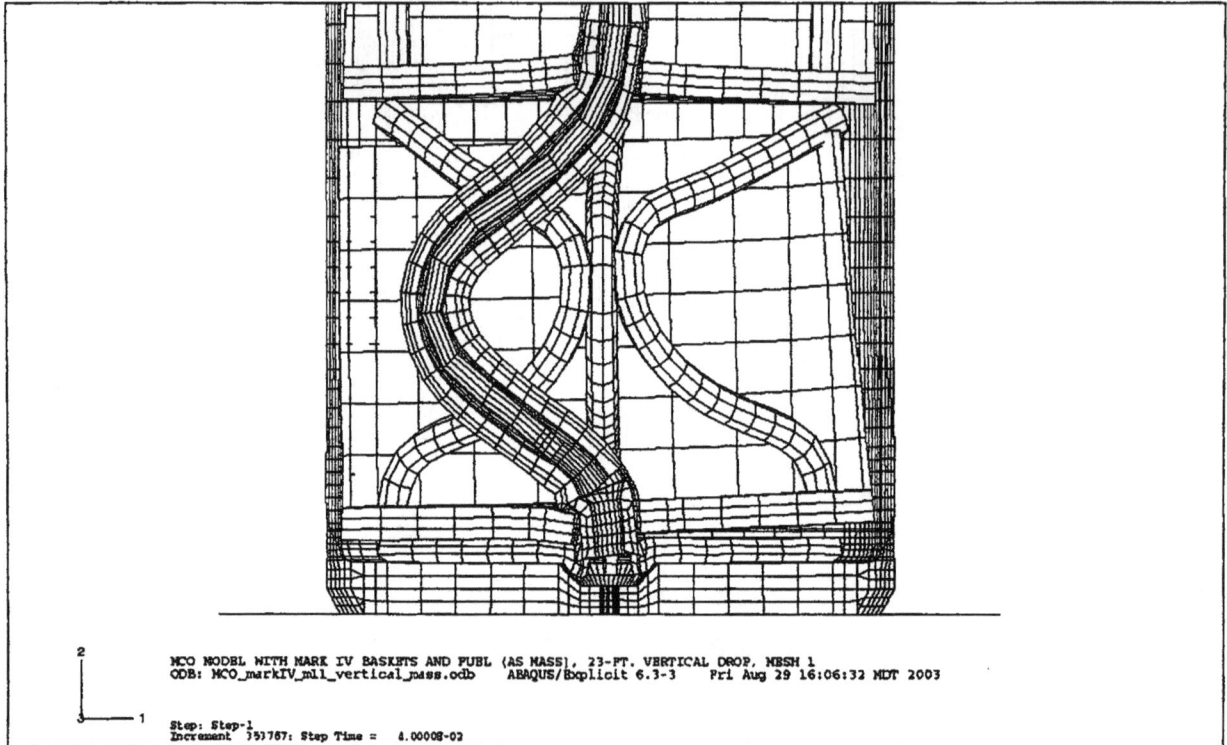

Fig. 68. MCO Mark IV, 23-foot (6.9-m) vertical drop – deformed shape (bottom end) (Snow 2003).

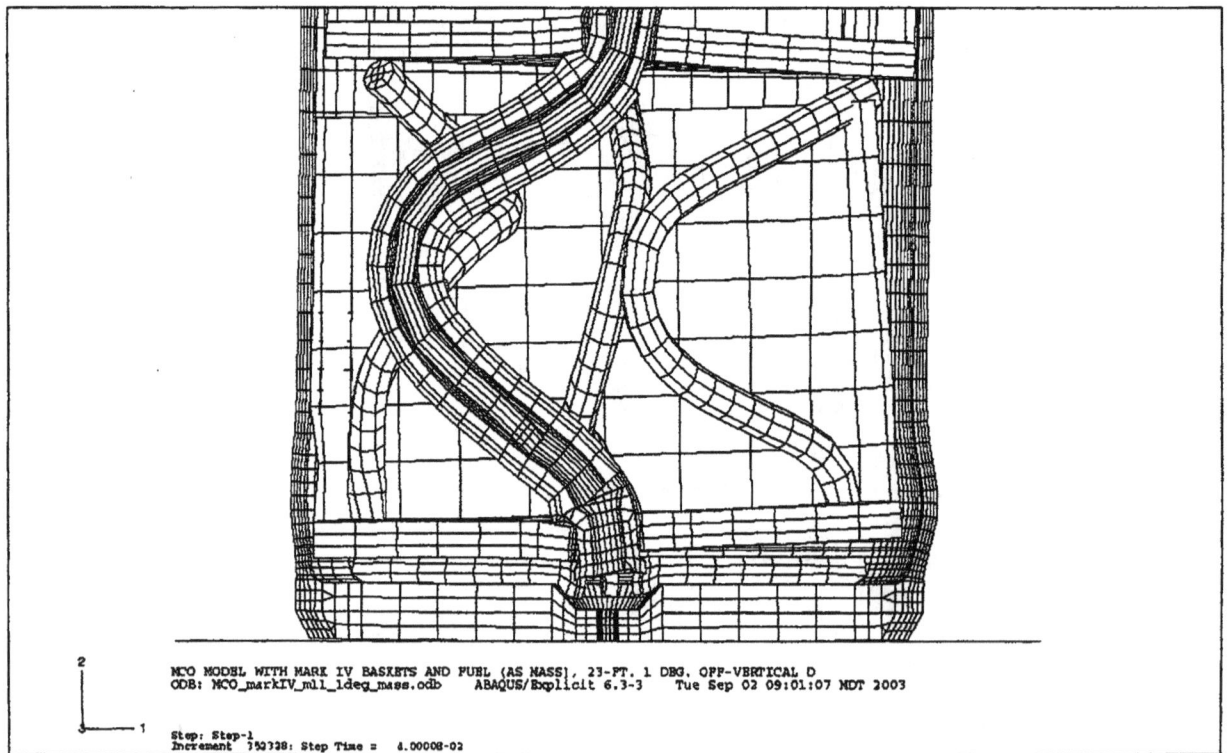

Fig. 69. MCO Mark IV, 23-foot (6.9-m) 1 deg. off-vertical drop – deformed shape (bottom end) (Snow 2003).

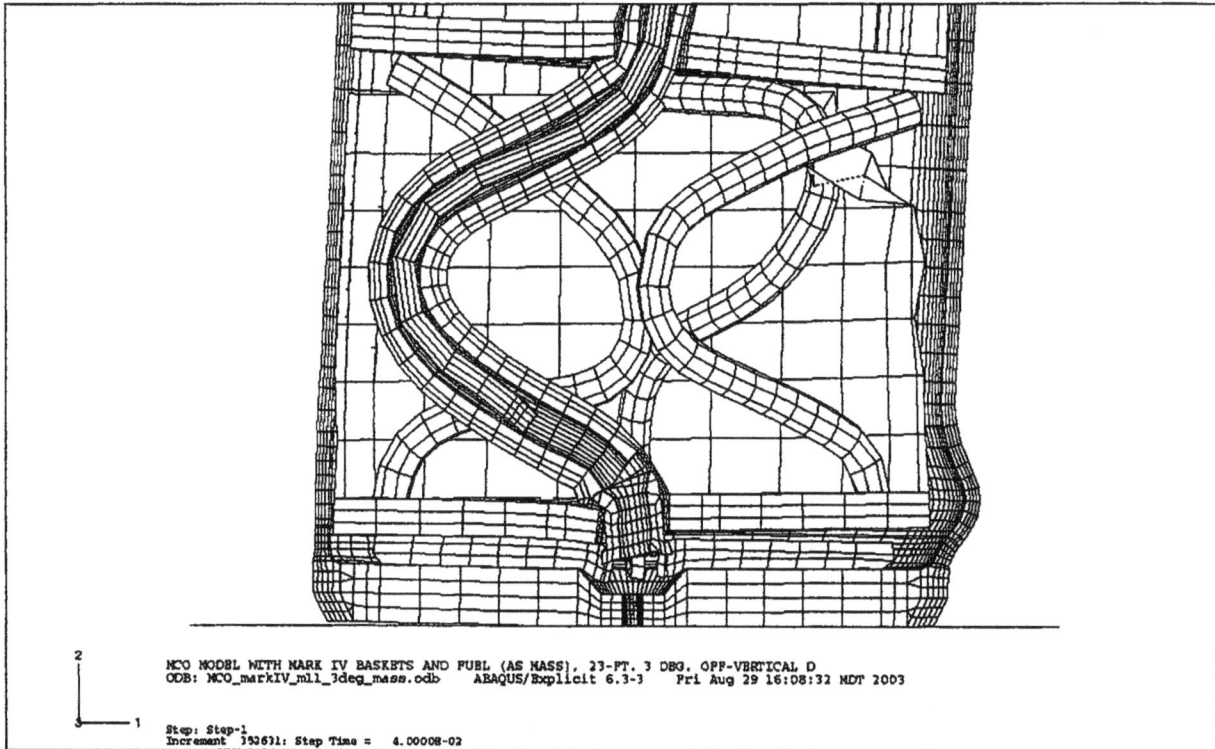

Fig. 70. MCO Mark IV, 23-foot (6.9-m) 3 deg. off-vertical drop – deformed shape (bottom end) (Snow 2003).

Vertical and near-vertical drop model plastic strains

During these drop events, the bulk of the kinetic energy at impact is transformed into plastic work. The best measure of that plastic work is the equivalent plastic strain, which is a cumulative strain measure that takes into account the entire deformation history. Table 18 lists the peak equivalent plastic strains occurring in the various containment boundary components of the MCO during the specified vertical and near-vertical drop events. Peak strains are listed for the outside surface of a component and inside surface. However, these peak strains did not necessarily occur at the same location on a component. Several conclusions may be drawn from the peak plastic strain data listed in Table 18:

1. The strains in the collar were negligible. Though not listed in the above table, the strains for the other remaining containment boundary component—the cover—were zero for these drop events.
2. The peak basket strains were quite different. The Mark 1A basket (a much more substantial basket in terms of the center post) experienced low peak strains whereas the Mark IV basket (with a comparably thin center post) resulted in peak surface strains approaching 100%. This straining was due to bending of the center post and perimeter bars. This indicated that damage to the fuels within the Mark IV baskets would likely be much greater than in the Mark 1A baskets for these drop events.
3. The peak strains in the MCO bottom at the flange and the main shell were similar for both basket configurations. This was expected because, in these vertical and near vertical drop events, the MCO must absorb the energy of its own structure, while the baskets must absorb their own drop energy. These peak strains were primarily due to the bending of the main wall and bottom flange during the impact. Those strain levels were much higher for the 1 and 3 degrees off-vertical drops than for the vertical drops.

127

Table 18. Peak plastic strains in the MCO for 23-foot (6.9-m) vertical and near vertical drop events (Snow 2003).

| MCO With Basket Configuration | Drop Angle From Vertical (deg.) | Peak Equivalent Plastic Strain (PEEQ) % | | | | | | |
| | | Bottom at Flange | | Main Shell | | Collar | | Lower Basket |
		Outside Surface	Inside Surface	Outside Surface	Inside Surface	Outside Surface	Inside Surface	Maximum Anywhere
Mark 1A (t = 0.012 sec)	0	4	5*	1	3	1	2	15
	1	18	7	5	13	<0.1	0.1	14
	3	34	13	13	28	0	0	14
Mark IV (t = 0.040 sec)	0	5	5**	1	3	1	2	99
	1	20	4	5	14	0.1	0.1	95
	3	35	14	13	29	0	0	93

(100% threshold nodal averaging, used extrapolated, averaged values - typical throughout this report.)
* For this model, the maximum strain in the bottom of 9% occurred under a basket support bar and was due to bearing loads.
** For this model, the maximum strain in the bottom of 8% occurred under a basket support bar and was due to bearing loads.

The maximum strain from the vertical drop was due to the baskets bearing on the MCO bottom through the basket support bars. This maximum strain was 9%, and was limited in depth. The other significant straining in this vertical drop occurred in the flange of the bottom and the main shell. Those strain levels were 5% or less through the thickness. This strain level was very low, especially when compared to the minimum fracture strain level of 118% or the minimum elongation of 47% listed in Table 15 for MCO containment components. It is possible that material damage short of fracture could cause leaking. In that case, a maximum allowable strain must be less than the fracture strain. Elongation, an "average" strain over a gage length at fracture, is much lower than the fracture strain and is considered acceptable for comparative purposes. Therefore, the MCO would most likely maintain containment during the vertical drop event.

The maximum strain from the near-vertical drop events (specifically, from the 3 degrees off-vertical drop) of 35% was also less than the minimum fracture strain level of 118%, or minimum elongation of 47% listed in Table 15 for MCO containment boundary components. Therefore, it was expected that the MCO would maintain containment during and after these near-vertical drop events as well.

Internal pressure

The MCO with the cover welded in place has a design pressure of 450 psig (3.10 MPa) (Lorenz et al. 1999). In order to provide insights into the effect of a 450 psig (3.10 MPa) maximum internal pressure on an MCO during a drop event, Snow (2003) evaluated the MCO with Mark 1A baskets and an internal pressure of 450 psig (3.10 MPa) under a 23-ft (6.9 m) vertical drop scenario. The pressure-stiffening effects provided by the 450 psig (3.10 MPa) internal pressure had a small—though beneficial effect on the MCO containment boundary components during the specified vertical drop event. Results for other drop scenarios were expected to be similar. Therefore, all other evaluations used zero internal pressure.

Friction

From a review of the data presented by Snow et al. (2001), it was expected that a coefficient of friction (COF) between the MCO and the rigid surface of 0.3 would produce valid results. For the drop angles considered, a range of COF values would be acceptable. This would indicate that the deformations and strains were not highly dependent on the COF value. To show this, the MCO with Mark 1A fuel baskets was also evaluated for the 3 degrees off-vertical drop with a COF of 0.1 and a COF of 0.5. All previous drop evaluations used a COF of 0.3 between the MCO and the rigid surface, as discussed earlier.

The calculated deformations and strains on the MCO containment boundary components for a 3 degrees off-vertical drop using a COF of 0.1 through 0.5 were consistent. Results for the vertical and 1 degree off-vertical drop scenarios were expected to be similar. In large part, because of the low drop height, the remaining 2-ft (6.9 m) drop scenarios were also expected to show similar results. Thus, using a COF of 0.3 for the MCO under the specified drop events was considered acceptable.

RESULTS OF VERTICAL EDGE 23 FT (6.9 M) DROP

Edge drop model energies

Figure 71 shows a plot of the energy history for this edge drop scenario. The plot shows model kinetic energy history (ALLKE), plastic dissipation history (ALLPD), frictional dissipation history (ALLFD), elastic energy history (ALLSE), and artificial energy history (ALLAE). An artificial energy total of 3%-5% for a drop evaluation is typical. This drop experienced a 2.2% total artificial energy, which was better than expected. Figure 71 shows that about 1/3 of the kinetic energy was expended in the edge impact. Most of that energy went into plastically deforming the collar in the impact area (at the diameter transition), with a small quantity going to frictional dissipation. As expected, an oscillation of the elastic energy occurred. From the standpoint of model energy, these FE model results were considered valid.

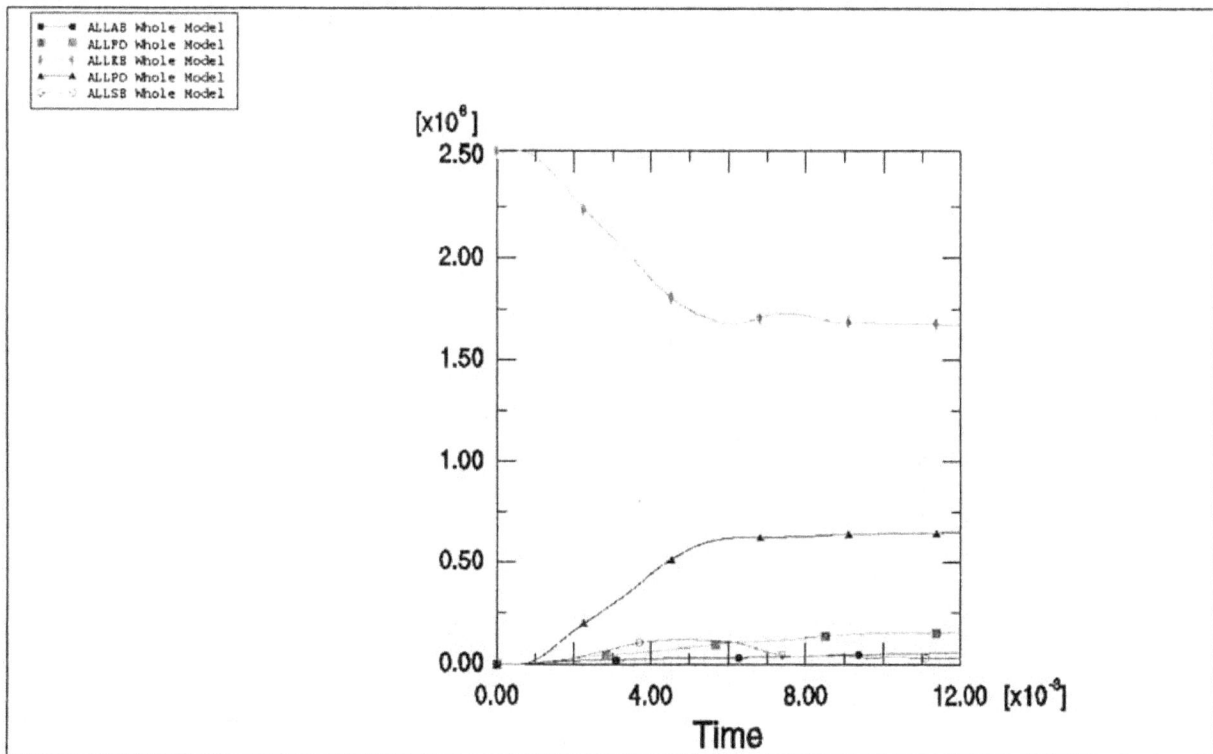

Fig. 71. MCO Mark 1A, 23-foot (6.9-m) vertical edge drop – model energy (Snow 2003).

Edge drop model deformations

Figure 72 shows the deformed shape of the MCO model for the edge impact for the drop. As expected, the collar at the diameter transition deformed significantly—pulling material at the outer diameter up and out. This resulted in some local bending of the collar wall as well.

Edge drop model plastic strains

The only significant material straining that occurred in the MCO during this edge impact event was in the collar. The equivalent plastic strains in the collar reached a peak of 130%. If this level of strain had occurred through the collar wall it would most likely have ruptured. However, this peak strain level was only on the surface. The outside half of the collar wall at this location was strained from 17% (midplane) to 130% (outside wall), where the inside half of the wall was strained from 6% (inside wall) to 9% (adjacent to the midplane). These strain levels were very low, especially when compared to the minimum fracture strain level of 118% or the minimum elongation of 47% listed in Table 15 for MCO containment components. This shows that most of the collar wall maintained its integrity during this event, and no rupture would have occurred. The secondary impact of this MCO onto a rigid surface (e.g., representing a waste package bottom) would not increase the damage in this area of the collar since the impact would be vertical or near-vertical (which was shown not to cause significant collar damage).

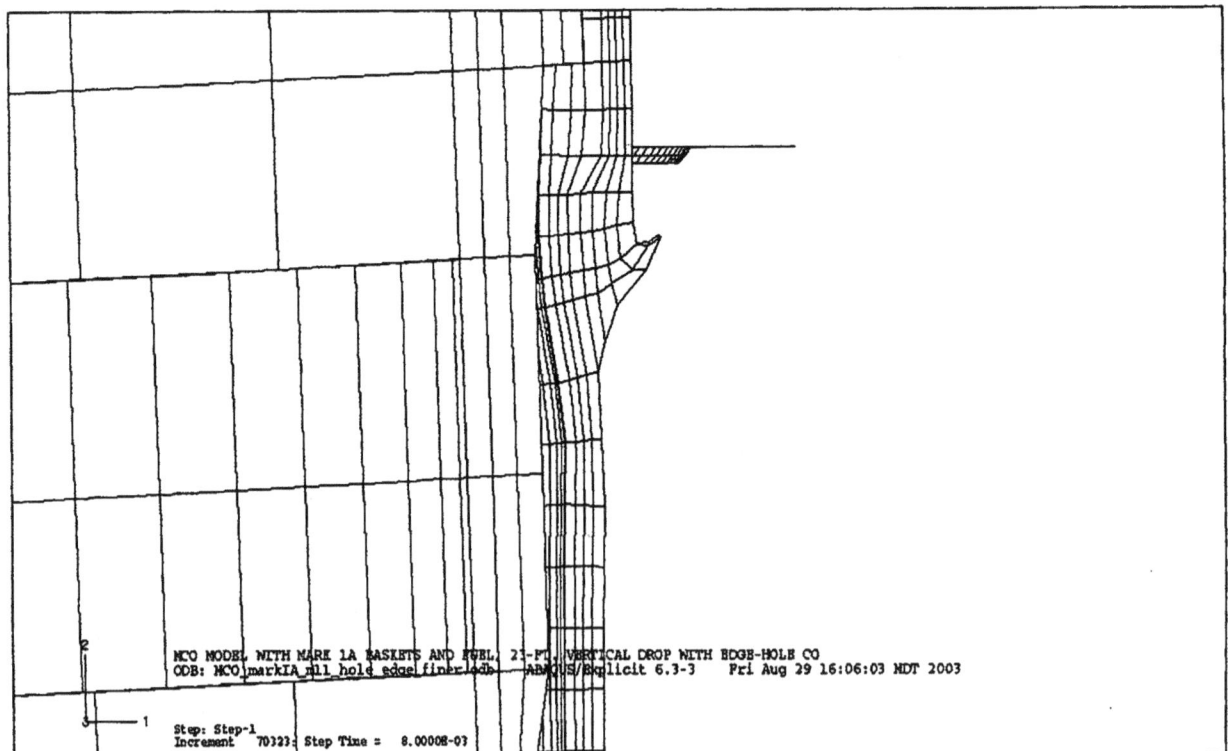

Fig. 72. MCO Mark 1A, 23-foot (6.9-m) vertical edge drop – deformed shape (Snow 2003).

WORST ORIENTATION DROP RESULTS AT 2 FT (0.6 M)

Worst orientation drop model energies

Figures 73 through 77 show plots of the energy history for each considered drop scenario. The plots show model kinetic energy history (ALLKE); plastic dissipation history (ALLPD); frictional dissipation history (ALLFD); elastic energy history (ALLSE); and artificial energy history (ALLAE). An artificial energy total of 3%-5% for a drop evaluation is typical. Figures 73 through 77 show that these drop scenarios were quite challenging from a numerical standpoint. When stretching, bending, or crushing occur in a FE model the numerical routines work considerably well. However, when the primary energy absorption mode is plastic-straining due to bearing (compression) loads, the numerical routines require more artificial energy to prevent numerical instabilities. In this present case, the top end of the MCO at the collar was quite rigid due to the locking ring and plug structures within. The thick bottom looked near-rigid, especially since the 2-ft (0.6 m) drop height did not provide enough drop energy to cause bending in the bottom. For the MCO with Mark 1A Baskets, the 60 degrees off-vertical drop required 0.4% artificial energy for the initial impact (and an additional 5.6% artificial energy for the secondary impact); while the 90 degrees off-vertical (horizontal) drop required 6.3% artificial energy. For the MCO with Mark IV Baskets, the 60 degrees off-vertical drop required 1.4% artificial energy for the initial impact (and an additional 6.3% artificial energy for the secondary impact); the 90 degrees off-vertical (horizontal) drop required 8.8% artificial energy; and the 115 degrees off-vertical drop required 2.9% artificial energy for the initial impact (and an additional 6.3% artificial energy for the secondary impact). The above artificial energies were higher than ideal, but were considered acceptable based on the numerical difficulties inherent to the specific conditions of these drop evaluations.

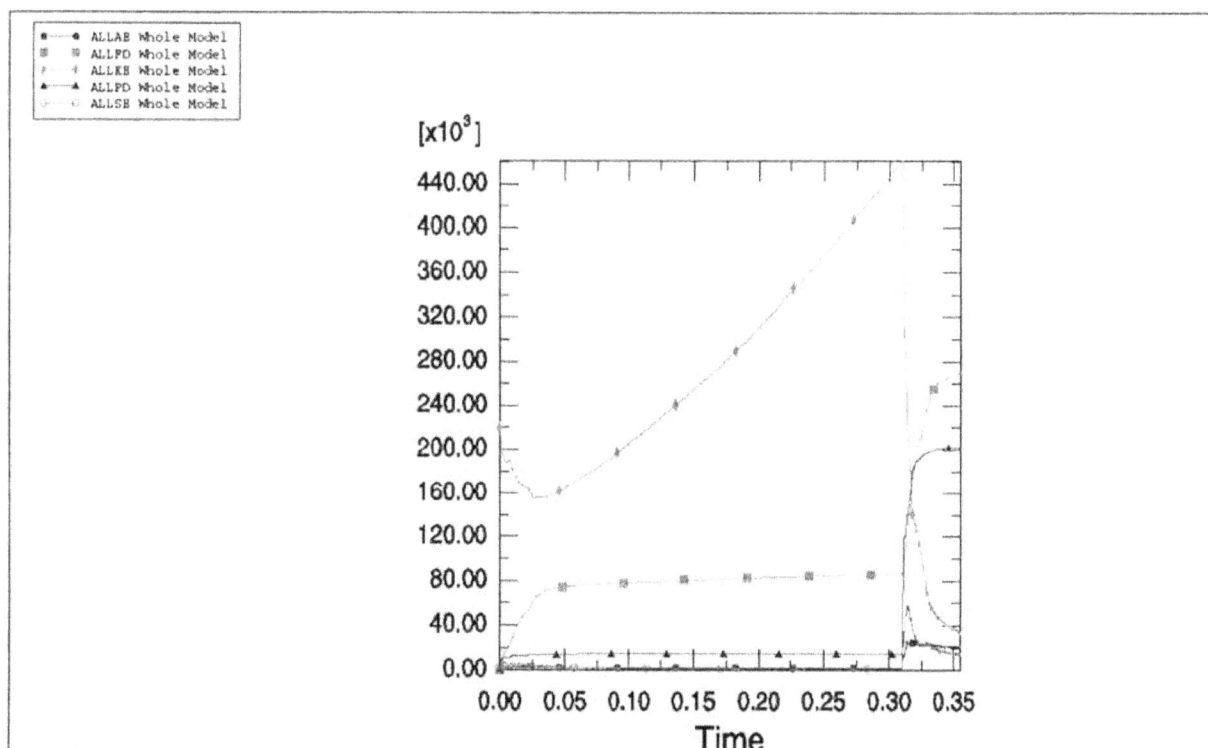

Fig. 73. MCO Mark 1A, 2-foot (0.6-m) 60 deg. off-vertical drop – model energy (Snow 2003).

[x10³]

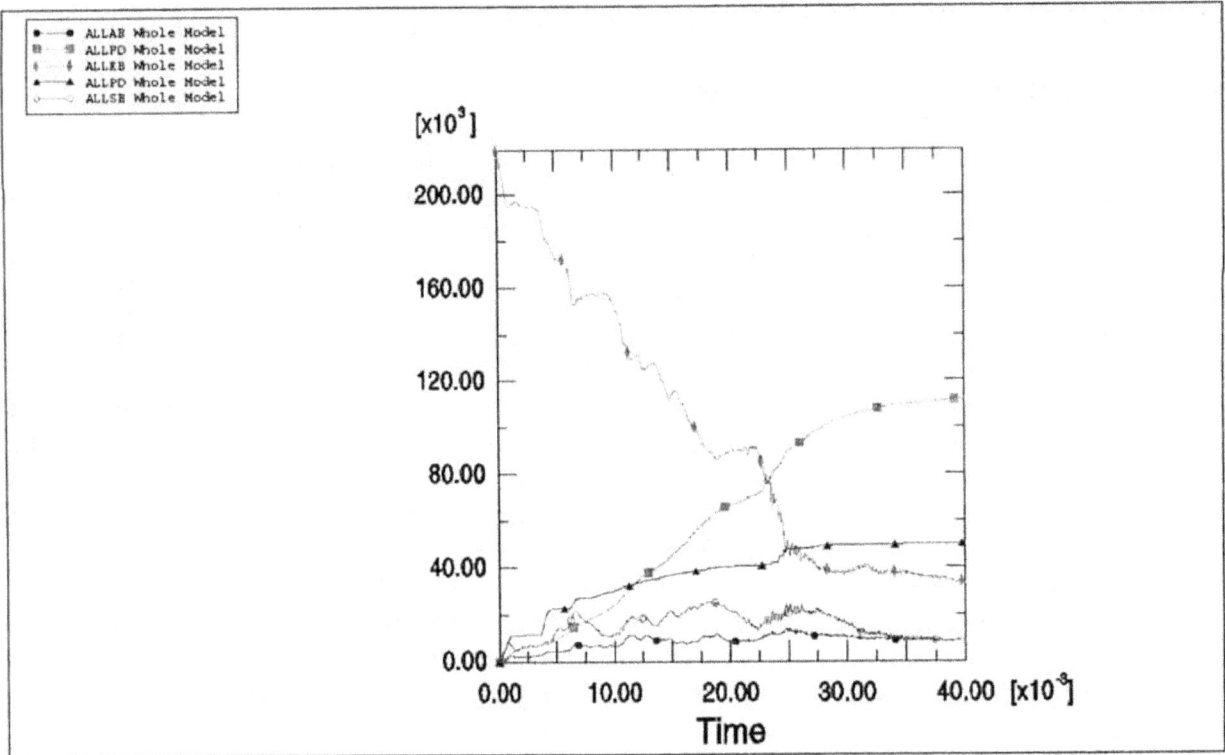

Fig. 74. MCO Mark 1A, 2-foot (0.6-m) 90 deg. off-vertical drop – model energy (Snow 2003).

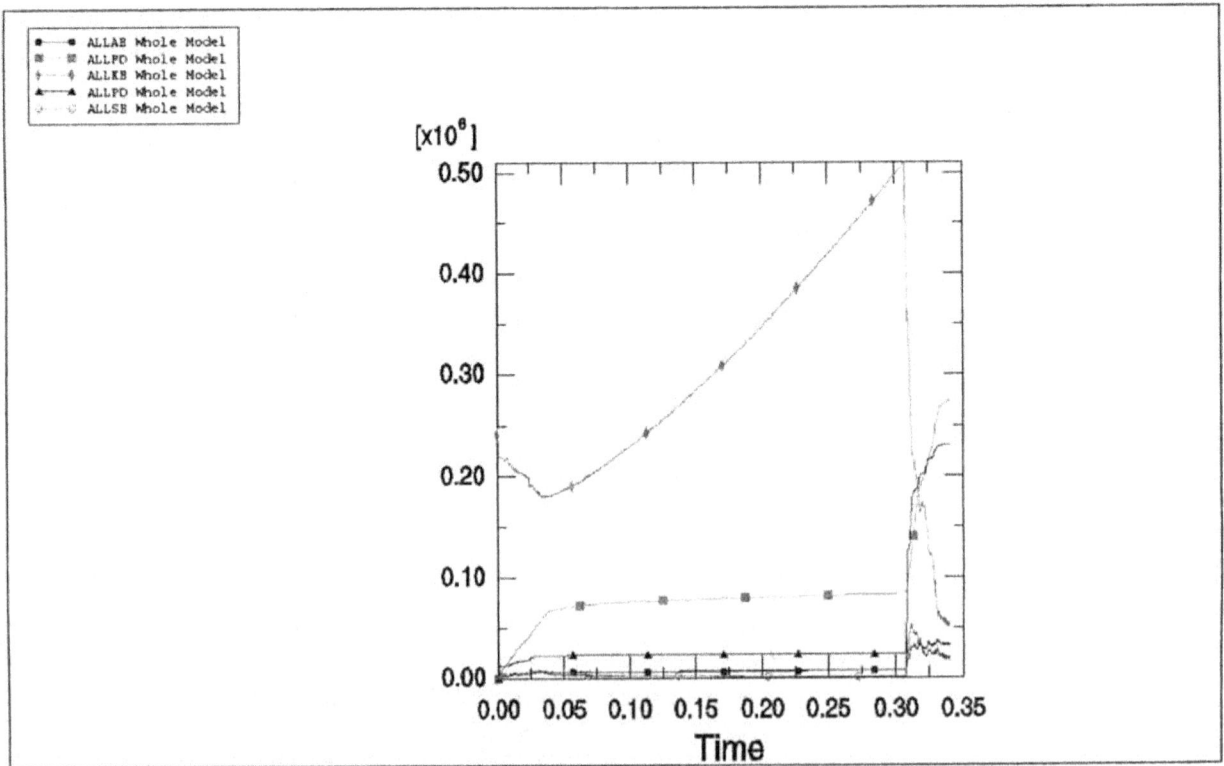

[x10⁶]

Fig. 75. MCO Mark IV, 2-foot (0.6-m) 60 deg. off-vertical drop – model energy (Snow 2003).

[x10³]
240.00
200.00
160.00
120.00
80.00
40.00
0.00

0.00   10.00   20.00   30.00   40.00 [x10⁻³]

Time

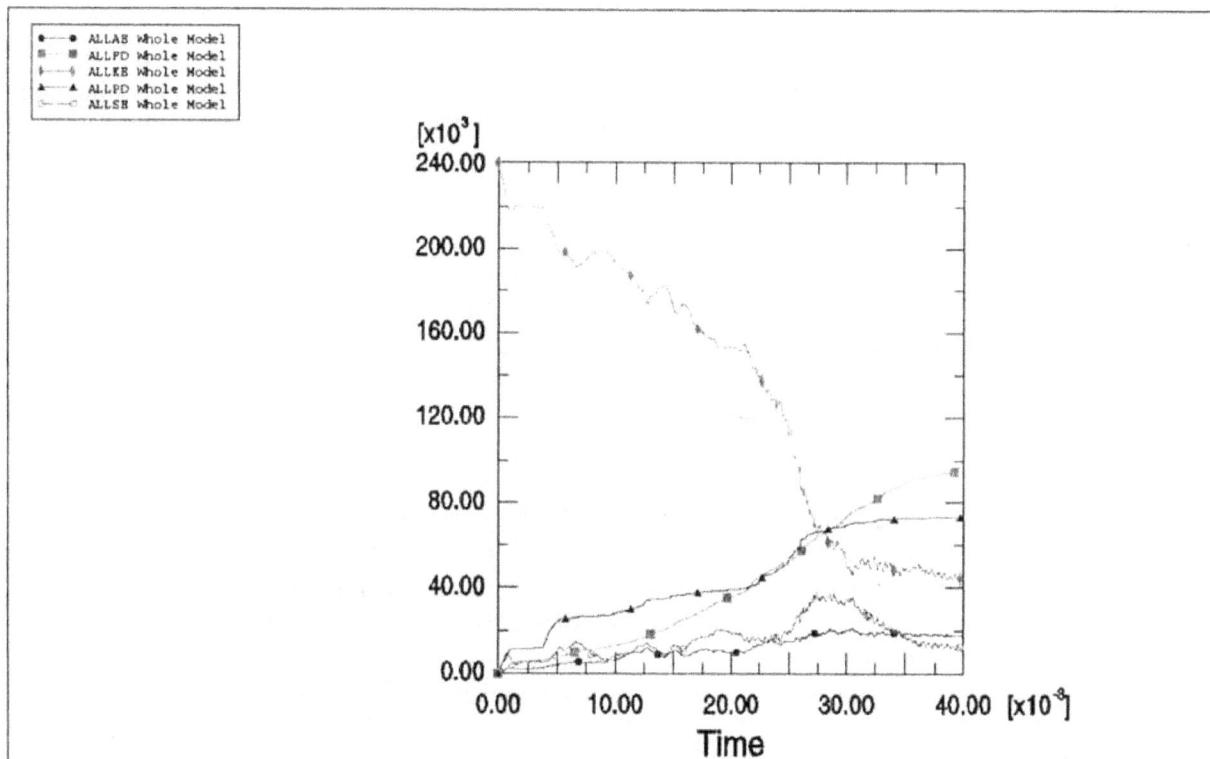

Fig. 76. MCO Mark IV, 2-foot (0.6-m) 90 deg. off-vertical drop – model energy (Snow 2003).

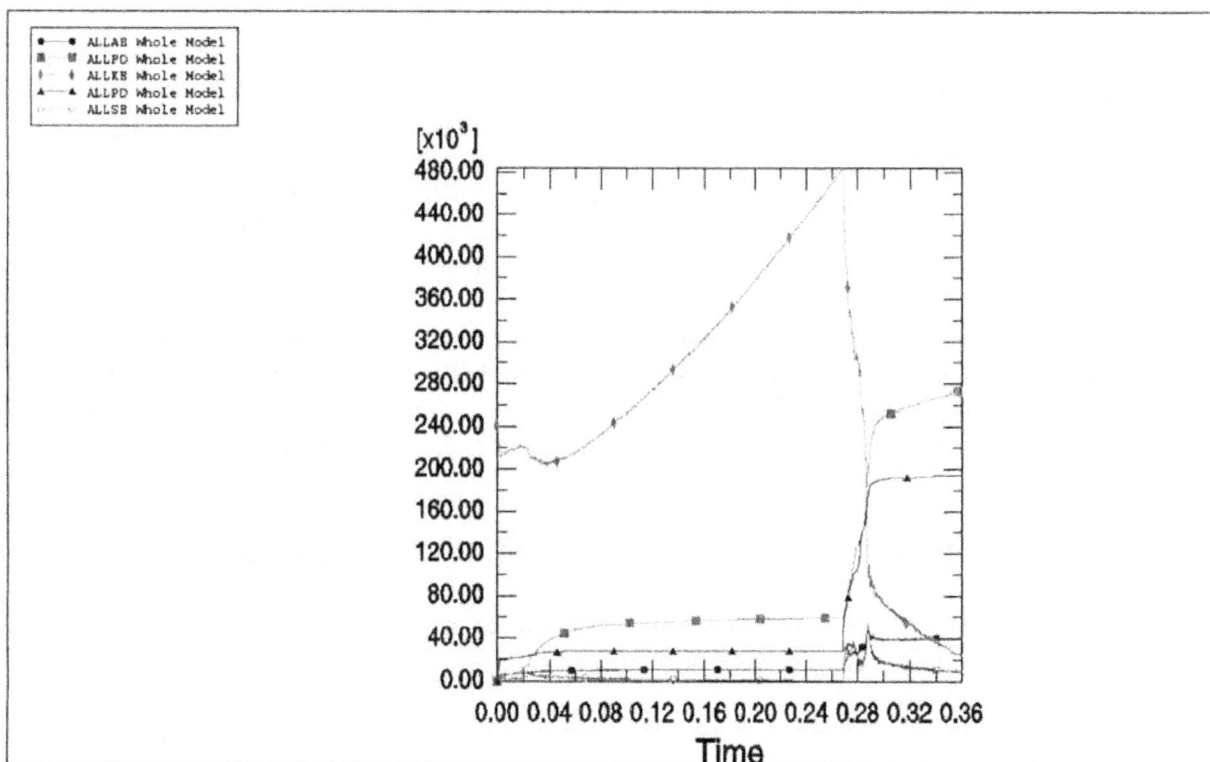

[x10³]
480.00
440.00
400.00
360.00
320.00
280.00
240.00
200.00
160.00
120.00
80.00
40.00
0.00

0.00 0.04 0.08 0.12 0.16 0.20 0.24 0.28 0.32 0.36

Time

Fig. 77. MCO Mark IV, 2-foot (0.6-m) 115 deg. off-vertical drop – model energy (Snow 2003).

133

Worst orientation drop model deformations

The deformations primarily took the form of compression deformation and small bending. No component buckling occurred. This was due to the low drop height (and resulting low drop energy) and the relative rigidity of the MCO end components.

Worst orientation drop model plastic strains

Table 19 lists the peak equivalent plastic strains occurring in the various containment boundary components and the baskets of the MCO during the specified drop events. Peak strains are listed for both the outside and inside surfaces of a component. However, these peak strains did not necessarily occur at the same location on a component. Several conclusions may be drawn from the peak plastic strain data listed in Table 19:

1. The 60 degrees off-vertical drops of the Mark 1A and Mark IV MCOs resulted in peak strains that were nearly identical in the containment boundary components. The peak strain in both MCOs (22% maximum strain) occurred where the bottom impacted the rigid surface, and was due to compressive loading. As expected, the strains in the Mark IV basket exceeded those of the Mark 1A basket due primarily to the less-substantial center post design.
2. The 90 degrees off-vertical (horizontal) drops with Mark 1A and Mark IV baskets both produced material straining in the MCO containment boundary components at about the same level everywhere except on the inside surface of the main shell. At that location, the Mark 1A MCO experienced 10% peak strain whereas the Mark IV MCO produced only 6% peak strain. As has been seen in other drop events evaluated herein, the Mark IV baskets saw higher peak strains (12% strain) than the Mark 1A baskets (1% strain) due to straining in the base of the less-substantial Mark IV center posts.
3. Comparing the Mark IV MCO 115 degrees off-vertical drop to the 60 degrees off-vertical drop, it appears that the main shell, cover, and collar were strained more in the 115 degrees drop than in the 60 degrees drop. However, the peak strain in the impact area of the bottom for the 60 degrees drop (22% strain) was a 7% strain increase over the 115 degrees drop (15% strain). Therefore, the 2-ft (0.6 m) 60 degrees off-vertical drop of the Mark IV MCO was more challenging to the containment boundary than the 2-ft (0.6 m) 115 degrees off-vertical drop overall, but the 115 degrees drop was more challenging to the containment boundary if the strains in the MCO bottom were not considered.

The maximum strain from the above-drop events (specifically, from the 60 degrees off-vertical drop of the Mark IV MCO) of 22% was less-than-the-minimum fracture strain level of 118% and the minimum elongation of 47% listed in Table 15 for MCO containment components. Therefore, it is expected that the MCO would maintain containment during and after these worst orientation 2-ft (0.6 m) drop events.

DROP RESULTS COMPARISON BETWEEN MCO AND DOE SNF CANISTER

Both the standardized DOE SNF canister and the MCO are 24-in (610 mm) nominal diameters with a nominal ½-in (13 mm) thick wall. The main differences are:

1. The standardized DOE SNF canister is 15-ft long, whereas the MCO is just under 14-ft long
2. Whereas the standardized DOE SNF canister has energy-absorbing skirts on each end, the MCO does not

3. The standardized DOE SNF canister uses dished heads while the MCO uses a thick, flat bottom
4. The MCO has an internal shield plug with a locking ring whereas the 24-in (610 mm) diameter canister does not
5. The MCO uses a cover to complete the containment whereas standardized DOE SNF canister uses the top head for that purpose
6. The MCO is about twice as heavy when fully loaded as compared to the standardized DOE SNF canister

Table 19. Peak plastic strains in the MCO for 2-foot (0.6-m) worst orientation drops (Snow 2003).

| Component* | Surface | Peak Equivalent Plastic Strain (PEEQ, %) | | | | |
| | | Mark 1A | | Mark IV | | |
		60 Deg. Off-Vertical Drop (t = 0.340 sec.)	90 Deg. Off-Vertical Drop (t = 0.040 sec.)	60 Deg. Off-Vertical Drop (t = 0.340 sec.)	90 Deg. Off-Vertical Drop (t = 0.040 sec.)	115 Deg. Off-Vertical Drop (t = 0.360 sec.)
flange	outside	11	6	11	7	6
Bottom	inside	6	4	7	4	5
bottom	outside	4	1	4	<1	1
	inside	6	7	6	7	11
impact area***	outside	20	10	22	10	15
Main Shell	outside	4	3	4	2	10
	inside	7	10	6	6	15
Collar	outside	9	4	6	4	3
	inside	12	6	11	6	20
Cover	outside	5	2	5	2	14
	inside	4	<1	4	<1	0
Baskets**	(max.)	8	1	14	12	20

* Components not listed in this table experienced zero or low plastic strains.
** Peak strains in the Mark 1A baskets occurred in the basket walls or the top of the center post. Peak strains in the Mark IV baskets occurred in the base of the center post.
*** This was the side of the MCO bottom where the impact occurred - which resulted in significant compressive strains.

Table 20 compares the peak strains resulting in the standardized DOE SNF canister and the MCO for equivalent drop conditions. From Table 20 it can be seen that the repository 23-ft (6.9 m) vertical drop caused similar peak strains in the containment boundary of both the standardized DOE SNF canister and the MCO. However, when the container was slightly off-vertical at impact, the Mark IV MCO experienced peak containment boundary strains of 35% (outside surface) and 14% (inside surface) for a 23-ft (6.9 m) 3 degrees off-vertical drop. In comparison, the standardized DOE SNF canister saw less than 1% peak containment boundary strains for the same drop. Therefore, when the standardized DOE SNF canister impacted slightly off-vertical, the containment boundary strains dropped because the skirt was effective in absorbing the drop energy. The opposite was the case for the MCO in an off-vertical impact orientation because the containment boundary (bottom flange and main shell) material buckled and bent to absorb the drop energy (no impact-absorbing skirt was available).

Table 20 also shows that the peak containment boundary strains for the repository 2-ft (0.6 m) worst angle drop were slightly higher, though comparable, for the standardized DOE SNF canister than for the MCO. Peak standardized DOE SNF canister containment strains for this drop occurred in the secondary impact region of the canister where the skirt was least effective.

Table 20. Peak strain comparison - MCO vs. 24-inch (610-mm) standardized DOE SNF canister (Snow 2003).

Drop Event	Peak Strain in 24-Inch (610-mm) Canister Containment Boundary			Peak Strain in MCO Containment Boundary	
	Outside	Middle	Inside	Outside	Inside
23-Foot (6.9-m) Vertical Drop	6	0.6	4	5	5
2-Foot (0.6-m) Worst Orientation Drop	23	15	16	22	11

CONCLUSIONS

The design of the MCO reflected a robust ability to survive a perfect vertical drop. However, there did appear to be a probability of the MCO experiencing significantly higher strains in the shell if the impact angle was slightly off vertical. Though experiencing much higher strains than occurred in the 23-ft (6.9 m) vertical drop event, it is expected that the MCO would still maintain containment during and after the 23-ft (6.9 m) near-vertical (1 and 3 degrees off-vertical) drop events as well. Damage to the baskets and fuels was similar to that experienced in the vertical drop.

Snow (2003) performed an edge-drop analysis for an MCO accidentally dropped from 23-ft onto another container (such as a repository waste package) and impacting the MCO collar. It is concluded that a capacity-loaded MCO, with either Mark 1A or Mark IV baskets, would maintain containment during and after such a drop event.

The evaluations considered by Snow (2003) suggest that the deformations of the MCO and the resulting material strains are not highly dependent on the coefficient of friction employed (i.e., a range of coefficients of friction produces essentially the same results).

INTERACTION OF SPENT NUCLEAR FUEL AND PACKAGING MATERIALS

INTRODUCTION

Maintaining the structural integrity of U.S. Department of Energy (DOE) spent nuclear fuel (SNF) canisters during interim storage, transport, and repository disposal is required to prevent the spread of radionuclides into the environment. Anderson (1998) performed a sensitivity analysis to determine if the fission products or any other materials present in DOE SNF are capable of undergoing chemical or physical changes that would allow them to interact with the SNF canister materials and significantly threaten their structural integrity during interim storage, transportation, or repository disposal. This sensitivity analysis evaluated the chemical reactions and possible physical interactions of all fuel materials with each other as well as with the canisters. Reactions and interactions considered included the fuel's structural materials; structural activation products; alloying metals; heavy metals and their decay products; and the fission products. Of greatest interest were components that may be in a physical form that will allow migration and contact with the inner surfaces of the canister. Since gaseous materials and liquids are the most mobile, they received primary attention. Solids were not ignored, however, because they can migrate by entrainment in the gaseous or liquid phases, or even by the subtle long-term effects of thermal gradients which will be present in any SNF fuel canister.

Anderson considered two types of SNF as follows:

1. zirconium-alloy-clad fuels, represented by N Reactor fuels
2. aluminum-alloy-clad fuels, represented by Advanced Test Reactor (ATR) fuels

As shown in Table 21, low enrichment N Reactor fuels make up the largest number of fuel elements; the greatest total mass; and the greatest mass of fissile and fertile heavy metals. High enrichment ATR fuels are important because they are among the highest burnup and, therefore, contain some of the greatest fission product inventories of the fuels that exist in large quantities. Although these two types of fuels do not represent the compositions and structures of all other SNFs in DOE's inventory, they represent a large fraction. Much of the information about these fuels will be applicable to other types of fuels.

Table 21. Contribution of N Reactor and ATR fuels to the USDOE's total SNF inventory projected for CY 2035 (Anderson 1998).

	Number of FHUs	Portion of total FHUs	Total Mass, kg	Portion of Total Mass	Mass of Heavy Metals, kg	Portion of Total Heavy Metals	Mass of U-235, kg	Portion of U-235
N Reactor Fuels	103,673	47.85%	3,524,882	40.22%	2,096,020	86.87%	25,165	23.27%
ATR Fuels	3,900**	1.80%	37,992	0.42%	3,399.8	0.14%	3,399.8	0.14%
Total of all USDOE SNF	216,644	100%	8,763,307	100%	2,412,774	100%	108,136	100%

** ATR fuels comprise approximately 11% of the approximately 35,600 aluminum clad fuel assemblies in this inventory.

137

The thermodynamic, chemical, and physical property data that provided the bases for the prediction of the feasibility of chemical reactions and physical changes were taken from the HSC Chemistry® (1997) database. The database was compiled from publicly-available chemistry, metallurgical, and engineering handbooks commonly used by universities and industry. Radiological and compositional data for SNF were taken from ORIGEN2 computer code calculations, and are identified in this document as "TBV" meaning, "to be verified."

DESCRIPTION OF SENSITIVITY ANALYSES

The ORIGEN2 data was tabulated, sorted, and combined chemical element-by-element to obtain the total amount of every chemical element for which significant quantities are present. The levels above which elements were deemed to be significant were arbitrarily designated as 0.000001 g per fuel assembly for ATR fuel, and 0.002 g per fuel assembly for N Reactor fuel. The much lower level of 0.000001 g was selected for ATR fuels because they are much smaller assemblies and contain less than 5% of the amount of uranium (BOL) that is present in the more massive N Reactor Mark IV assemblies. In addition, ATR fuels have greater irradiation histories (340 MWd/kg of U compared to 2.9 MWd/kg of U for N Reactor fuels). Therefore, they contain proportionally larger inventories of fission products. Consequently, the inclusion of all chemical elements present at levels above 0.000001 g per fuel assembly provides a paradigm for future work with extremely high burnup fuels.

The quantities of fuel materials assumed to be present for these assessments were the total amount of the combined activation products (including all structural material in fuels); actinides and progeny; and fission products. Since both fuels contain uranium in elemental form (as a metal or metal alloy at zero oxidation state), it is reasonable to assume that most of the fission products would be in the zero oxidation state, or at least they would have been initially. In the alloy matrix, some reactive fission products would react with each other (e.g. cesium and iodine) and have their respective oxidation states changed. But as minimal migration occurs inside the matrix, most of the less-reactive fission and activation products as well as their decay products are assumed to be in their original oxidation state which is predominantly zero. This condition could continue throughout the normal irradiation process, defueling, and water storage periods until the fuel is dried and placed into a storage canister—assuming that the fuel cladding maintains its integrity.

Design and structure of storage canisters

The canisters in which the fuels will be stored will be fabricated from stainless steel, most likely Type 316L with American of Society and Testing Materials (ASTM) Standard Specification SA-312. Type 316L stainless steel is a face-centered austenitic stainless steel. Stainless steels are resistant to chemical attack by virtue of a protective film of chromium oxide (Cr_2O_3) that forms when the elemental chromium in the alloy is exposed to oxygen in the air. That layer is continuously replenished in an air environment; however, it is not replenished when oxygen is absent, as is the case with the blanket of inert gas inside a fuel canister. Hence, the present assessment considers the chromium in stainless steel to be present as elemental chromium and not (in significant quantities) as the oxide.

The currently-proposed canister for ATR fuel is the DOE Standardized Canister that is expected to be fabricated from 18-in (457 mm) outside-diameter tube with 3/8-in (9.5 mm) thick walls, either 10 or 15 ft long. The DOE Standardized Canisters are proposed to have dished aid caps that are ½-in (13 mm) thick and are welded in place using a weld rod that in this study is assumed also to be Type 316L stainless steel. The shorter 10-ft (3 m) long canister is considered to be the more likely model. It would have an inner surface area of approximately 45,000 cm² or 6,970 in². The total mass of stainless steel in the canister itself would be approximately 358 kg or 788 lbs.

The Multi-Canister Overpack (MCO) for N Reactor fuels is expected to be a 165-in (4.19 m) long 24-in (609 mm) outside diameter cylinder with ½-in (13 mm) thick walls and relatively thicker flat-disk end plates. It would have an inner surface area of approximately 81,000 cm² or 12,600 in². The total mass would be approximately 875 kg or 1,920 lbs. This MCO design will accommodate up to:

1. four fuel baskets (each containing 54 Mark IV assemblies) and one scrap basket
2. four fuel baskets (each containing 48 Mark IA assemblies) and two scrap baskets

Case 1 was selected for this assessment because it is the more conservative case with approximately 6,000 kg of fuel material compared to approximately 4,500 kg of fuel material in case 2. The larger amount of fuel material is considered to be more conservative in that it provides more material for any potential chemical reactions with the canister. The total mass fuel material present in case 1, including scrap, is equivalent to 243 Mark IV outer-fuel assemblies. This number of assemblies would contain a maximum of 68.9 kg of ^{235}U, based on a maximum pre-irradiation enrichment of 1.25% for Mark IA fuels (Mark IV are 0.947% ^{235}U), which is well-below the recommended repository limit of 200 kg of ^{235}U per canister for very low-enriched uranium (VLEU) (DOE 1996).

Assumed conditions for stored SNFs

In addition to the characteristics described above for ATR and N Reactor fuels and for the ATR and MCO canisters, the conditions listed below are assumed for the sensitivity analysis (Anderson 1998):

1. The fuels are aged 90 years out of the reactor, nominally representing the storage of 1995 fuels until year 2035 plus an additional 50 years in a geologic repository.
2. The fuels are dry with only a negligible amount of water associated with oxide and hydroxide hydrates.
3. The canisters maintain their integrity and hold all gaseous and liquid fuel components in addition to a blanket of helium gas that is placed in the canister at one atmosphere pressure at ambient temperature (25°C) when it is sealed.
4. The fuel cladding fails, making all fuel components inside the canister available to react with all other fuel components and with the canister. The assumption that all materials in the structure of the fuel cladding are available to react is extremely conservative. Even if the cladding cracks and releases the fuel meat, the majority of the materials in the structure of the cladding would remain in place, because it is not credible for all of the cladding to release all of its individual elemental chemical constituents at the maximum temperatures expected for SNF.
5. The temperatures after 50 years in the geologic repository are nominally 200°C, but could be as high as 350°C if placed near commercial nuclear fuel in the same repository. For conservatism, temperatures as high as 400°C are considered for some cases where a temperature profile of reactivity is used.
6. The cylindrical canisters are situated horizontally in the repository, meaning that any liquid or fluid-entrained solids from the fuel could accumulate along the length of the bottom inner surface of the canisters. However, it is unlikely that most canisters will be exactly horizontal, so the collection of fluids over a small surface area at one end of the canister can be assumed.

Thermodynamic feasibility calculations

The potential reactivities of the fuel and canister materials were evaluated using the HSC Chemistry® (1997) for Windows Chemical Reaction and Equilibrium Software on a 166 MHz Dell OptiPlex GXi personal computer with a Microsoft Windows 95 operating system. The software was used to determine the thermodynamic feasibility of chemical reactions occurring between any and all of the individual fuel components and the inner surfaces of the canisters. The results of such calculations cannot predict that the indicated chemical reactions will definitely occur, but that such reactions are thermodynamically feasible and are limited only by chemical kinetics. The evaluations were comprehensive in that they considered both direct and indirect reactivities between fuel and canister materials. Indirect reactivities are based on the properties of secondary species which may occur after the original subject materials have inter-reacted to form intermediate materials.

The software cannot handle equilibrium analyses for all 69 chemical elements of interest at the same time. It was, therefore, necessary to approach such calculations systematically. First, a series of screening calculations was performed for each type of fuel to confirm the relative chemical reactions of the elements that were expected to be the most reactive. Then, after the most reactive elements had been identified and confirmed, some elements with similar properties were combined (or substituted) and entered as input data as if they were single elements. For example, the number of moles of the rare gases xenon and krypton were combined and then calculated only as moles of krypton. Another example is combining the moles of noble metals ruthenium, palladium, and rhodium, and then entering them all as ruthenium. Such combinations (or substitutions) were used only when there was a sound technical basis for doing so. The calculations for each group of chemical elements were based on the properties of the most reactive element in the group (e.g., Ru in the group of Ru, Pd, and Rh) or the element with the largest mass (e.g., Cs in the group of Cs, Na and Rb). In a few cases, the substitution was driven by the availability of reaction product data in the software data base. Whenever a group of combined (or substituted) elements was shown to be reactive with the canister or other fuel materials, additional calculations were performed in order to delineate the behaviors of the individual elements in the combined group. Some elements that occur only in small amounts were simply combined with the most chemically-similar elements present, like americium (an actinide) in ATR fuel being combined with neodymium (not an actinide) and other chemically-similar lanthanides. The primary sources of such information for the current study are Nebergall and Schmidt (1959); Cotton and Wilkinson (1962); Daniels and Alberty (1962); Sargent-Welch (1979); Walker et al. (1989); Lange (1961); and Peckner and Bernstein (1977).

Equilibrium calculations were performed for each fuel at 100, 200 and 350°C. The analysis at 100°C was included for comparison and to obtain a spectrum of data to make it easier to identify trends. Sometimes 400°C was used for the same reason. Only the two middle (200 and 350°C) temperatures are pertinent to actual long-term conditions in a geologic repository.

RESULTS OF SENSITIVITY ANALYSES

The original fuel matrices of both types of fuel selected for this assessment are uranium metal (zero oxidation state) or uranium that is alloyed with other metals—primarily aluminum. These metallic compositions contribute to the predominantly chemical-reducing environments inside both types of fuels and later inside the canisters. Because the fission products formed during irradiation in the reactor also are predominantly metals, the chemical nature of the fuel matrices of irradiated fuels is at least as strongly-reducing as were the unirradiated fuels. Any strongly-oxidizing species that form in the fuel matrices of ATR or N Reactor fuels (e.g., fission product iodine as I_2) would readily react with metals to reach their reduced equilibrium oxidation states. Indeed, most such species would be completely reacted and no longer in elemental form well before the time of the fuel-cladding failure assumed for the current sensitivity analysis. Such chemical reactions between fuel components would occur only to the degree that there are reactants present. Hence, the extent of such reactions and the composition of the reaction products are controlled by the amounts of the reacting species present in the least molar (or "equivalent") quantities. When the reaction products are ultimately released into the canister environment, they would be in the form of chemical compounds or ionic species, and not as the pure chemical elements themselves.

Note that both types of fuels selected for this assessment contain metal uranium fuel matrices in contrast to many other fuels that contain uranium as U(IV), commonly known as the ceramic form 1702. The chemical environments inside such oxide fuels would not be as strongly-reducing as are those in ATR and N Reactor fuels.

Fission product gases

Significant amounts of fission product gases will accumulate inside SNF canisters when they are released from the SNF. An unknown portion of the fission product gases may have already escaped the cladding while in the reactor and during the subsequent storage period before entering the canister. Nevertheless, this assessment assumes that the entire inventory is still present when the fuels are sealed inside the SNF canisters. The entire inventory of the fission product gases krypton and xenon, combined with the original blanket of helium, would fill the canisters for each type of fuel. The ORIGEN2 code also indicates that a small amount of neon would be present in ATR fuel at a level that is negligible compared to krypton and xenon. The resulting pressures from inert gases alone could exceed the specified pressure loadings in the Disposability Interface Specification for DOE SNF canisters. The possibility of metal hydrides releasing gaseous hydrogen at storage temperatures and the vapor pressures of any liquid metals could also contribute to the total pressure inside stored SNF canisters, thereby exceeding the specified pressure loadings even further.

Since these noble gases are all chemically-inert with respect to all other materials present under the physical and chemical conditions of these SNF canisters, the entire quantity of such gases that escape the fuel matrix would simply collect inside the sealed canisters. They would not corrosively attack the stainless steel fuel canisters nor significantly affect them other than to contribute to the pressure and facilitate the transfer of heat from the fuel to the canister walls. Although no material, including Type 316L stainless steel, is completely impermeable to gases—especially gases with small diameter molecules like mono-atomic helium—the rate of such permeation has been shown to be extremely low (Aderl and Nagata 1989). Based on the reported rates of permeation of helium through stainless steel, it would take around 10×10^{40} years for the helium to escape through the 3/8-in (9.5 mm) thick stainless steel wall of a DOE standardized SNF canister. Krypton and xenon, with atomic radii of 1.03×10^{-10} m and 1.24×10^{-10} m, respectively, would escape even slower than helium atoms with an atomic radius of 0.49×10^{-10} m. Hence, inert gases will not escape by permeating the canister walls. That also means that the quality of the welds and seals will become the factors that govern the integrity of SNF canisters with respect to gases.

Atmospheric gases

The major gases in the atmosphere (diatomic nitrogen and oxygen and triatomic water vapor) will not permeate the stainless steel canister from the outside for the same reasons presented above for fission product gases (as long as the canister is sealed and maintains its integrity. In addition, according to Fick's law, the ingress of gases from the outside would be inhibited by the presence of a positive internal concentration gradient which in the case of gases, is the same as a pressure gradient (Reed-Hill 1973).

Hydrogen and the formation of hydrides

The behavior of hydrogen in this assessment is based solely on the amount of hydrogen indicated by the ORIGEN2 calculations. This hydrogen is solely from nuclear sources such as ternary fission products. The potential effects of larger amounts of hydrogen on SNFs and stainless steel canisters have been evaluated by DOE (2000a). Although the amount of hydrogen present from nuclear sources may be small compared to the amount that could be present in hydroxides, hydrated oxides, and occluded water, there are still more moles of hydrogen (as H_2) than krypton in both ATR and N Reactor fuels. If the total amount of hydrogen associated with both types of fuel were present inside the canisters as elemental diatomic hydrogen gas, it would essentially double the internal gas pressures.

It is thermodynamically favorable for hydrogen to react with cerium, barium, praseodymium, neodymium, zirconium, plutonium, and uranium (in the order listed) to form metal hydrides at both 200 and 350°C. The formation of such metal hydrides in the current scenarios depends upon the availability of the component materials; that is, upon the availability of both the metals and the hydrogen as discussed below.

When hydrogen atoms interstitially penetrate a metal, the resulting hydride is essentially an "intermetallic" compound or alloy in which both the metal and hydrogen remain in the zero oxidation state. As the assumed maximum amount of hydrogen available in the current scenario is 19.2 moles for ATR fuel and 10.5 moles for N Reactor fuel, the greatest amounts of metal hydrides that would be expected to form with these metals are 9.6 and 5.3 moles, respectively, based on the stoichiometry associated with MH_2. Conversely, the capability of forming hydrides of each of those metals under the current scenarios with only hydrogen from nuclear sources depends upon the availability of the metals which may be governed by the following two factors:

1. Hydrides that have not already reacted with something else to form more stable compounds: a likely limiting factor in the fuel meat
2. The physical accessability of hydrides: a likely limiting factor in fuel cladding where a more tenuous mechanism such as permeation through the alloy is required for the hydrogen to reach the reactive metal.

For ATR fuel, the predominant species of hydrides at both 200 and 350°C will be NdH_2 followed by much smaller amounts of BaH_2 and CeH_2. More reactive cerium is not the leading former of hydrides because it is less available, having already formed more stable intermetallic compounds with aluminum, magnesium, and the cesium-rubidium solution. Less ZrH_2 is expected to form in ATR fuel than in the N Reactor fuel because there is much less zirconium available in ATR fuel. Also, a portion of the zirconium in ATR fuel would preferentially form silicides and would not be available to react with hydrogen.

For N Reactor fuel, the equilibrium inventory of metal hydrides at 200°C is predicted to consist of nearly 50% BaH_2, a slightly lesser amount of ZrH_2, and a significantly smaller amount of NdH_2. At 350°C the predominant species is ZrH_2, followed by significantly lesser amounts of BaH_2 and NdH_2. Blumenthal (1958) indicates that ZrH_2 may not form at a "substantial rate" below temperatures of at least 250°C. However, the long periods of time associated with repository storage justify the assumption that the equilibrium states are achieved. Although zirconium is thermodynamically less likely to form hydrides than are barium and

142

neodymium, ZrH_2 is predicted to be the predominant hydride at 350°C. The reason is that there is so much more of it than there is of the other major hydride-forming metals. However, most of the zirconium hydride is expected to be formed with fission product zirconium and not with structural zirconium from the cladding for two reasons:

1. Fission product zirconium is formed in the fuel matrix where it is immediately available to react with any ternary fission product hydrogen generated there or even by the "n,p" reaction in which neutrons replace protons in the nuclei of atoms exposed to the high neutron flux from the fissioning uranium (and plutonium) in the fuel matrix.
2. Structural zirconium in Zircaloy-2 is physically protected by the zirconium oxide film that forms when it is manufactured and is reinforced by exposure to water at high temperatures in the reactor. Blumenthal (1958) reported that the presence of a thick layer of ZrO_2 inhibits but does not prevent the penetration of hydrogen into zirconium alloys where it can form hydrides—mostly ZrH_2.

The formation of significant amounts of metal hydrides, especially zirconium hydrides, is recognized as undesirable for two reasons:

1. The intrusion of air could cause rapid combustion of metal hydrides, in the case of ZrH_2 yielding $Zr(OH)_2$ and water, which in turn could then react with alkali metals.
2. Hydrides can release gaseous hydrogen when heated and generate high pressures inside the fuel canisters. The presence of hydrogen gas at high pressures and temperatures also increases the potential for hydrogen damage of the SNF canisters. At 350°C, only a small portion (less than 10%) of the hydrogen from BaH_2, NdH_2, and ZrH_2 would be released into the canister environment. It would require much higher temperatures to release the majority of it as hydrogen gas. For example, 50% of the hydrogen associated with NdH_2 and ZrH_2 is not released until it is heated to 950°C. The BaH_2 must be heated to nearly 1,150°C to release 50% of its hydrogen.

Based on the limited amounts of materials available in the scenarios considered in this assessment, it appears unlikely that a large amount of the hydrogen would ever be released inside the canister as gaseous hydrogen. Conversely, it must be assumed that at least some of the hydrogen would be released, especially at 350°C. Any released hydrogen gas would contribute to the internal gas pressure of the fuel canister. Its contact with the canister would be equally distributed over the entire 45,000 cm² or 81,000 cm², respectively, of the inner surfaces of the DOE standardized SNF canisters for ATR fuel or MCOs for N Reactor fuel.

Theoretically, the following two phenomena are possible when gaseous hydrogen is present under the conditions of the current scenarios:

1. It may penetrate the stainless steel canister via the mechanism of disassociated mono-atomic hydrogen atoms diffusing between the interstitial spaces inside the crystalline structure of the stainless alloy to form hydrides (Reed-Hill 1973)
2. There could be an extremely slow but continual penetration of the stainless steel canister at a rate governed by the crystalline structure of the alloy and expressed by the diffusivity constant of hydrogen in stainless steel (Aderl and Nagata 1989; Redd-Hill 1973)

In either case, once hydrogen atoms are inside stainless steel—including the welds and their heat affected zones—hydrogen embrittlement can occur. In the present scenarios with dry fuels in which only ternary fission product hydrogen is considered, the potential for a significant amount of damaging hydrogen embrittlement to occur is obviously less than when hydrogen from other sources is present.

Vapor pressure of liquid materials

The inert gases and hydrogen are not the only species in 90-year-cooled ATR and N Reactor fuels that would contribute to the pressure inside sealed SNF canisters. All materials in the gaseous state at any given temperature would contribute to the total pressure inside the canister, as would any liquids through their vapor pressures (more specifically, through the partial pressures of their miscible or dissolved components). For example, cesium metal, which is a liquid between 29 and 685°C, would contribute 5.9 mm of Hg (or 0.0078 atmospheres) of vapor pressure if present in the SNF canister at 350°C. Rubidium, the only other alkali metal fission product likely to be present in a significant amount, is a liquid between 30 and 701°C, and has a vapor pressure of 4.1 mm of Hg (or 0.0053 atmospheres) at 350°C. These metals are very reactive and at least a portion of them would react with other SNF materials.

Halogens

Although elemental bromine and iodine would both be gases at the canister temperatures of 200 and 350°C, neither would be present as such in either of the SNF canister environments. All elemental halogens would be chemically reduced by any of the alkali or alkaline earth metals present. ATR and N Reactor fuels contain 40-fold and 30-fold excesses, respectively, of the molar amount of each of these metals required to reduce the total inventories of halides present to alkali halide and alkaline earth halide salts like $CsBr$, RbI, $SrBr_2$ and BaI_2. All bromide and iodide salts of cesium, rubidium, barium, and strontium have melting points greater than 400°C; therefore, these halide salts would remain in solid form and would not corrosively attack the inner surfaces of the stainless steel canisters. This is true when water is absent, which is an assumed condition in this assessment. The limited amount of hydrogen precludes the formation of water from inside the canisters. Any water in quantities great enough to significantly contribute to halide corrosion of the stainless steel canister would have to come from external sources. Such water would react with the excess alkali metals (that is, alkali metals not already oxidized by a halogen) to form hydroxides before it could accumulate in an amount large enough to form an aqueous halide system that would be corrosive to the stainless steel canister. If such a large amount of water were to intrude, excess water (after the alkali metals have all reacted) could accumulate and provide a medium in which metal salts could dissolve and form solutions that would be potentially corrosive to the stainless steel canisters. The entrainment of such salts in liquid cesium or other non-polar (non-aqueous) liquids would not be very likely to set up the conditions required for a chemical attack on the stainless steel canister.

Oxygen

The small but significant amount of oxygen for ATR fuels is associated with the aluminum 6061 alloy cladding as a surface layer of alumina (Al_2O_3). This oxygen is already in the reacted (reduced) form of a metal oxide and would not be available to react with and degrade the iron, nickel, chromium, or manganese in the stainless steel canisters. There are numerous fuel components that would preferentially react with or hold oxygen more strongly than the iron, nickel, or chromium in stainless steel at 350°C and 3.4 atmospheres pressure. Specifically, calcium, plutonium, neodymium, magnesium (a significant component of aluminum alloy 6061), cerium, strontium, zirconium, and aluminum (in that order) are all thermodynamically more likely to hold the oxygen as oxides than is the chromium in stainless steel. Oxygen is thermodynamically more likely to associate with these fuel materials than with the stainless steel components (iron, nickel, and chromium) of the canister. Hence, it would be thermodynamically favorable for oxygen to be driven from the stainless steel to the fuel and not vice versa.

The data for N Reactor fuel in these studies do not include oxygen. However, it is believed that the Zircaloy-2 cladding of Mark IV fuel would have an oxide coating similar to that encountered on the surfaces of zirconium fuels processed at the Idaho Chemical Processing Plant and confirmed by autoclave tests (Hillner 1985). In addition, oxygen levels of 1,000 to 1,600 parts per million are typical in Zircaloy. The presence of such levels of oxygen contributes to the hardness desired for nuclear fuel cladding. Oxygen—along with alloyed tin, iron, chromium, and nickel—is an integral part of the structure of the alloy that is stable at temperatures up to at least 790°C. Similar to the oxygen associated with the structural materials in ATR fuels, oxygen in N Reactor fuels would not be released from the structural Zircaloy-2 and, therefore, would not be available to react with other materials.

Alkali liquid metals

The alkali metals are extremely reactive in elemental form. In the closed environment of a SNF canister, a portion of the alkali metals cesium and rubidium would be oxidized by the halogen fission products bromine and iodine. They would also react with oxygen or water that might contact them. Some unreacted excess cesium and rubidium (beyond what will be oxidized by halogens or other species) will be present in both ATR and N Reactor SNF canisters. Sodium, if present in the same state, would accompany the cesium and rubidium. Sodium and lithium are present in much lesser amounts than cesium and rubidium and, therefore, need not be considered separately from cesium and rubidium. If a greater amount of lithium was present, it might need to be considered separately because its miscibility with a cesium-rubidium mixture is very slight (Massalski 1986).

The elemental alkali metals cesium and rubidium in the canister environment would be entirely miscible with each other. The larger molar quantity of cesium (compared to rubidium) for both ATR and N Reactor fuels, dictates that the solutions of these metals in SNF canisters would likely consist of 60 to 90 wt % cesium with a melting point of approximately 15 to 20°C. The behavior of such a liquid metal system in the SNF canister environment would be similar for both the 200 and 350°C scenarios considered in this assessment, although the Cs-Rb vapor pressure would be substantially greater at the higher temperature. At both temperatures the bimetallic vapor would penetrate the entire closed system and be adsorbed onto virtually all exposed surfaces. There are two possible scenarios with the condensed cesium-rubidium metal at this point:

1. It is likely that some of the liquid cesium/rubidium would remain in place on the exposed surfaces as it forms a thin film of intermetallic alloys with metals on or near the surface of the various fuel materials.
2. It is less likely but possible that the metal combinations or alloys that form during contact with the condensed cesium-rubidium alloy would have even lower melting temperatures and, thereby, increase the amount of liquid phase material inside the canister. Eventually, at least a portion of the liquid metal could condense and collect on the bottom inner surface of the canister. The most significant difference between the 200 and 350°C scenarios is that the lower temperature model takes longer, due to the lower vapor pressure of the cesium/rubidium combination at the lower temperature.

Austenitic stainless steels (including Types 304L, 316 and Nitronic 50) offer excellent resistance to liquid sodium and potassium at 350°C (Peckner and Bernstein 1977). Chemically-similar cesium-rubidium would behave the same way with respect to stainless steel corrosion (Peckner and Bernstein 1977; Reed-Hill 1973). Lithium, however, is different and would attack stainless steel if it were present in a large enough quantity.

Non-alkali liquid metals

Other metals that melt at temperatures below 350°C, including tin, lead, cadmium, and indium, are likely to dissolve in or alloy with cesium-rubidium due to their predicted miscibility and/or solubility. Any such dissolved tin or lead would likely be of fission product or actinide decay product origin in the fuel matrix, and not from the aluminum alloy 6061 cladding of ATR fuel or Zircaloy cladding of N Reactor fuel. The majority of the tin (83% in ATR fuel and 98% in N Reactor fuel) and virtually all of the lead (>99.9%) in both types of fuels are in the cladding and would not be released into the canister environment. In both types of fuel, most of the fission product tin would be melted at 350°C (Lange 1961). At 200°C the tin would not melt, but it could dissolve in or form inter-metallic compounds with rubidium and cesium. Cadmium, which is present in greater quantities in N Reactor fuel than in ATR fuel, can dissolve in the cesium/rubidium system and may form $CsCd_{13}$ or other inter-metallic species. Only a few grams of indium would be present in an MCO filled with N Reactor fuel, and even less would be present in a DOE standardized canister filled with ATR fuel. However, the presence of liquid indium dissolved in cesium/rubidium is a distinct possibility.

The small amount of selenium in both types of fuel would most likely not be released from the fuel matrix in elemental form. Thermodynamically, selenium (along with the chemically-similar tellurium) is expected to form selenides (and tellurides) in compounds that would be analogous to metal sulfides. Selenium is more reactive than tellurium. Because it melts at a lower temperature (221°C), it would almost certainly react and be consumed before tellurium which is still a solid at 350°C. Selenium would preferentially react with magnesium to form MgSe, if magnesium were to become available from the structural aluminum alloy 6061 in ATR fuel. If the magnesium in the structural material of ATR fuel is unavailable (physically inaccessible to the potential reactant), or if it is absent as is the case with N Reactor fuel, selenium would react with the fission products cerium, strontium, and cesium—in that order. It would form the selenides CeSe, SrSe, and Cs_2Se, respectively. Another reason that reaction with magnesium is unlikely is the fact that the fission products selenium and tellurium are both generated by fissioning uranium inside the fuel matrix where they are physically in contact with cerium, strontium, and cesium (which also are present only as fission products). The more abundant and less mobile solid phase tellurium also would react with magnesium (if it were available), and then with the remaining strontium and cesium to form SrTe and Cs_2Te. It is less likely to form CeTe. Neither selenium nor tellurium is thermodynamically likely to react with barium, which was combined with strontium as a surrogate in the earlier chemical equilibrium calculations for N Reactor fuel.

Metals contacting liquid cesium and rubidium

It is extremely difficult to predict all of the elements and intermetallic compounds that could dissolve in or form additional alloys with condensed cesium and rubidium. Most of the metals present in both ATR and N Reactor fuels have the ability to form various alloys that melt at higher temperatures than do the individual component metals. In addition, many can also form solid solutions with eutectic phases that may liquefy at lower temperatures than do the individual component metals. However, it appears that only a very small amount of the elements present in the complex assortment of fission products in these 90-year cooled fuels would actually dissolve in a CsRb solution (Massalski 1986). Based on the limited solubilities or immiscibilities of these materials in cesium, very little additional material will dissolve in the liquid cesium-rubidium. There is a strong case for predicting that there will be an accumulation of some liquid cesium-rubidium on the bottom inside surface of the canisters with N Reactor fuel, up to a maximum of several hundred grams. While the liquid would be composed predominantly of cesium and rubidium (probably 60 to 90% cesium), it could contain a small amount of most of the other chemical elements, either dissolved in the cesium-rubidium solution or in adjacent droplets of immiscible liquids (Massalski 1986).

146

It is not thermodynamically favorable for liquid or condensed cesium and rubidium or their vapors to react with or corrode Type 304L stainless steel. The addition molybdenum metal to the equilibrium calculation does not change the results, indicating that Type 316L stainless steel (which contains molybdenum) is an acceptable material of construction with respect to compatibility with the ATR and N Reactor fuel materials. The addition of nitrogen, vanadium, and niobium to the assessment yields similar results for Nitronic 50 stainless steel. The chemical reactions that would comprise corrosion of the stainless steel are not feasible.

The diffusion of extraneous materials into stainless steel can potentially change its physical properties and structural integrity. Two primary types (mechanisms) of diffusion of material into or through a metal occur by the formation of substitutional solid solutions or interstitial solid solutions (Reed-Hill 1973). Substitutional solid solutions occur when atoms of the intruding or solute material actually replace (or substitute) atoms of the solvent or major constituents (i.e., it actually occupies the place of the original atom in the structure of the alloy). This mechanism occurs only between solute and solvent atoms of similar size. Interstitial solid solutions are formed when the solute atoms are physically small enough to fit between the interstices of the crystalline solvent atoms. Interstitial division occurs only when the solute atoms have an apparent diameter that is smaller than 59% the diameter of the solvent atoms. Cesium and rubidium atoms are clearly much larger than the primary atoms in stainless steels; hence, neither substitutional nor interstitial diffusion pose a potential threat to the integrity of the SNF canisters. The relatively high melting range of stainless steel— approximately 1,000°C higher than the 350°C maximum temperature considered in this assessment— also contributes to the physical integrity of its micro crystals. This, in turn, promotes resistance to substitutional diffusion by similar sized atoms (Peckner and Bernstein 1977).

Liquid metals in contact with stainless steel

Intergranular corrosion (i.e., the penetration and corrosion of materials between the grains of the stainless steel in the canister) is not a credible failure mechanism because the selected materials of construction do not sensitize (and become susceptible to intergranular corrosion) at temperatures below 500°C. Type 316L stainless steel planned for DOE Standardized SNF canisters and possibly for MCOs would not be susceptible to such attack at 350°C.

Technically, Liquid Metal Embrittlement (LME) is not corrosion (degradation that results from the chemical interaction of fuel and canister materials), but a degradation of mechanical properties of the solid alloy. Alloys that experience LME become brittle and less ductile and, therefore, are more likely to crack in stressed locations. Stresses produced by welding alone have been reported to produce cracking of austenitic stainless steel exposed to liquefied zinc.

The LME occurs if there is a prolonged intimate contact of the solid alloy with a liquid metal. Such contact can be prevented by a thin film layer of oxide several Angstroms thick. This is the reason that stainless steels are less susceptible to LME than are carbon steels and other alloys that are not protected by an oxide surface film. However, austenitic stainless steels like the Type 316L stainless steel planned for SNF canisters can and do experience LME when exposed to certain liquid metals. In addition, the replacement of an air atmosphere with a helium atmosphere means that air would not be available to replenish the Cr_2O_3 protective film on the inner surfaces of the canisters.

Aluminum, silver, cadmium, copper, indium, lead, antimony, selenium, tin, tellurium and zinc (of the metals present in SNF) are capable of causing liquid metal embrittlement of ferritic and/or austenitic stainless steels. The high melting temperatures of elemental aluminum, silver, copper, antimony, tellurium, and zinc preclude those materials from being present as pure liquids. However, the alkali metals rubidium and cesium which may be present as liquids inside the canister, can indirectly contribute to LME of stainless steel by transporting dissolved or entrained metals to the stainless steel surface. Based on the liquid metal solubilities and miscibilities, some aluminum; some cadmium; some indium; some lanthanides; any available lead; and some tin could dissolve in liquid rubidium/cesium and therefore reach the stainless steel surface. The possibility of LME degradation of the stainless steel fuel canisters from exposure to these metals in liquid form cannot be ruled-out.

The melting temperatures of the remaining metals known to cause LME in stainless steel, cadmium, indium, lead, selenium, and tin are low enough that they could exist as elemental liquids at 350°C. However, selenium is unlikely to be present as a liquid metal because (along with chemically-similar tellurium) it will likely have reacted with the alkali and alkaline earth metals to form selenide (and telluride) analogs of the sulfides. The possibility of LME of the stainless steel from exposure to liquid cadmium, indium, lead, selenium, tellurium, and tin cannot be ruled-out, but the likelihood is small in comparison to those metals listed in the previous paragraph.

Molten salts

Molten salt corrosion is unlikely for two reasons:

1. In the strongly-reducing chemical environment of the canister, few salts form.
2. The salts that are known to be present in significant quantities have high melting temperatures that keep them in the solid state at 350°C (Lange 1961).

The solubility of the salts most likely to be present (alkali halides) in liquid cesium-rubidium is not known; however, based on the general insolubility of polar compounds in non-polar liquids (e.g., molten cesium-rubidium), such dissolved salts can be eliminated as a credible source of chemical attack on the stainless steel canisters.

Perhaps the best-known corrosion mechanism associated with liquid state materials is one that is commonly encountered in aqueous media: the galvanic cell. The components that are required to form a galvanic cell are a cathode, an anode, and an electrolyte. Theoretically, any two metals can form the anode and cathode since no two elements have exactly the same electric potentials. The conditions assumed for this sensitivity analysis provide the presence of approximately fifty elements that exist as metals in the zero oxidation state providing numerous cathodes and anodes. This system, however, lacks an electrolyte. While the liquid cesium-rubidium will conduct a flow of electrons, it does not provide a medium in which iron, nickel, chromium, or manganese ions can dissolve and flow to a cathode. Therefore, any corrosion that occurs with such a liquid metal system would not be an electrochemical reaction of the type traditionally known as galvanic corrosion.

None of the liquid materials identified in this assessment would be expected to contribute significantly to the chemical degradation of stainless steel fuel canisters or cause them to lose their integrity under the defined conditions. However, the possibility of physical degradation of the stainless steel canister by LME cannot be ruled-out.

Solid SNF materials

Solid state materials are less mobile than are liquid or gaseous state materials. However, over long periods of time, and especially at elevated temperatures, components of materials in the solid state can and do migrate. Both physical and chemical changes can occur when different metals contact each other. Such contact is assumed between uranium metal and the inner wall of the stainless steel SNF canister following failure of fuel cladding. Uranium forms low-melting eutectics with elements from the group chromium, cobalt, iron, manganese, and nickel. "Low-melting" in this case means eutectics with liquid phases in the low 800 or high 700°C range, compared to approximately 300°C higher for uranium metal and another 200 to 300°C higher for uranium-aluminum alloys. The use of alloys containing these elements (Cr, Fe, Mn, and Ni) for cladding metallic uranium is, therefore, limited to temperatures below approximately 700°C.

None of the solid materials identified in this assessment would be expected to contribute significantly to the degradation of stainless steel fuel canisters or cause them to lose their integrity under the assumed packaging conditions and storage temperatures.

CONCLUSIONS

The chemical properties and potential reactivities of the materials expected to be present in dry, 90-year cooled, ATR and N Reactor SNF stored in a helium atmosphere inside stainless steel canisters were assessed to determine the likelihood of the canisters degrading enough to lose their structural integrity. With the possible exceptions of exceeding the specified pressure loadings for the canisters; hydrogen embrittlement of the canisters; and liquid metal embrittlement, no significant mechanisms of degradation were determined by which DOE Standardized Canisters or Multi-Canister Overpacks would be likely to fail at nominal repository storage temperatures of 350°C. This assessment indicates that the foregoing potential mechanisms of structural failure of Type 316L stainless steel canisters could not be ruled out, even though they are deemed to be very unlikely under the conditions studied. Resolution of these potential problems will be required before this mode of disposition can be approved for spent nuclear fuels. These results, based on the assessment of two typical and representative fuels, may be applied to most of the fuels in the DOE's SNF inventory.

HYDROGEN DAMAGE IN SPENT NUCLEAR FUEL PACKAGES

INTRODUCTION

The spent nuclear fuel (SNF) standard canisters of the U.S. Department of Energy (DOE) are expected to safely store the SNF for 40 years in interim storage, and meet the U.S. Department of Transportation (DOT) and U.S. Nuclear Regulatory Commission (USNRC) shipping requirements. Free water carryover into the canister; hydrated oxides on the fuel and cladding; and potentially-organic materials may provide sources of hydrogen within the canisters. DOE (2000a) performed a study regarding the potential for hydrogen damage to the canister and internal structural components from hydrogen over-pressurization, embrittlement, or other hydrogen damage mechanisms. An extensive analysis of the variety of DOE SNF types was not considered necessary for this study. Most DOE SNF has historically been stored in water that provides a readily-available source of hydrogen within the package. Damaged SNF and aluminum-clad SNF are of primary concern because of their ability to hold significant quantities of water in surface films (oxides or hydroxides) or corrosion products. Therefore, DOE (2000a) focused on the following representative types of DOE fuel to analyze the potential for hydrogen damage:

1. Aluminum-clad research reactor fuels such as the Materials Test Reactor (MTR) type. This fuel has been stored under poor water quality conditions causing cladding corrosion that results in bound water in surface hydroxides.
2. Zircaloy-clad N-Reactor fuel stored in the K-Basins at the Hanford Site. This fuel represents the largest mass of DOE fuel, and a significant percentage of the fuel incurred mechanical handling damage after irradiation, which has been exacerbated by water storage.

Standardized canisters

Most SNF, other than N-Reactor SNF being considered for movement to interim storage, transportation, and final disposal, is expected to go into standard canisters. Two standard canister sizes are currently specified: a 457-mm (18-in) and 610-mm (24-in) nominal outer diameter (OD), with lengths and other dimensions as listed in the preliminary specification (DOE 1999).

High-integrity cans

Some SNF that contains badly-damaged fuel, particulates, fuel samples, sections, or other miscellaneous small lots may require packaging in a sealed container within the canister referred to as the high-integrity can (HIC). This is a 137-mm (5.4-in) OD can designed for ease of packaging, structural integrity, and criticality safety. The SNF that is placed in the can is assumed to be adequately characterized for storage, transportation, and final disposal.

Multi-canister overpacks

The Multi-Canister Overpack (MCO) is a specialized 610-mm (24-in) canister designed for packaging N-Reactor fuel from the K-Basins at Hanford, and holding the fuel in the Canister Storage Building for interim storage. There has been considerable interest in using the MCO as a disposable canister to minimize handling of the SNF. It is discussed here because of the large percentage of the DOE inventory that will be placed in the container.

Pressure limits

The following specified pressures are listed for all of the DOE SNF standardized canister geometries:

1. The Maximum Normal In-Plant Handling Pressure (MNIP) is the maximum pressure that would develop in a DOE SNF canister during initial handling, interim storage, transportation, or initial repository handling or disposal container loading prior to actual emplacement in a repository drift under the most severe conditions of normal in-plant handling operations. The DOE SNF canister shall be designed for an MNIP not to exceed 344.8 kPa (50 psig).
2. The Maximum Normal Operating Pressure (MNOP) is the maximum pressure that would develop in a DOE SNF canister during initial handling, interim storage, transportation, or initial repository handling or disposal container loading prior to actual emplacement in a repository drift without venting. The DOE SNF canister shall be designed for an MNOP not to exceed 151.7 kPa (22 psig).

The pressure limit for the MCO has been set at 450 psig (3.10 MPa) to ensure that the canister is capable of handling the internal pressure that may be generated through radiolysis of water within the canister during interim storage. It is expected that the MCO would need to be vented or pressure tested prior to shipment to the repository.

The HIC pressure limit has been set at 250 psig (1.72 MPa) to handle potential saturated water conditions within the can. It can be shown, however, that failure of the HIC within the standardized canister will not result in overpressurization of the standardized canister.

Temperature

The DOE SNF containers will be designed to maintain confinement of radioactive material from -40°C to 343°C (DOE 1999). The nominal operating temperatures for the DOE SNF canisters in interim storage are expected to vary according to the specific system in use and the relative decay heat of the SNF being stored. For Hanford N-Reactor SNF in the Canister Storage Building, interim storage temperatures are expected to reach a maximum of 170°C.

No temperature requirement has been placed on the repository surface facility where canisters must be capable of withstanding a 24-ft (7.2 m) drop. Repository waste packages within the sealed drift (disposal location) may reach a temperature of 200°C, about 50-100 years after emplacement if they are placed adjacent to fully-burned commercial SNF packages.

Container alloys

The material of construction for the DOE spent nuclear fuel standard canister is 316L stainless steel. The MCO is designed using 304L stainless steel. These low-carbon austenitic alloys were chosen for their corrosion resistance, mechanical properties, and their resistance to sensitization, increasing their resistance to localized corrosion and environmental cracking, including hydrogen cracking. Alloy 22 is a nickel-chromium-molybdenum alloy that has been chosen as the corrosion-resistant barrier alloy for the Yucca Mountain Waste Package. It is also the choice for the HIC. This alloy series was chosen because of its exceptional resistance to general and pitting corrosion. The chemistry and mechanical properties of these alloys are listed in Tables 22 and 23.

Table 22. Alloy composition (% wt) (DOE 2000a).

Element	304L	316L	Alloy 22
C	0.03 max	0.03 max	0.010 max
Cr	18.0-20.0	16.0-18.0	22 (nom.)
Ni	8.0-12.0	10.0-14.0	Balance
Mo	-	2.0-3.0	13 (nom.)
Fe	Balance	Balance	3 (nom.)
P	0.045 max	0.045 max	0.02 max
S	0.03 max	0.03 max	0.010 max
N	0.10	0.10	-
W	-	-	3 (nom.)
Co	-	-	2.5 max
Si	1.00 max	1.00 max	0.08 max
V	-	-	0.35 max
Hf	-	-	
Nb	-	-	
Sn	-	-	
Zr	-	-	

Table 23. Room temperature mechanical properties (DOE 2000a).

Alloy	0.2% Yield Strength		Ultimate Tensile Strength		Elongation (%)
	(10^3 psi)	(MPa)	(10^3 psi)	(MPa)	
304L (Min. per ASTM A 473)	25	172	65	448	40
316L (Min. per ASTM A 473)	25	172	65	448	40
Alloy 22 (Typical)	54	372	114	786	62

SOURCES OF HYDROGEN

Due to the presence of water or organic compounds, the anticipated sources of hydrogen inside SNF canisters are the chemical or electrochemical reactions. It is difficult to ascertain how much residual water will actually be present in a given container after undergoing a drying regimen. The water could be present as (Lessing 1998):

1. water of hydration of a corrosion product
2. physically adsorbed water on corrosion products
3. water inclusions (water logging)
4. water vapor

Assuming water is present, hydrogen could be generated by the following mechanisms (Sridhar et al. 1995):

1. **Galvanic Corrosion:** The aluminum SNF cladding will be in contact with the stainless steel or Hastelloy C-22 canister material. The stainless steel or Hastelloy C-22 would then be cathodic to the aluminum if there is an aqueous environment in the canister.

2. **Crevice Corrosion:** The Idaho National Engineering and Environmental Laboratory (INEEL) SNF corrosion inspection program has reported crevice corrosion of the cladding of aluminum SNF. The micro-environment inside these crevices can become acidic and evolve hydrogen.

3. **Microbially Influenced Corrosion (MIC) Effects:** Sampling of aluminum-clad SNF in wet and dry storage has identified microbes on the cladding surface. Hydrogen could evolve from MIC of the cladding, or the microbes could naturally evolve hydrogen without participating in a corrosion reaction.

4. **Radiolysis Production through Interaction with the Canister Environment:** Radiolysis of any free and bound water within the canister can form hydrogen-free radicals that could be absorbed by the canister material. This gaseous hydrogen could also result in an overpressurized condition.

DOE (2000a) evaluated the effect of hydrogen on the materials and design of the canisters by looking at the possibility of an overpressurized condition due to hydrogen generation and the effect of hydrogen on materials properties.

OVERPRESSURE DUE TO HYDROGEN

Hydrogen pressure calculation for INEEL SNF

The Advanced Test Reactor (ATR) fuel elements have not been characterized after drying for remaining fuel cladding thickness; oxide film thickness on aluminum cladding surface; localized corrosion; and adherent corrosion products or sludge. To assess the potential for hydrogen pressurization, the following storage conditions for ATR fuel elements were assumed:

1. The fuel has been stored under water for 20 years prior to drying and canning
2. The surface of the aluminum cladding will have oxidized and formed hydrated forms of Al_2O_3
3. The fuel has been dried below 400°C prior to canning
4. The canister is sealed and backfilled with helium
5. There are 20 ATR elements in the canister with a surface area of 86.3 m²
6. The canister volume is 387 L
7. The fuel will be in interim storage for 40 years

Wefers and Misra (1987) indicate that after drying at less than 400°C, the hydrated alumina will be in the form of boehmite (AlOOH, equivalent to $Al_2O_3 \cdot H_2O$). This is used as the basis for estimating water content and hydrogen formation from radiolysis. Sindelar and Basden (1997) report that the technical performance requirement for a dry storage system for aluminum fuels is that the fuel/cladding consumption limit is less than 3 mils (0.076 mm). This is taken as an upper limit for the depth of aluminum that has been converted to hydrated oxide. While the specifications for the ATR aluminum cladding are for a thickness of 15 mils (0.379 mm) for the bulk of the plates, the actual thickness ranges from 6 to 12 mils

(0.152 to 0.304 mm). Based on the estimate of 30 percent wall thickness corrosion loss and conversion to surface oxide, this would correlate to approximately 1.8 to 3.6 mils (0.046 to 0.091 mm) of aluminum converted to hydrated oxide. Considering an approximate average cladding thickness of 9.5 mils (0.240 mm), the average conversion thickness of aluminum would be 2.9 mils (0.072 mm). This is at the upper limit of the technical specification that the Savannah River Site (SRS) has established for receipt, but it is a very approximate estimate.

Measurements in the INEEL CPP-666 Fuel Storage Basin show corrosion rates of 0.029 and 0.048 mils per year (0.0007 to 0.0012 mm/y). An extrapolation for 20 years of exposure then gives a total metal loss of 0.58 to 0.96 mil (0.015 to 0.024 mm). The maximum observed long-term aluminum corrosion at Argone National Laboratory (ANL) is 2 mils (0.051 mm), with 0.4 to 0.8 mil (0.01 to 0.02 mm) being typical.

Lam et al. (1997) have measured aluminum corrosion rates in humid air at up to 100% relative humidity. The Arrhenius equation derived for 6061 aluminum in the saturated water vapor environment is:

Wt. Gain ($\mu g/dm^2$) = 3.30×10^6 exp[-6820(cal/mol)/RT] $t_{hr}^{0.47}$

The aluminum consumed in mils is 1.19×10^{-6} × Wt. Gain ($\mu g/dm^2$), and the oxide film thickness (boehmite) in nm is 0.0533 x Wt. Gain ($\mu g/dm^2$). This corresponds to less than 0.3 mil (0.008 mm) penetration loss of aluminum in 20 years at near-ambient temperature. It may be considered a lower limit value.

It appears that the expected depth of aluminum that may be oxidized is between 0.7 and 2 mils (0.018 to 0.051 mm). The central value of 1.4 mils (0.035 mm) is used for a "best" estimate, with an uncertainty of ± 0.7 mil (0.018 mm) expected to bound the range.

For the density of aluminum of 2.7 g cm^{-3}, this corresponds to 0.0096 ± 0.0048 g Al/cm^2 surface area. The total surface area of fuel in a canister is 86.3 m^2, so the total quantity of aluminum converted to the hydrated oxide is 8,290 ± 4,140 g or 307 ± 154 atom g. This corresponds to 154 ± 77 moles water of hydration after drying to boehmite. As free water, this would correspond to 2.8 ± 1.4 L. Note that the water is, in fact, bound and not free.

The canister internal volume is 387 L. Considering that the fuel cladding is on the average approximately 5.5 mils (0.139 mm) less in thickness than the specification limit (but that the aluminum side plates are at full thickness), a calculation of the fuel volume results in 65.2 L. Therefore, the net free volume in the canister is 387 - 65 = 322 L. Assuming that the waters of hydration become completely dissociated over time by radiolysis to H_2, the perfect gas calculation of hydrogen pressure (expressed at 25°C) is 12 ± 6 atm. If the canister were vented, the pressure would not build up. But if it is sealed, the hydrogen pressure, exclusive of any original gas, will accumulate.

The rate of radiolysis could result in substantially-less degradation of the water than the 100% assumed above. A more accurate analysis would consist of the following:

1. An evaluation of the radiation field intensity
2. An estimation of the fraction of radiation energy absorbed
3. Determination of G-values (the decomposition rate of a chemical as a consequent of radiation exposure expressed as molecules per 100 eV) for H_2 and O_2 production
4. Incorporation of back reactions and steady-state considerations

The radiation intensity is based on the expected ^{235}U depletion and the age of the fuel. Typical ^{235}U depletion is 35%, and the maximum for ATR fuels is 50%. Therefore, 50% depletion is used to bracket all fuels. Because alpha and beta rays would be stopped by the cladding, only gamma energy is considered for H_2O adsorbed on the surface of intact fuel.

Tromp (1991) calculated the fission and actinide product compositions of ATR fuel that had been depleted by a maximum of 50%. The results were combined with the gamma power (W/Ci) values given for individual isotopes by Pajunen (1997), as well as the fuel composition of 1075 g BOL ^{235}U per assembly to derive the gamma power as a function of time for the 20 assemblies in a canister. At 20 years of cooling (when the fuel is assumed to be packaged), the gamma power in a canister is 62 W. After an additional 40 years of storage, it has decayed to 24 W.

The fraction of the gamma energy adsorbed by the canister contents was estimated from the following sources. Based on the Green (1994) estimates of the fraction of gamma energy absorbed by the contents of packages of various geometries and sizes, it results that approximately 80% would be absorbed in the standardized fuel canister. Bibler (1978) measured gamma radiation absorbed in drums of concreted waste to be approximately 90%. Therefore, DOE (2000a) uses 85%.

Of the 85% of the gamma energy absorbed by the canister contents, the fraction absorbed by the AlOOH is calculated by multiplying it by the fraction of the canister contents that is AlOOH. This follows the procedure of Pajunen (1997).

Using the models of Pajunen (1997) and Garibov et al. (1987), the computed values of the fraction of absorbed AlOOH are similar (0.101 Pajunen model versus 0.113 Garibov et al. model). DOE (2000a) uses the value of 0.107 for the nominal aluminum corrosion thickness of 1.4 mils (0.035 mm). For the range ends of 0.7 and 2 mils (0.018 to 0.053 mm), the value is 0.0545 and 0.155, respectively.

Pajunen (1997) has tabulated G values from a number of sources for the production of H_2 from water in a solid matrix as a function of percent H_2O. For the 15 wt% H_2O in AlOOH, the bounding initial G value for H_2 production of 1.2 molecules/100 eV of energy is absorbed by AlOOH, neglecting any recombination effects. Similarly, for the production of O_2, Pajunen (1997) indicates an experimental bound of 0.08 molecules/100 eV of energy absorbed. Thus, O_2 production, relative to H_2, may be neglected.

Using this information, it results that the radiolysis of the entire hydration water will be converted to H_2 in 7.5, 7.7, and 8.0 years, respectively (for the minimum, nominal, and maximum thickness of boehmite modeled), producing H_2 pressures (at 25°C) of 5.9, 12, and 18 atm.

These results were obtained using the initial G_{H2} value without regard for a decrease in the value over time and without regard for back reactions. Thus, the accumulation of pressure may be at a lesser rate and, in fact, the final pressure may be limited by the establishment of steady-state conditions as a result of back reactions. Garibov et al. (1982) observed a rapid drop-off in the initial G values with a general leveling off of accumulated H_2. Bibler (1978) found that concreted waste in a 2 by 10 ft (0.6 by 3 m) cylindrical canister reached a steady-state H_2 pressure of approximately 2 atm from gamma radiation. Oxygen initially present in the canister decreased, indicating a negative G_{O2}. The primary back reaction that leads to the removal of H_2 and establishment of a steady-state is considered to be:

$$OH + H_2 \rightarrow H_2O + H$$

Fairly conservative calculations indicate that all the water in a canister may undergo radiolysis during the interim storage life and pressurize to levels of H_2, exceeding the canister design limit of 50 psig (4.4 atm absolute). However, experimental evidence exists that actual pressure may not exceed 2 atm (14.7 psig) due to the establishment of a steady-state in the presence of back reactions.

Hanford site analysis

Duncan and Ball (1997) and Pajunen (1997) have performed a detailed analysis on radiolysis effects in SNF storage and expected water content of the MCOs that will store the zircaloy clad N-Reactor fuel. Duncan and Ball (1997) generated a range of expected water content for the MCO and calculated the resulting pressures, gas concentrations, and temperatures for the following three MCO conditions:

1. **Bounding:** An MCO whose contents are not reasonably expected to exceed the particulate matter quantities derived by using the bounding parameter value. This MCO would be used for the design basis accident (DBA).

2. **Nominal or Best Estimate:** An MCO that represents the average or expected MCO.

3. **Low:** An MCO that could occasionally exist.

The results listed in Table 24 show for the "bounding" MCO a pressure of 148 psia (10.1 atm) when all the water inventory is decomposed.

Table 24. MCO pressure estimates (DOE 2000a).

Source	Bounding	Best-Estimate	Low
Initial Pressure	1.5	1.3	1.1
Backfill gas heatup	0.75	0.65	0.55
Uranium oxide hydrates	3.27[a]	0.47[a,c]	0.0[d]
Al(OH)$_3$	4.55[b]	1.20[b]	0.0[d]
Total Pressure	10.1 atm	3.62 atm	1.65 atm
	148 psia	53 psia	24 psia
Pressure Increase	133 psig	38 psig	9 psig

[a] No hydrogen gettering
[b] No hydrogen gettering, but oxygen gettering
[c] 75% thermal decomposition
[d] Gettering assumed

HYDROGEN EMBRITTLEMENT

When applied to metals, hydrogen embrittlement occurs when H_2 diffuses into the metal. The hydrogen is evolved from the reduction of water in the following electrochemical reaction:

$H_2O + e^- = H_{ads} + OH^-$

157

When it comes in contact with the metal surface, it dissociates and diffuses into the bulk material:

$$H_2 = 2\underline{H}$$

where $\underline{H}$ is the dissociated hydrogen atom in the bulk of the metal. $\underline{H}$ is found by dividing the permeability (Φ) by the diffusivity (D). Both Φ and D are described by the Arrhenius equation:

$$D = D_0\, e^{-Q/RT}$$

$$\Phi = A\, e^{-Q/RT}$$

where A and D_0 are pre-exponential constants; Q is the activation energy; R is the gas constant; and T is temperature in K.

From Perng and Alstetter (1986), it results that:

$$\underline{H} = \frac{3.381\cdot 10^{-3}\cdot \exp(-\dfrac{1075.1}{RT})}{P^{\frac{1}{2}}} \qquad \frac{cm^3\,(STP)}{cm^3}$$

where R is expressed in cal/mole and the pressure (P) is expressed in Pa. To perform the calculations, the following assumptions were made:

1. The partial pressure of hydrogen produced as a result of radiolysis is a nominal value of one atmosphere (about 101,000 Pa).
2. The hydrogen produced as a result of radiolysis is diatomic. The hydrogen will enter the metal as mono-atomic hydrogen.
3. All pressure-retaining austenitic stainless steel (SST) components of the spent fuel cask behave similarly when exposed to the hydrogen produced as a result of radiolysis.
4. The hydrogen concentration calculated in this manner includes only the mobile, diffusable hydrogen and none of the irreversibly trapped hydrogen.

For T = 295 K and P = 101,121 Pa we get:

$$\underline{H} = 2.69\cdot 10^{19}\ \frac{cm^3\,(STP)}{cm^3} \;=\; 4.6\cdot 10^{18}\ \frac{molecules}{cm^3}$$

Singh and Altstetter (1982) used a fracture mechanics approach to show that if the hydrogen level is kept below 4-5 wppm (ppm by weight) for one 304 alloy composition and 7 wppm for another 304 composition, there would be no slow-crack growth at a constant load. For this calculation, DOE (2000a) chose the value of 3.7 wppm. At 273 K, Type 304 stainless steel has a density of 7.9 g/cm³. Therefore, 3.7 ppm by weight in Type 304 stainless steel is $2.92 \cdot 10^{-5}$ g (H_2)/cm³ = $1.75 \cdot 10^{19}$ molecules/cm³. This is about 4 times the concentration that would result from a 1-atm partial pressure of H_2 gas. Therefore, it is not expected that the concentration of mono-atomic hydrogen in the stainless steel matrix (due to 1 atmosphere of H_2) will cause any significant decrease in mechanical properties.

Assuming that the pressure is at the bounding condition of 10.1 atm for the Hanford Multi-Canister Over-pack (MCO), as shown in Table 24, and that all other factors stay the same, the above result needs to be multiplied by $P^{1/2}$. The molecules of mono-atomic hydrogen in the metal at 10.1 atm would then be $1.46 \cdot 10^{19}$ molecules/cm^3. These values show that the bounding condition pressure (10.1 atm) reported for the Hanford Site MCO may result in a reduction in the ductility of the specified 304L construction material.

The above calculations are based on theoretical thermodynamic and diffusion relations. The calculated hydrogen quantity has been related to reported property changes in the alloys of interest. However, various other effects that are determined by the surface conditions and metallurgical condition of the metal can alter the final amount and distribution of hydrogen in a metal lattice. Passive oxide films such as those encountered in stainless steel and nickel-chromium-molybdenum alloys may slow the absorption and permeation of hydrogen. The hydrogen may interact with secondary phase particles; grain boundary impurities; dislocations; and microstructure defects. The amount and distribution of hydrogen may be less than calculated using these relationships.

MECHANICAL PROPERTIES DEGRADATION DUE TO HYDROGEN

Stainless steel

Studies of the mechanical properties of the stainless steels with low-bulk hydrogen levels have shown that the yield strengths changed very little, but the elongation (ductility) can be reduced. The loss of ductility could have a deleterious effect on an accidental drop of the canister (Figs. 78 - 80). The samples were cathodically charged in 5% sulfuric acid saturated with carbon disulfide. These figures show a large reduction in the elongation of the alloys with smaller reductions in the yield and tensile strengths. These data show that the elongation of 316 stainless steel is affected less by hydrogen than 304 stainless steel.

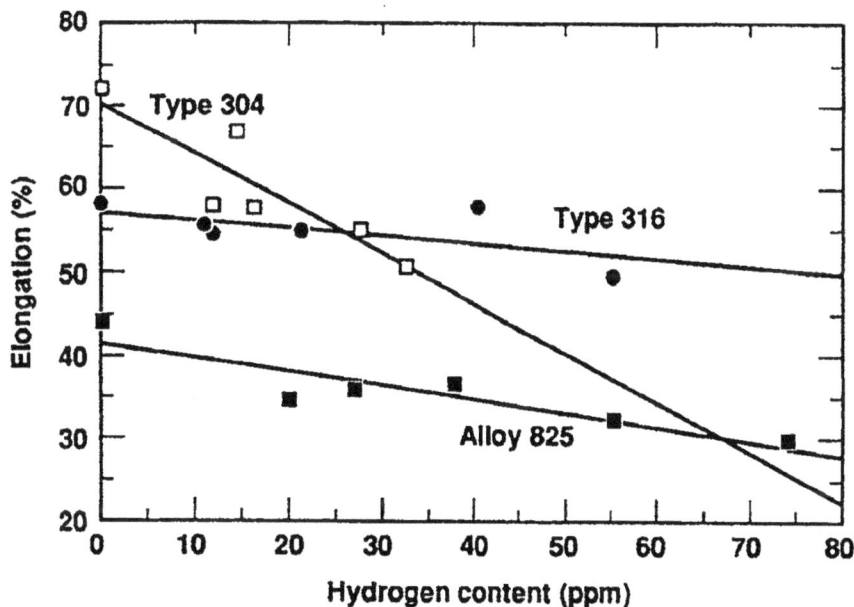

Fig. 78. Elongation vs. hydrogen concentration for Types 304 and 316 stainless steel and Alloy 825 (DOE 2000a).

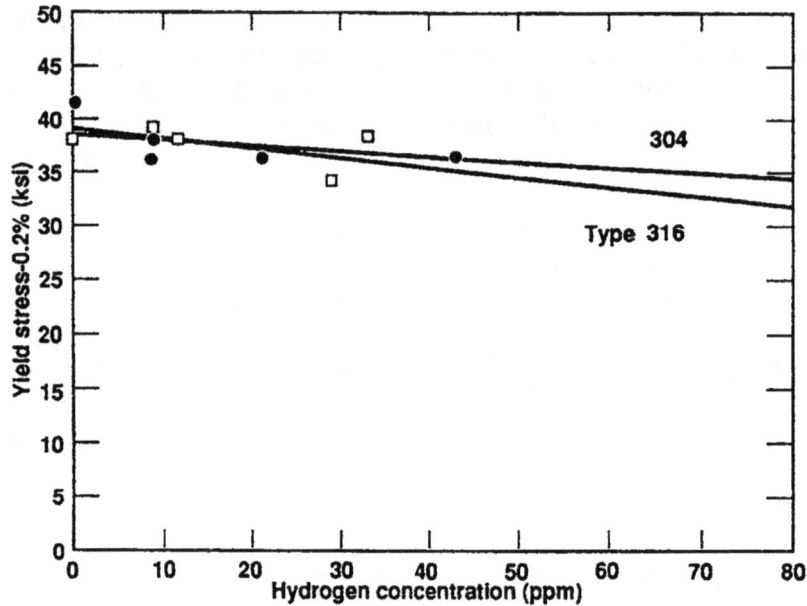

Fig. 79. Yield strength vs. hydrogen concentration for Types 304 and 316 stainless steel (DOE 2000a).

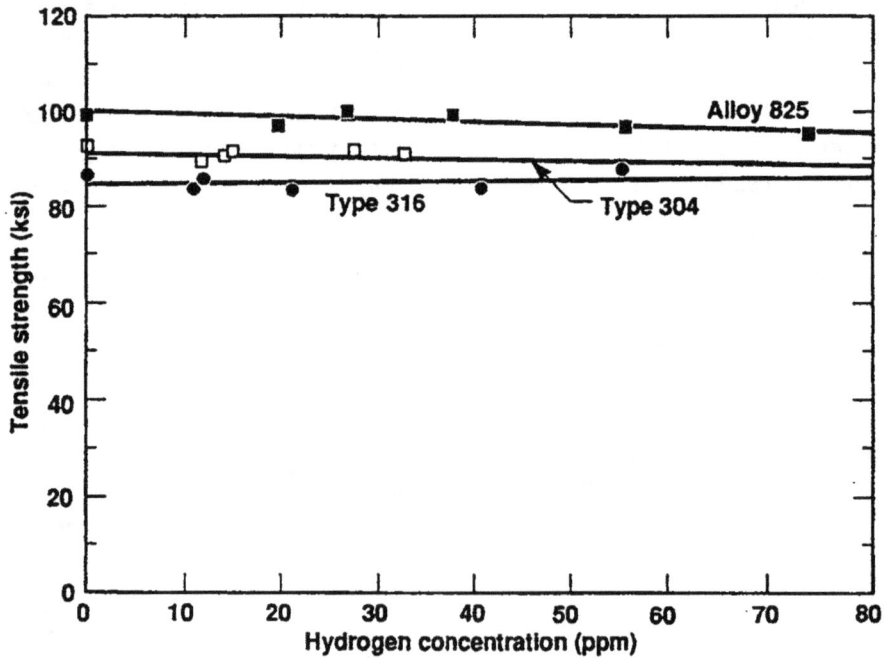

Fig. 80. Tensile strength vs. hydrogen concentration (DOE 2000a).

Hazarabedian and Overjero-Garcia (1990) directly compared 304 and 316L stainless steels using a slow-strain rate test with rates of 4×10^{-5} s^{-1} and 5×10^{-6} s^{-1} in an aqueous environment of H$_2$S saturated NACE solution consisting of 5% NaCl and 0.5% acetic acid at one atmosphere. The coupons were annealed or given a sensitizing heat treatment at 1023 K for 0.25 h, 1 h or 24 h (Table 25). The results show that 304 is much more susceptible to reduction of ductility (elongation) and ultimate tensile strength than 316L.

Table 25. 304 and 316L SSRT results (DOE 2000a).

Alloy	Alloy Condition	Test Medium	Strain Rate (s^{-1})	Ultimate Tensile Strength (MPa)	Elongation (%)
316L	annealed	air	4×10^{-5}	710	46
316L	annealed	H_2S	4×10^{-5}	710	46
304	annealed	air	4×10^{-5}	790	60
304	annealed	H_2S	4×10^{-5}	678	37
316L	1023K-1 h	air	4×10^{-5}	710	46
316L	1023K-1 h	H_2S	4×10^{-5}	710	46
304	1023K-0.25 h	air	4×10^{-5}	778	59
304	1023K-0.25 h	H_2S	4×10^{-5}	593	30
316L	1023K-24 h	air	4×10^{-5}	700	45
316L	1023K-24 h	H_2S	4×10^{-5}	678	45
304	1023K-24 h	air	4×10^{-5}	789	54
304	1023K-24 h	H_2S	4×10^{-5}	589	19
304	annealed	air	5×10^{-6}	756	60
304	annealed	H_2S	5×10^{-6}	534	35
304	1023K-0.25 h	air	5×10^{-6}	760	57
304	1023K-0.25 h	H_2S	5×10^{-6}	494	20
304	1023K-24 h	air	5×10^{-6}	778	53
304	1023K-24 h	H_2S	5×10^{-6}	494	14

The austenitic stainless steels are susceptible to a strain-induced martensitic transformation from an austenite (γ) fcc phase to a martensite of bcc (α) or hcp (ε) phase. This transformation is affected by strain rate, alloy chemistry, and temperature (Hecker et al. 1982; Huang et al. 1989). The evidence is mixed on the effect martensite has on the susceptibility to hydrogen-reduced ductility effects in stainless steels, but there is evidence that the presence of hydrogen can result in the transformation from austenite to martensite at lower stress levels (Rozenak et al. 1990).

Caskey (1977) has measured the change in ductility with test temperature for stainless steel (Fig. 81).

Alloy 22 (Hastelloy C-22)

The use of Alloy 22 is considered for the high integrity container (HIC) that is proposed to store miscellaneous SNF. It is also the present prime candidate for the corrosion barrier in the National Repository due to its excellent resistance to various forms of general and localized corrosion.

The resistance of Hastelloy C-22 to hydrogen embrittlement has been studied by Sridhar et al. (1991; 1995) and Kesavan (1991). The tests were performed on annealed materials using the slow strain-rate technique (4×10^{-6} s^{-1}) on samples that were given a 24-hour hydrogen precharge with a continuous charge during the test. Their data shows a loss of ductility (elongation) at temperatures in the 25 to 85°C range as shown in Table 26 and plotted in Fig. 82.

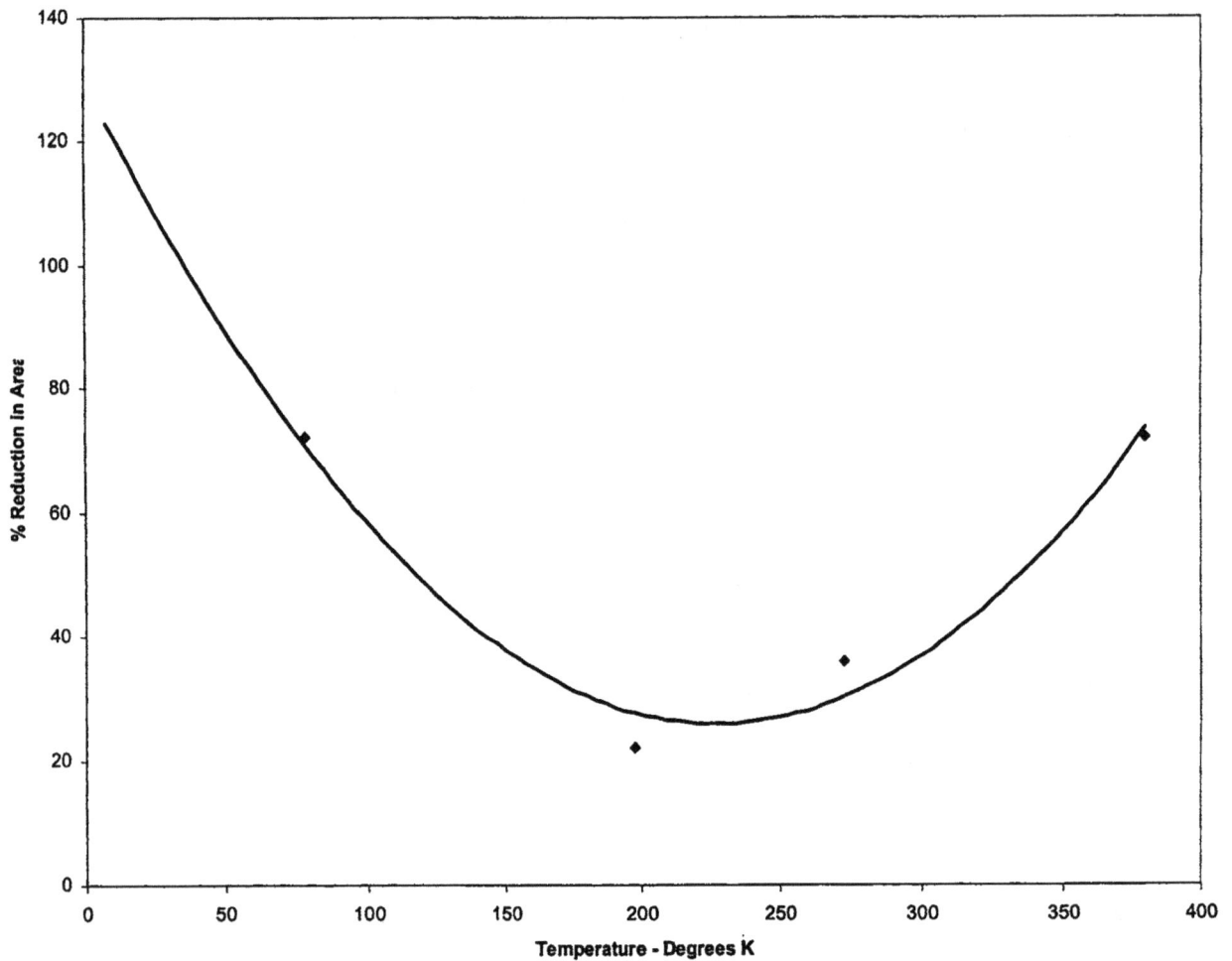

Fig. 81. Percent reduction in area vs. test temperature for Type 304L stainless steel at a bulk hydrogen concentration of $2'10^{-4}$ mol/cm^3 (DOE 2000a).

Table 26. Variation of ductility with temperature for Hastelloy C-22 (DOE 2000a).

Temp (°C)	I_c (mA/cm^2)	Time to failure (hours)	Max. Load (ksi)	(MPa)	% Reduction in Area
25	0	165	112.58	776	74.62
25	20	118.4	103.5	714	42.88
50	20	135.25	105.24	726	47.54
85	20	159.5	111.77	771	64.20

162

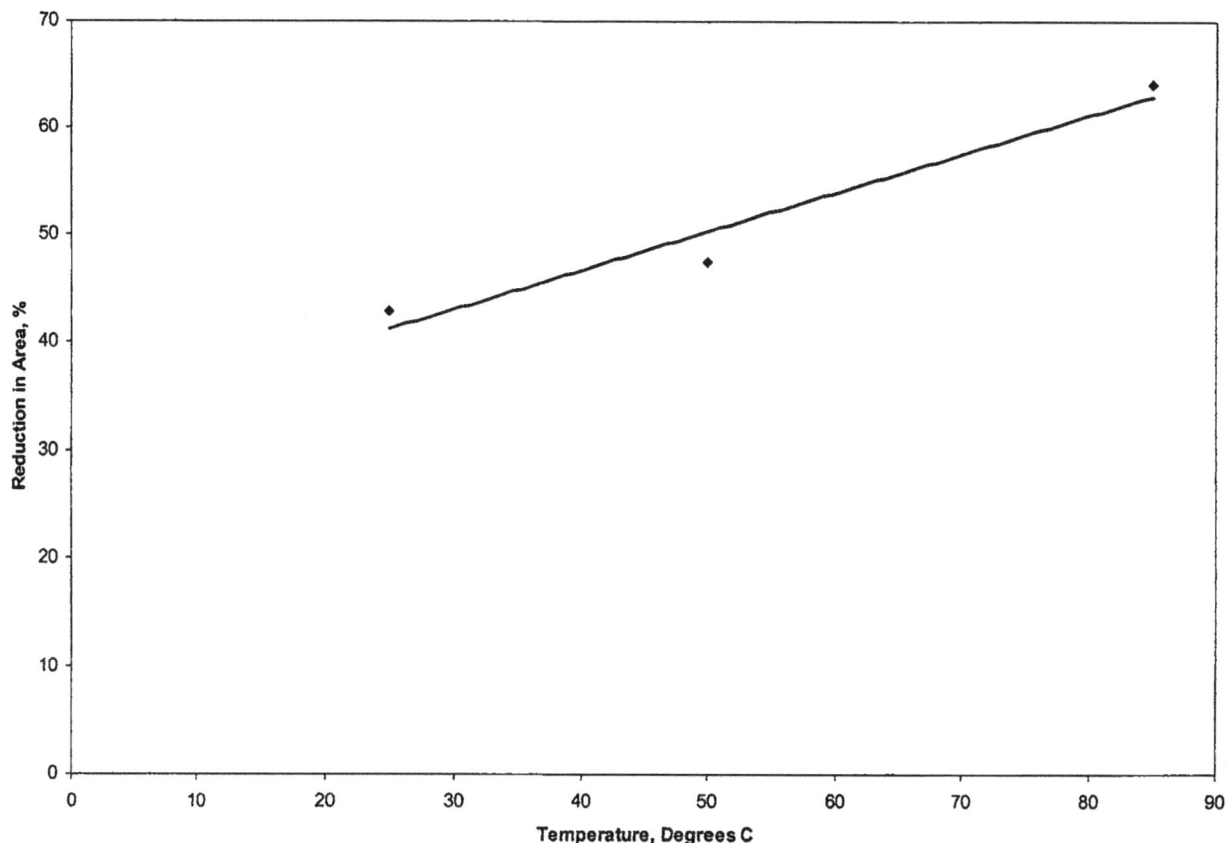

Fig. 82. Variation of ductility with temperature for hydrogen charged C-22 (DOE 2000a).

OTHER EFFECTS

Surface finish

Stainless steels derive their corrosion resistance from a thin, invisible surface layer, which is formed during a reaction between the metal and oxygen present in the ambient environment. This oxide layer drastically decreases the corrosion rate of the material. When this layer is formed, the material is said to be passivated. Other materials that can form this passive film are the various nickel alloys that contain chromium, titanium, and aluminum. Very slow consumption of the metal, in the form of uniform, corrosion will occur in the passive state as the passive layer is slowly dissolved, but it will reform by oxidation of the underlying metal.

The passive layer on stainless steels consists mainly of chromium oxide, and it forms spontaneously in environments containing enough oxidants. The metal surface exposed by mechanical damage (e.g., scratches) is also spontaneously repassivated. The oxygen content of air, and also of most aerated aqueous solutions, is enough for both the creation and maintenance of the passive layer of stainless steels. In general, any oxide-formed surface film that is likely to form on a stainless steel surface during oxidation or corrosion will have a lower permeability than the alloy itself (Caskey 1976). The permeability of a stainless steel membrane with an oxide film formed in a strongly-oxidizing solution can be a factor of .01 to .0001 smaller than for a laboratory-cleaned surface where all barriers to hydrogen absorption are removed. Permeation through membranes with electropolished surfaces has a 4 to 10 fold reduction in permeation rate compared to machined and mechanically polished surfaces.

The canister fabrication process would benefit from a "passivating" chemical treatment to form a thicker surface film to take advantage of the decreased hydrogen solubility, diffusivity, and permeability relative to the base alloy. However, this technique might be impractical in a shop fabrication environment due to difficulties in handling and disposing of the passivating acids.

Welded vs wrought SST

Most of the hydrogen embrittlement test results on stainless steels were developed with wrought products that were tested in the annealed or sensitized condition. Table 27 shows that while hydrogen does embrittle stainless steel plate and welds, they still do possess some ductility (at least 10% elongation) even after charging to very high concentrations. Hydrogen-charged Type 304 stainless steel plate possesses more ductility than uncharged Type 304 stainless steel welded with Type 308.

Pressure

Most of the data for the effect of gaseous hydrogen was developed under high pressure and temperature conditions: 473K (392°F) and 24 MPa (240 atm) of H_2 gas. This is much higher than the postulated pressures and temperatures postulated for the standardized canisters. Drying the fuel will also mitigate the hydrogen embrittlement problem from another perspective. It was mentioned above that the concentration of H in the stainless steel matrix was proportional to $P^{1/2}$. If P could be kept down by decreasing the amount of water (i.e., drying) on the fuel, the concentration of $\underline{H}$ in the stainless steel matrix could be kept below that which causes undesirable embrittlement.

Temperature

The data for both the austenitic alloys (304L, 316L) and the nickel-based alloys show that the ductility of the alloys is affected most by hydrogen in the 25 to 85°C temperature range. This could be significant if the canister wall temperature is in this range when shipped from interim storage to the National Repository.

Table 27. Comparison of wrought 304 vs. 308 weld properties in hydrogen (DOE 2000a).

Type of Material	Charging Conditions	Yield Stress (psi)	Yield Stress (MPa)	Ultimate Stress (psi)	Ultimate Stress (MPa)	% Elongation	cc/cm³ (atom ppm)
Plate	Uncharged	31,000	214	87,000	600	73	None
Plate	10,000 psi (69 MPa) at 212°C	32,000	221	77,000	531	33	5.33 (3376)
Weld	Uncharged	57,500	396	89,800	619	23.2	None
Weld	25,000 psi (172 MPa) at 72°F	62,300	430	98,600	680	12.4	5.25 (3331)

Strain rate

One of the considerations for the material choice for the spent fuel canisters is the National Repository Surface Facility acceptance drop-test requirement. This will induce a high strain rate in the canister material that could be many orders of magnitude higher than the data presented earlier, which was obtained from slow strain tests. DOE (2000a) could not find any data on tensile tests using a high strain rate on hydrogen-charged austenitic stainless steels or nickel-based alloys.

One possible scenario is that the hydrogen-charged stainless steel will suffer small cracks immediately upon fast plastic deformation (from the drop test). However, the cracks would not propagate due to inadequate time for the hydrogen to diffuse to the crack tips, or be transported by diffusion or dislocation movement. This would cause the cracks to blunt and stop growing. The ductility loss would then be dependent on the hydrogen in the lattice at the time of any high strain rate event. The data in Fig. 83 show that the ductility measurement is dependent on strain rate for AISI 4340 steel with the higher strain rates giving higher values. The reason given for this behavior is that the decreasing strain rates allow a greater net-hydrogen accumulation at a greater depth in the specimen.

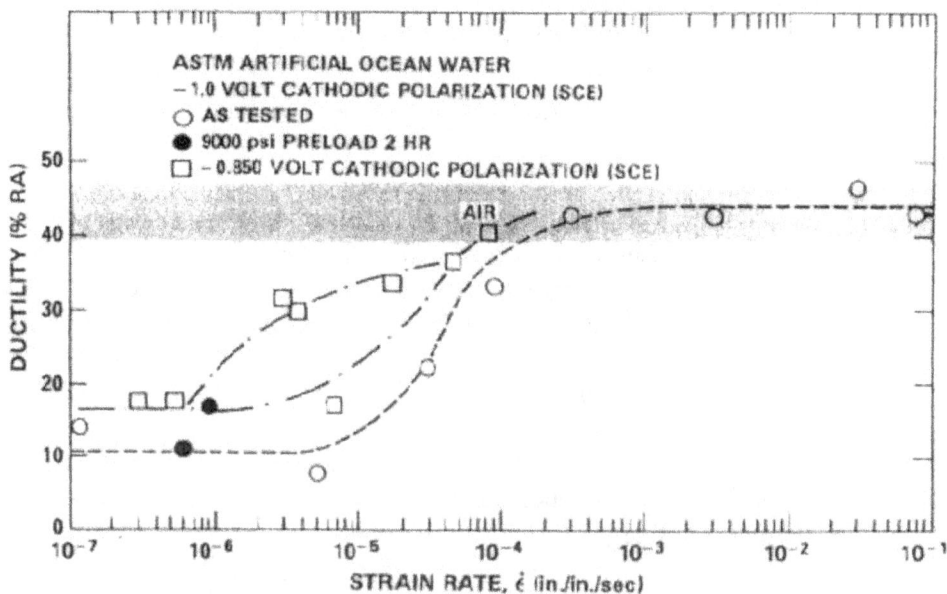

Fig. 83. Effect of strain rate on ductility of AISI 4340 steel under cathodic polarization in ASTM artificial ocean water (DOE 2000a).

There are similarities in material performance in a hydrogen environment between carbon steels, stainless steels, and nickel-based alloys. Toh and Baldwin (1956) showed that the properties of hydrogen-charged carbon steel approach that of uncharged carbon steel as the strain rate increases.

The only datum found by DOE (2000a) for a material in the same alloy family of Hastelloy C-22 is for Hastelloy C-276 as reported by Coyle et al. (1981). Figure 84 shows that the ductility goes to normal, annealed material values at the higher strain rates.

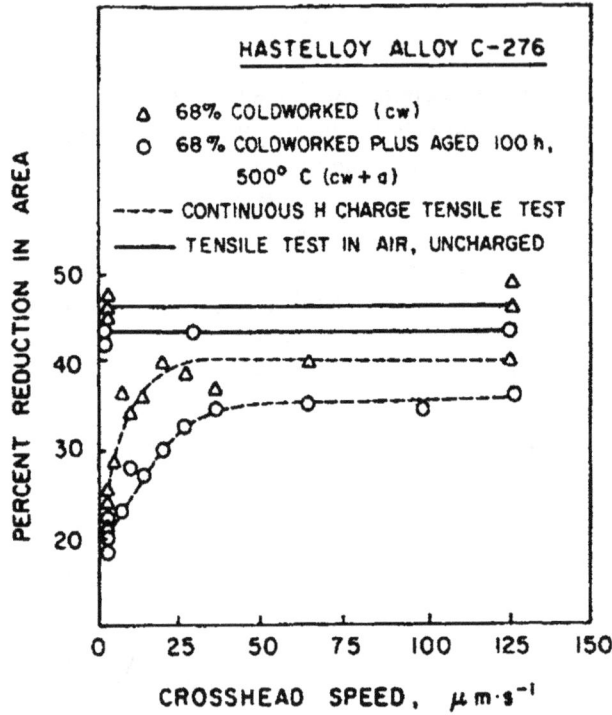

Fig. 84. Effect of crosshead speed (strain rate) on embrittlement of hydrogen charged tensile samples of cold worked and cold worked and aged Alloy C-276 (DOE 2000a).

Hyzak et al.(1981) tested modified Charpy-V samples of annealed and high-energy rate forged (HERF) Type 304L stainless steel that were thermally charged with hydrogen by exposure to 18 MPa or 30 MPa hydrogen gas at 478 K. The hydrogen charge was calculated to be approximately 2350 appm for the annealed 304L, and 3090 appm for the HERF material. This treatment supersaturated the austenitic lattice. This level of hydrogen in the metal would probably not be achieved during interim storage of DOE SNF. The test results in Table 28 show that the absorbed hydrogen reduces the impact toughness of both materials at both temperatures. The results show that strain-induced hydrogen redistribution through dislocation transport is not needed for toughness reduction. The hydrogen present in the lattice is sufficient. The conclusion from this test is that toughness will be reduced during application of a high strain rate. However, the reductions in fracture energy are not large.

Table 28. Hydrogen effects on impact behavior of 304L stainless steel (DOE 2000a).

Sample Condition	Temperature (K)	Fracture Energy (Joules), Annealed	Fracture Energy (Joules), HERF
No exposure	297	194,194	199
Exposed*	297	188,182	152
No exposure	77	159,172	160
Exposed*	77	107,115	95

* Annealed 304L: 1000 hours in 18 MPa H_2 gas at 473 K (2350 appm H_2)
* HERF 304L: 1300 hours in 30 MPa H_2 gas at 473 K (2350 appm H_2)

166

Charpy impact testing

To provide additional information on the issue of the strain rate effect, Charpy Impact tests were performed. The specimen size was 5 mm × 10 mm × 55 mm. The samples were cut in the longitudinal direction. The 304L and 316L samples were tested at a room temperature (23°C) that would approximate the temperature of an accidental drop of the canister. As shown in Table 29, the average absorbed energy decreased by 10.11% as compared to the unexposed control specimens. There was considerable scatter in the 304L data with the hydrogen-charged specimens, showing a difference of 43 ft-lb (58.4 J). The control specimens showed a difference of 52 ft-lb (70.6 J) between the high and low values.

The data for the 316L specimens (Table 30) show that the control specimens had a slightly lower (4.5%) average of absorbed energy than the hydrogen-charged specimens. The 316L data show a smaller scatter in the absorbed energy data with a variance of 4-ft-lb (5.4 J) for the hydrogen-charged specimens and 9 ft-lb (12.2 J) for the control specimens between the high and low values.

Table 29. ASTM E 23 Charpy impact energy test results for Alloy 304L (DOE 2000a).

Specimen Identification	Notch Location	Absorbed Energy		Lateral Expansion			Percent Shear
		(J)	(ft-lb)	(mm)	(mils)	(in.)	
A*	Plate	107.2	79.0	1.75	68.9	0.069	100
B*	Plate	74.6	55.0	1.59	62.6	0.063	100
C*	Plate	115.3	85.0	1.68	66.1	0.066	100
D*	Plate	80.1	59.0	1.60	63.0	0.063	100
E*	Plate	133.0	98.0	1.50	59.1	0.059	100
F	Plate	131.6	97.0	1.49	58.7	0.059	100
G	Plate	150.6	111.0	1.36	53.5	0.054	100
H	Plate	84.1	62.0	1.44	56.7	0.057	100
I	Plate	115.3	85.0	1.56	61.4	0.061	100
J	Plate	80.1	59.0	1.53	60.2	0.060	100

* Specimens A-E: 1000 hours in 18 MPa H$_2$ gas at 473 K (2350 appm H$_2$), Specimens F-J are control samples
Test Temperature: 23°C

Table 30. ASTM E 23 Charpy impact energy test results for Alloy 316L (DOE 2000a).

Specimen Identification	Notch Location	Absorbed Energy		Lateral Expansion			Percent Shear
		(J)	(ft-lb)	(mm)	(mils)	(in.)	
1*	Plate	105.8	78.0	1.48	58.3	0.058	100
2*	Plate	109.9	81.0	1.40	55.1	0.055	100
3*	Plate	108.5	80.0	1.44	56.7	0.057	100
4*	Plate	111.3	82.0	1.45	57.1	0.057	100
5*	Plate	108.5	80.0	1.47	57.9	0.058	100
6	Plate	97.7	72.0	1.58	62.2	0.062	100
7	Plate	101.8	75.0	1.64	64.6	0.065	100
8	Plate	103.1	76.0	1.50	59.1	0.059	100
9	Plate	107.2	79.0	1.55	61.0	0.061	100
10	Plate	109.9	81.0	1.54	60.6	0.061	100

* Specimens 1-5: 1000 hours in 18 MPa H$_2$ gas at 473 K (2350 appm H$_2$), Specimens 5-10 are control samples
Test Temperature: 23°C

Photomicrographs of the fracture faces of the 304L and 316 samples show no apparent difference in the fracture surfaces. All samples failed in a ductile manner by a dimpled fracture caused by micro-void formation and coalescence. There was no evidence of a faceted structure on the fracture faces that indicate a brittle fracture mechanism.

Fracture toughness of welded 304 and 316 stainless steels

Mills (1997) investigated the failure potentials for components fabricated by welding from 304 and 316 stainless steels by analyzing their fracture toughness. He concluded the following:

1. The fracture of 304 and 316 base metals is high because of a ductile, dimple rupture fracture mechanism
2. There is a large heat to heat variability in measured values
3. The fracture toughness is dependent on the welding process
4. The recommended processes are Gas Tungsten Arc, Electron Beam, and Gas Metal Arc

These results have relevance to the hydrogen embrittlement of stainless steels for the following reasons:

1. Hydrogen can affect the ductile rupture mechanism and cause reductions in fracture toughness
2. The reported, hydrogen influenced, mechanical property values in the literature may be affected by the heat to heat variability of the stainless steels
3. There are no welding procedure restrictions in the present specification for the standardized canister

CONCLUSIONS

The properties of wrought and welded 304L stainless steel are of concern in a hydrogen environment because:

1. The ductility of austenitic stainless steels is affected by hydrogen
2. Type 304L suffers a greater reported loss of ductility than 316
3. The ductility of the uncharged Type 304L/308 welds starts out low and is further reduced after charging with hydrogen

The choice of 316L stainless steel is appropriate for the standardized canister due to better resistance to hydrogen embrittlement and localized corrosion. Wrought Alloy 22 shows a hydrogen-induced ductility decrease from 25 to 85°C, which is consistent with other nickel-based alloys. The quantity of free and bound water remaining in the DOE SNF canister after drying needs to be quantified through measurement or characterization of the SNF, sludges, corrosion products, etc. The mechanical properties data for hydrogen embrittlement of stainless steels and nickel-based alloys was generated using the slow strain rate technique and may not be applicable to the high strain rates observed under drop-test conditions. If the pressures generated by radiolysis of water are substantially higher than the design pressure of the spent fuel canister, failure could occur whether the material is hydrogen-embrittled or not. The results of the high strain rate testing show that annealed 316L stainless steel is unaffected by the test hydrogen exposure.

FUEL CANISTER STRESS CORROSION CRACKING SUSCEPTIBILITY; EXPERIMENTAL RESULTS

INTRODUCTION

The condition of spent nuclear fuel (SNF) owned by the U.S. Department of Energy (DOE) must be assessed prior to transportation and disposal in the repository. Because SNF will be in temporary storage for as much as 50 years, verification that no significant degradation has occurred to the canister is required to preclude repackaging all the fuel. Many canisters are removed from wet storage, vacuum-dried (hot or cold), and then placed into dry storage. This process could potentially impact the canister's integrity.

In order to increase the confidence of an intact canister after interim storage, a series of experiments were conducted to evaluate the stress corrosion cracking (SCC) susceptibility of two candidate canister materials (DOE 2003a). These experiments were designed to subject the candidate materials to a range of possible environmental conditions including temperature and solution composition. SCC is the brittle fracture of a normally-ductile material. It occurs as result of a combination of factors including a susceptible material; a corrosive environment; and the presence of tensile stresses. Corrosion of the material acts in a synergistic way with low stresses (e.g., residual stresses from welding) to cause crack initiation. Once the crack has been initiated, propagation occurs based on the condition at the crack tip including the metallurgical condition of the material (e.g., grain size, secondary phases). SCC of austenitic steels is typically associated with chloride containing solutions. However, it has also been seen with caustic solutions such as sodium hydroxide (Jones 1992). Both trans-crystalline and inter-crystalline cracking have been identified. Formation of carbide precipitation has been suggested as an area of weakness susceptible to inter-crystalline SCC. This occurs near welds where the temperature is optimum for carbon to react with chromium and precipitate at the grain boundaries. Welding is also a source of residual stresses. The difficulty in predicting SCC lies in the lack of visible indicators. Cracking can occur on a microscopic scale and then quickly propagate to an unrecoverable extent. It does appear, however, that catastrophic failure in austenitic steels is not common. Rather, the cracks are small enough to cause leaking without sudden large fractures (Logan 1966).

TEST MATERIALS

Two austenitic stainless steels are currently proposed for SNF canister material. Type 304L stainless steel is the material of construction for the Multi-Canister Overpack (MCO) being used at Hanford for N-Reactor fuel. Type 316L is the current material choice for the standardized canister. Therefore, both metals were included in the test project. The "L" designation indicates low carbon (less than 0.03% compared to less than 0.08% for normal grades) requirements in the material. The lower-carbon content reduces the potential for sensitization. Carbon can react with the chromium in the alloy matrix at temperatures between 1,100°F (593°C) and 1,300°F (704°C), forming a chromium carbide precipitate at the grain boundaries. The resulting chromium depletion in the matrix increases the susceptibility of the material to corrosion (known as sensitization). Areas near welding operations are exposed to the optimum temperature for sensitization for the longest time period. Increased time at the temperature increases the carbon-chromium extent of reaction and, thus, the susceptibility to corrosion.

TEST METHODOLOGY

The test methods selected by DOE (2003a) were crevice, U-bend, and slow strain rate tests (SSRTs). Standard practices are provided by ASTM and were used in these experiments. For the crevice tests followed ASTM G 78-95 (ASTM 1995a), the U-bend tests followed ASTM G 36-94 (ASTM 1994), and the SSRTs followed ASTM G 129-95 (ASTM 1995b). Coupon preparation and evaluation was according to ASTM G1-90 (ASTM 1990).

Sample preparations

Most of the samples were welded using the gas tungsten arc weld (GTAW) process. Welds were positioned across the U-bend coupons at the point of maximum bend. Tensile samples for the SSRT were machined with two-weld configurations. One set of experiments was performed with a weld longitudinally through the tension coupon, and one set was performed with a weld transversely across the gage area. In addition, at least one unwelded coupon was used in each test method to evaluate the effects of welding on cracking susceptibility.

Crevice

Crevice tests consist of a section of the candidate material (coupon) in close contact with another sample, typically a washer of either the same material or Teflon. A bolt assembly was used to maintain the configuration. Close contact between two materials forms a very small gap called a crevice. Solution can seep into the crevice, but free mixing with the bulk solution does not occur. Thus, stagnant conditions are created, and the solution chemistry can change causing accelerated localized corrosion, typically in the form of pitting. Although these tests are not specifically designed for SCC, they do indicate the susceptibility of material to localized corrosion. In addition, pitting in a material can be an initiator to cracking.

The coupon assembly was immersed in the selected solution at a specified temperature for the test duration. Test vessels, made of Teflon with full-reflux condensers, were placed in water baths to maintain a constant temperature. After 42 days, the crevice coupons were removed from the solution, disassembled, and examined with a stereo-microscope at 70X for initiation of pitting or cracking. Results from crevice tests are reported as number and depth of pits identified, weight changes, and any additional changes in the visual appearance of the coupon.

U-bend

A strip of material bent in the shape of a U and held in place by a bolt arrangement to induce a specific stress is called a U-bend coupon. Residual tensile stress is induced on the outer side of the coupon from the bending action. The inner side is in compression, but only the tensile stresses contribute to the increased risk of SCC. The bolts are hung from a Teflon hanger that is attached to the vessel lid. This immerses the bent portion of the coupon, but the ends are exposed to the vapor space of the vessel. As discussed earlier, SCC is a function of material stress and the environment. Increasing the stress in the material by bending the coupon results in more severe conditions and accelerates the onset of crack initiation. It may also accelerate crack propagation once the initiation has occurred.

As with the crevice method, the coupon was suspended in the solution at the designated temperature. The same equipment used in the crevice tests was used for the U-bend tests. Approximately once per week, the

U-bend coupons were removed and examined with a low-magnification microscope (10X to 70X) for indications of cracking (without removing the bolt assembly). If no cracking was observed, the coupons were returned to the test solution. The tests continued for approximately 6 months. Any cracks or other visual changes identified on the coupon were reported in the results.

SSRT

SSRTs require a tensile instrument to pull a tension specimen very slowly. Strain (units of extension divided by the gage length per time) is applied at a slow constant extension rate (units of extension per time). A strain rate is typically 10^{-4} to 10^{-8} s^{-1}. This rate is slow enough to allow the corrosive environment to act on the material but fast enough to have results in a reasonable time. For the purposes of these tests, the strain rate was 10^{-6} s^{-1}. The test ended when the coupon fractured. Similar to the U-bend test, the SSRT method increases the stress on a material resulting in an accelerated test.

A baseline or control stress-strain curve was generated by exposing a coupon to deionized water at a designated test temperature. Subsequent tests exposed samples to the desired solution at a designated temperature while being stressed. The resulting curves were then compared to the baseline. Accelerated breaking times indicated that the environment is increasing the tendency for that material to crack. As no definitive values were determined in this test method, technical judgment was required to interpret the results when comparing the stress-strain curves.

Environmental conditions

The solution chemistry was based on the thermodynamic assessment of two representative fuel canister configurations reported by Anderson (1998). Concentrations were selected based on best technical judgment, assuming the conservative side of a realistic value. Initially, two solution chemical conditions were chosen and are shown in Table 31. Bromine and iodine, also present in the Advanced Test Reactor (ATR) and N-reactor fuel elements, readily oxidize rubidium and cesium and were, therefore, included in the solution chemistries. Once the molar ratios of the relevant components in the solution were defined, dilutions to 40 and 20 wt% caustic (rubidium and cesium hydroxide) in deionized water were made. Both concentrations were used in the test matrix. A third solution without halides was added to investigate the role of the bromine and iodine. In addition, control tests were performed with deionized water. Triplicate samples were tested in the crevice and U-bend tests, but only a single sample for each configuration was used in the SSRTs. Table 32 has the complete test matrix.

Table 31. Chemical composition of the test solutions (DOE 2003a).

Solution	#1 20% OH (w/ Bromine and Iodine)	#2 40% OH (w/ Bromine and Iodine)	#3 40% OH (w/o Bromine and Iodine)
CsOH	0.57 M	1.14 M	1.14 M
RbOH	1.12 M	2.24 M	2.24 M
NaBr	0.004 M	0.004 M	
NaI	0.015 M	0.015 M	

Table 32. Stress corrosion cracking test matrix (DOE 2003a).

Test Solutions	316L No weld	316L T weld	316L L weld	304L No weld	304L T weld	304L L weld
90°C Deionized water		Crevice U-bend SSRT	SSRT		Crevice U-bend SSRT	SSRT
90°C 20% caustic (w/ Bromine and Iodine)		Crevice U-bend SSRT	SSRT		Crevice U-bend SSRT	SSRT
90°C 40% caustic (w/ Bromine and Iodine)	U-bend	Crevice U-bend SSRT	SSRT	U-bend	Crevice U-bend SSRT	SSRT
90°C 40% caustic (w/o Bromine and Iodine)		Crevice U-bend			Crevice U-bend	

SSRT = Slow strain rate test

EXPERIMENTAL CONDITIONS

Type 316LW crevice tests

Welded Type 316L (316LW) crevice experiments at 90°C were tested for 42 days. The results listed in Table 33 indicate that the general corrosion rates were extremely low. The general industry standard considers corrosion rates <2 mils per year (mpy) (0.51 µm/y) to be acceptable. Corrosion rates in the halide free solution are not significantly different with respect to those from the solution with iodine and bromine, so corrosion inhibition from the halides is not indicated. No indications of pitting or cracking were seen in any of the tests. Figures 85 through 88 show the final appearance of the coupons from the test solutions. The color of the coupons tested in deionized water (Fig. 85) has remained unchanged except for an occasional oxide (rust) spot. As a result of the chemical attack, the coupons tested in the hydroxide solutions have turned a brown color. Examination by scanning electron microscopy found the layer too thin for chemical analysis estimated at less than a few nm thick.

The coupons tested in the 40-wt% hydroxide solutions (Figs. 87 and 88) are darker than the ones tested in the 20-wt% hydroxide solution (Fig. 86). Different photography techniques have made this color contrast less apparent in the pictures. Coupons from the 20-wt% hydroxide solution and 40-wt% hydroxide solution without halides show a tinted "footprint" where the crevice washer had been, but very little localized corrosion has occurred. Interferometry techniques detected no discernible depth to the crevice area. The surface roughness of the corrosion site appeared to be the same magnitude as the rest of the coupon. The welds are etched in all the hydroxide solutions tested. Neither the etched welds nor the minor accelerated crevice corrosion are severe enough to indicate an early failure mechanism.

172

Table 33. Type 316LW crevice test corrosion rates at 90°C (DOE 2003a).

Coupon ID	Alloy	Solution Tested	Corrosion Rate	
			(mpy)	(μm/y)
07	316LW	Deionized water	(-0.002)	(-0.051)
08	316LW	Deionized water	0.002	0.051
09	316LW	Deionized water	0.008	0.202
	(Disregard (-) rate)	Average Rate =	0.005	0.127
W4223	316LW	20% OH (w/ Bromine and Iodine)	0.02	0.51
W4224	316LW	20% OH (w/ Bromine and Iodine)	0.04	1.01
W4225	316LW	20% OH (w/ Bromine and Iodine)	0.03	0.76
		Average Rate =	0.03	0.76
W4275	316LW	40% OH (w/ Bromine and Iodine)	0.08	2.02
W4276	316LW	40% OH (w/ Bromine and Iodine)	0.04	1.01
W4277	316LW	40% OH (w/ Bromine and Iodine)	0.03	0.76
		Average Rate =	0.05	1.27
10	316LW	40% OH (w/ Bromine and Iodine)	0.07	1.77
11	316LW	40% OH (w/ Bromine and Iodine)	0.05	1.27
12	316LW	40% OH (w/ Bromine and Iodine)	0.05	1.27
	316LW	Average Rate =	0.06	1.52

Fig. 85. Type 316LW tested in deionized water at 90°C (DOE 2003a).

173

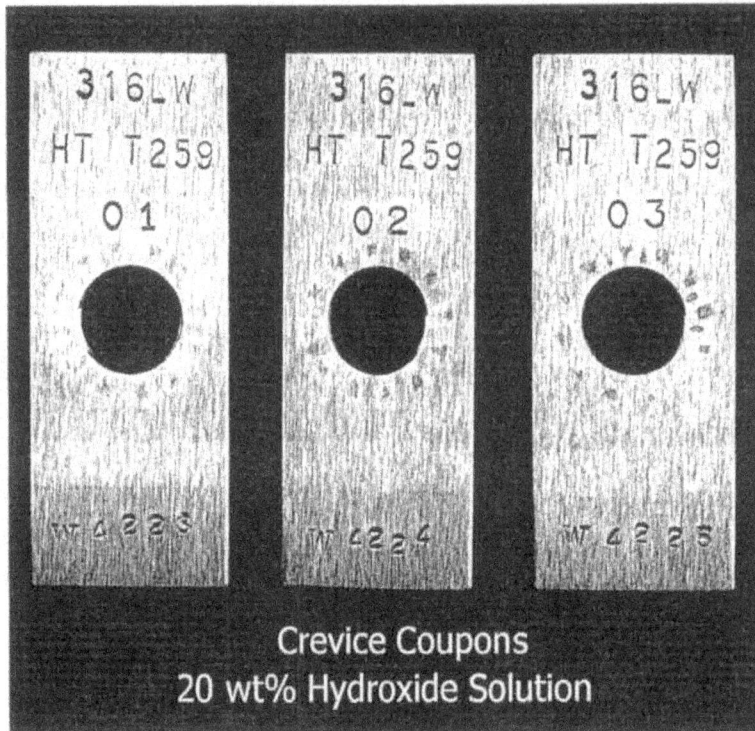

Fig. 86. Type 316LW tested in 20-wt% hydroxide solution with halides at 90°C (DOE 2003a).

Fig. 87. Type 316LW tested in 40-wt% hydroxide solution with halides at 90°C (DOE 2003a).

174

Fig. 88. Type 316LW tested in 40-wt% hydroxide solution without halides at 90°C (DOE 2003a).

Type 316LW U-bend tests

Type 316LW U-bend tests in the deionized water were immersed for 6 months at 90°C. None of the three test coupons have shown any clear indication of cracking. A very small weight-loss trend indicated that a small level of general corrosion was occurring, and this correlates with the results of the crevice tests. The coupons did not darken, but some small rusty spots were randomly visible across the surface. Figure 89 shows the post-test appearance of the coupons.

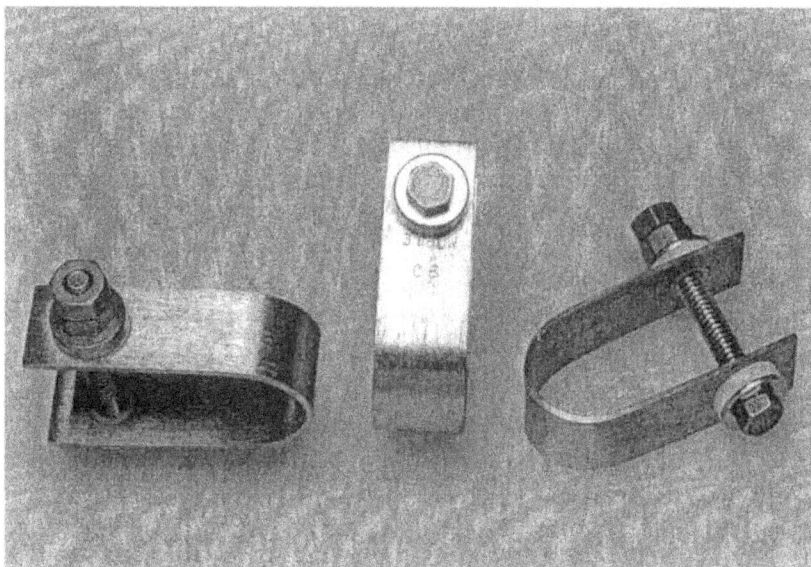

Fig. 89. Type 316LW tested in deionized water at 90°C (DOE 2003a).

None of the 316LW U-bend test coupons had any clear indications of cracking after 6 months exposure to the three hydroxide solutions. Microscopic examination of coupon numbers 2, 5, and 6 during the first week showed an opening that appeared to be deformation damage. No changes were seen throughout the test period, and they were assumed to have occurred during the bending of the coupon. Although the primary analysis method with U-bend tests is visual, some weight measurements were made before and after exposure. Small weight losses (less than 0.1 g) indicate some general corrosion was occurring, which correlates with the results from the crevice tests. Figures 90 through 92 show the condition of the U-bend coupons in 20% and 40% cesium/rubidium/hydroxide with iodine and bromine. The hydroxide solution turned the coupons dark brown as seen in Figs. 90 and 92. The general corrosion appears worse in the area exposed to the vapor. Figure 91 details this area showing the discoloration.

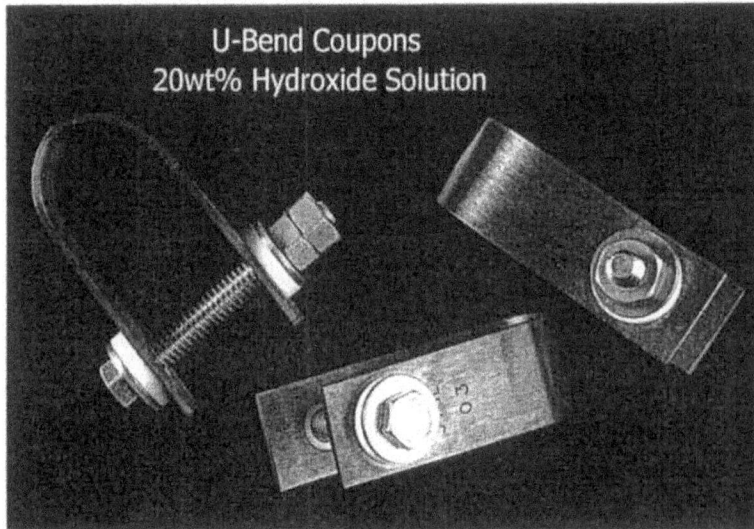

Fig. 90. Type 316LW tested in 20-wt% hydroxide solution at 90°C (DOE 2003a).

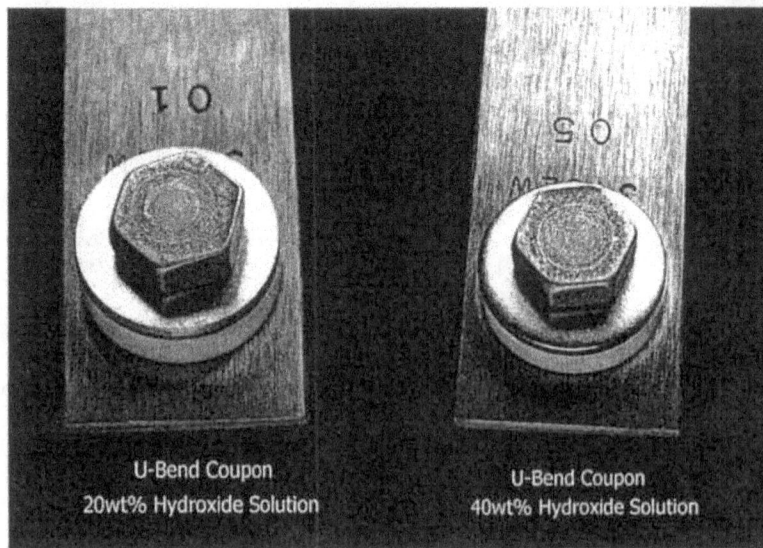

Fig. 91. Type 316LW U-bend coupons showing liquid/vapor interface area (DOE 2003a).

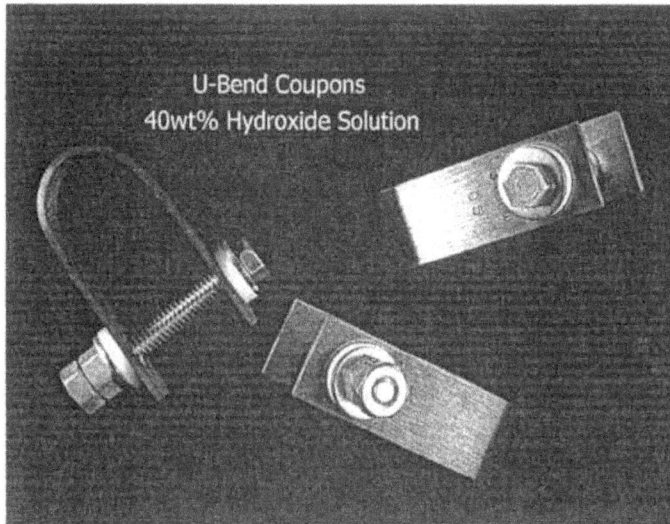

Fig. 92. Type 316LW tested in 40-wt% hydroxide solution with halides at 90°C (DOE 2003a).

Results from the 316LW U-bend tests in the 40-wt% hydroxide solution with halides were similar. After 6 months of exposure, no cracking was identified on the three coupons. Microscopic examination of coupon numbers 5 and 6 revealed openings that also looked like deformation damage caused by the initial bending of the coupon. This damage did not change throughout the test period and is assumed to have occurred during the bending of the coupon. A small weight-loss trend indicated that a small level of general corrosion was occurring, which correlated to the results of the crevice tests. Overall, the solution has turned the coupons a dark brown as can be seen in Fig. 92. The general corrosion was worse in the vapor space portion of the coupon as opposed to the immersed section in the same fashion as the coupons tested with the 20-wt% solution. Figure 91 also details the interface area of the coupon tested with the 40-wt% hydroxide solution.

The 316LW U-bend tests in the 40-wt% hydroxide solution without halides were immersed for 5 months at 90°C, and no indications of cracking were identified. Discoloration was more pronounced without the iodine and bromine (Fig. 93). However, microscopic examination and weight-loss measurements were similar to the tests with halides included in the test solution.

Fig. 93. Type 316LW tested in 40-wt% hydroxide solution without halides at 90°C (DOE 2003a).

Similar results were seen with 316L, unwelded coupons in 40-wt% hydroxide solution with halides at 90°C after 5.5 months (Fig. 94). General corrosion was worse in the vapor space portion of the coupon as opposed to the immersed section, as seen in the other U-bend tests. No difference was seen in the results as compared to the welded coupons. Heat input from welding can cause a region around the weld that is more susceptible to localized and general corrosion than the bulk material. Weld areas were not preferentially attacked in the U-bend experiments, thus no increased susceptibility to cracking due to welding was indicated.

Type 316L slow strain rate tests

Two weld configurations were investigated in the SSRTs: 1) a transverse weld across the gage area; and 2) a weld longitudinally down the test coupon. Each weld configuration had a baseline test in deionized water and was then tested with the 20-wt% and 40-wt% hydroxide solutions containing halides. Triplicate test coupons were not used for this test method because of the time duration of each test. The time to failure of each of the tests is shown in Table 34.

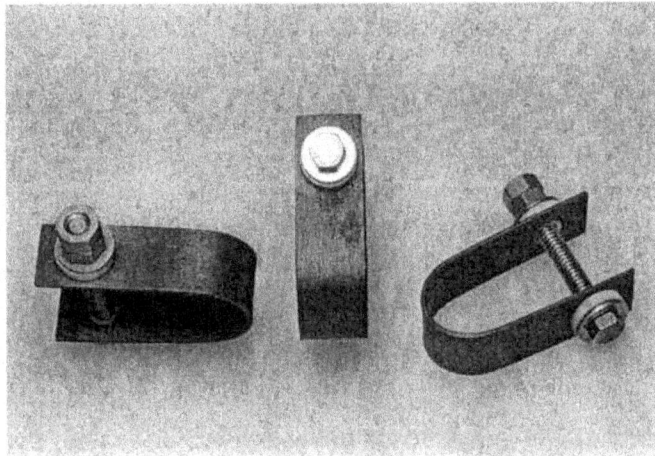

Fig. 94. Type 316L (unwelded) tested in 40-wt% hydroxide solution with halides at 90°C (DOE 2003a).

Table 34. SSRT results of Type 316LW testing (DOE 2003a).

Slow Strain Rate Test Results of Type 316L Transverse Welded Coupons	
Test Solution	Time to Failure (hours)
Baseline deionized water	168
20-wt% hydroxide, (w/ Bromine and Iodine)	160
40-wt% hydroxide, (w/ Bromine and Iodine)	160
Slow Strain Rate Test Results of Type 316L Longitudinally Welded Coupons	
Test Solution	Time to Failure (hours)
Baseline deionized water	161
20-wt% hydroxide, (w/ Bromine and Iodine)	156
40-wt% hydroxide, (w/ Bromine and Iodine)	155

The results of the SSRTs indicate that there is little probability for these solutions to cause a stress crack in the Type 316LW stainless steel material. Times to failure of the test solutions deviated less than 5% from the baseline tests. This is within the repeatability of the test equipment. In addition, visual observations under a microscope showed no indications of cracking. The failures appeared to be ductile in all coupons tested. Selected samples were examined with a scanning electron microscope to verify the appearance of a ductile fracture. Brittle fractures typically have no plastic deformation such as necking and side peaks (Fig. 95). Chevron (V-shaped) lines are also often seen on the fracture face in a brittle fracture; however, a dimpled surface indicative of ductile failures was found in these tests.

The combined results of the crevice, U-bend, and SSRT methods indicate that the tested Type 316LW and 316L (unwelded) material most likely will not be susceptible to cracking from potential cesium and rubidium hydroxide solutions. Welding changes the metallurgical structure near the weld, but it had no effect on the SCC susceptibility in these tests. Corrosion rates from those coupons tested with the hydroxide solutions did not significantly differ from the control coupons tested in deionized water. Welded coupons had similar results to the unwelded samples, indicating no increased susceptibility from the heat input. In addition, the corrosion rates from the hydroxide solution without halides did not differ significantly from those reported for the hydroxide solution containing the halides, indicating that iodine and bromine do not have a significant effect on the corrosion susceptibility of the Type 316L stainless steel. No visual indications of cracking or pitting were seen from the crevice and U-bend experiments, and no decrease in failure time was found in the SSRTs. In conclusion, the 316L stainless steel standardized canisters should not prematurely fail because of the presence of cesium and rubidium.

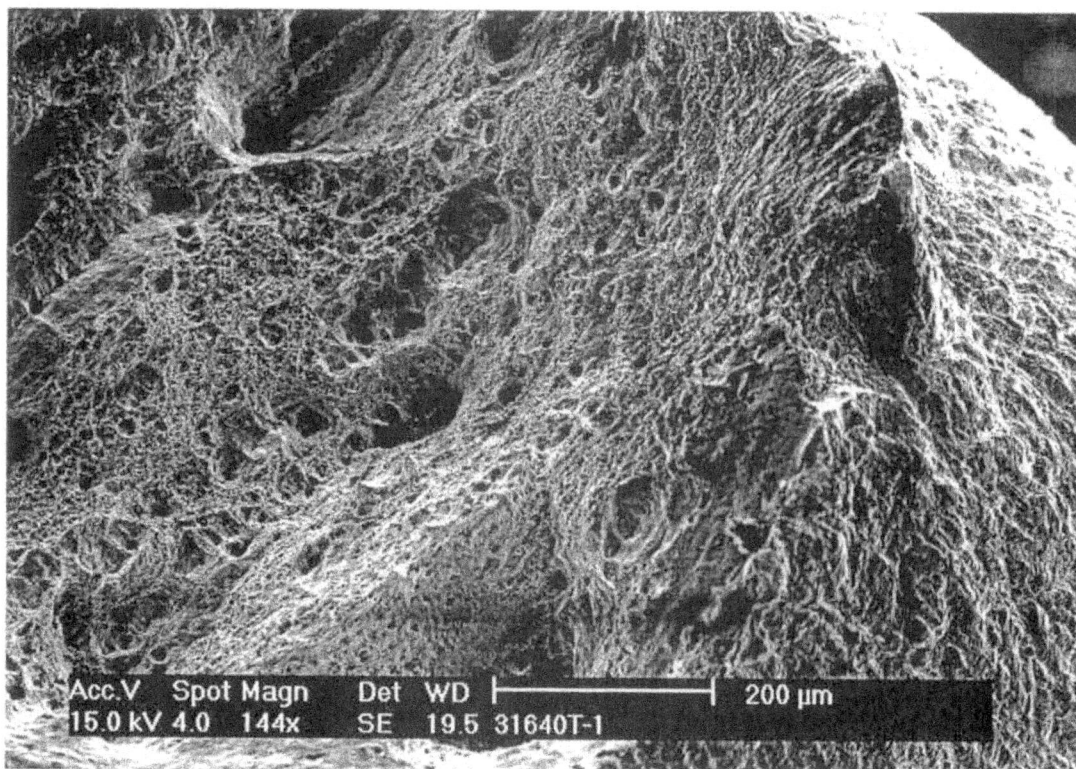

Fig. 95. SEM photo of the fracture surface of type 316L, transverse weld, slow strain rate test sample exposed to 40% caustic solution (DOE 2003a).

Type 304LW crevice tests

The results listed in Table 35 indicate that the general corrosion rates for Type 304L stainless steel were typically low, which were similar to the results for Type 316L. Coupon number 9, which was exposed to the 40-wt% solution without halides, is the exception with a corrosion rate of 2.1 mpy (0.53 μm/y). This higher rate was not repeated with the other two crevice coupons in the same test vessel. Visual examination of the coupon did not reveal corrosion attack any worse than seen on the other samples. The final weight was repeated, and a review of weights and dimensions compared to the other coupons was performed. No indication of measurement errors was found. Thus, the higher corrosion rate was included in the average rate. No indications of pitting or cracking were seen in any of the tests. Minor crevice corrosion was seen on the coupons tested in both 40-wt% hydroxide solutions. White light interferometer measurements were taken from these corrosion sites, but the surface roughness did not differ from the other parts of the corrosion coupon. Figures 96 through 99 show the final condition of the coupons from all the test solutions. The color of the coupons tested in deionized water remained unchanged except for a few oxide (rust) spots.

Table 35. Type 304LW crevice test corrosion rates (DOE 2003a).

Coupon ID	Alloy	Solution Tested	Corrosion Rate	
			(mpy)	(μm/y)
01	304LW	Deionized water	0.000	0.000
02	304LW	Deionized water	0.0008	0.020
03	304LW	Deionized water	(-0.008)	(-0.202)
	(Disregard (-) rate)	Average Rate =	0.004	0.101
10	304LW	20% OH (w/ Bromine and Iodine)	0.03	0.76
11	304LW	20% OH (w/ Bromine and Iodine)	0.02	0.51
12	304LW	20% OH (w/ Bromine and Iodine)	0.008	0.202
		Average Rate =	0.02	0.51
04	304LW	40% OH (w/ Bromine and Iodine)	0.05	1.27
05	304LW	40% OH (w/ Bromine and Iodine)	0.05	1.27
06	304LW	40% OH (w/ Bromine and Iodine)	0.07	1.77
		Average Rate =	0.06	1.52
07	304LW	40% OH (w/o Bromine and Iodine)	0.06	1.52
08	304LW	40% OH (w/o Bromine and Iodine)	0.07	1.77
09	304LW	40% OH (w/o Bromine and Iodine)	2.1	53.1
		Average Rate =	0.74	18.72

Fig. 96. Type 304LW tested in deionized water at 90°C (DOE 2003a).

Fig. 97. Type 304LW tested in 20-wt% hydroxide with halides at 90°C (DOE 2003a).

Fig. 98. Type 304LW tested in 40-wt% hydroxide solution with halides at 90°C (DOE 2003a).

Fig. 99. Type 304LW tested in 40-wt% hydroxide solution without halides at 90°C (DOE 2003a).

As a result of the chemical attack, the coupons tested in the hydroxide solutions turned a brown color. The coupons tested in the 40-wt% solution with halides were darker than the ones tested in the 20-wt% hydroxide solution. The coupons tested in the 40-wt% hydroxide solution without halides were the darkest. As discussed in the Type 316L experimental results, the discoloration is less than a few nm thick. All coupons from the hydroxide solutions showed a tinted print where the crevice washer had been placed, but only minor localized corrosion has occurred on the coupons tested in the 40-wt% hydroxide solutions. All the coupons tested in the hydroxide solutions had the welds etched.

Type 304LW U-bend tests

None of the three 304LW U-bend test coupons showed any clear indication of cracking after 6 months in deionized water at 90°C. Weight losses of the U-bend coupons were measured as a general indication of corrosion. A very small weight-loss trend (<0.1 g) indicated a small level of general corrosion was occurring, and this correlates with the results of the crevice tests. The coupons did not darken, but some small oxide (rust) spots were visible. Figure 100 shows the post-test appearance of the coupons.

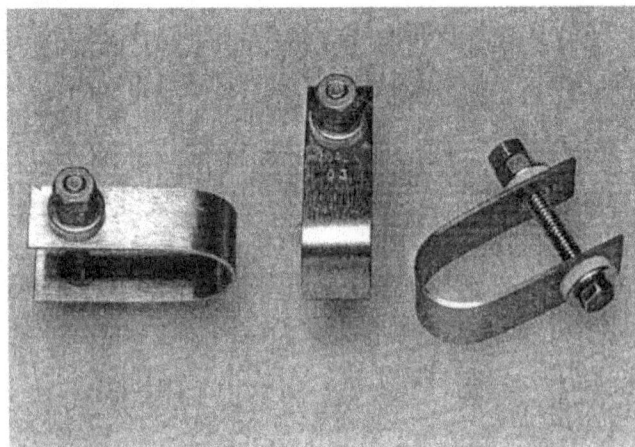

Fig. 100. Type 304LW tested in deionized water at 90°C (DOE 2003a).

The 304LW U-bend tests in the 20-wt% and 40-wt% hydroxide solutions with halides were immersed for 5 months at 90°C. A small weight-loss trend indicated that a small level of general corrosion was occurring, but did differ from the losses seen in deionized water. The corrosion appears more prevalent in the vapor space areas. Microscopic examination of the triplicate coupons revealed no localized corrosion (e.g., cracking) was occurring. Figure 101 shows the post-test appearance of the coupons exposed to 20-wt% hydroxide solution. Figure 102 shows the coupons from the 40-wt% hydroxide tests.

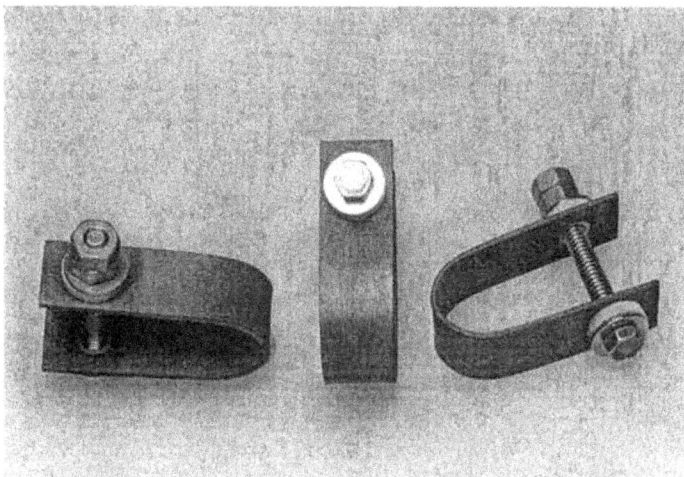

Fig. 101. Type 304LW tested in 20-wt% hydroxide solution at 90°C (DOE 2003a).

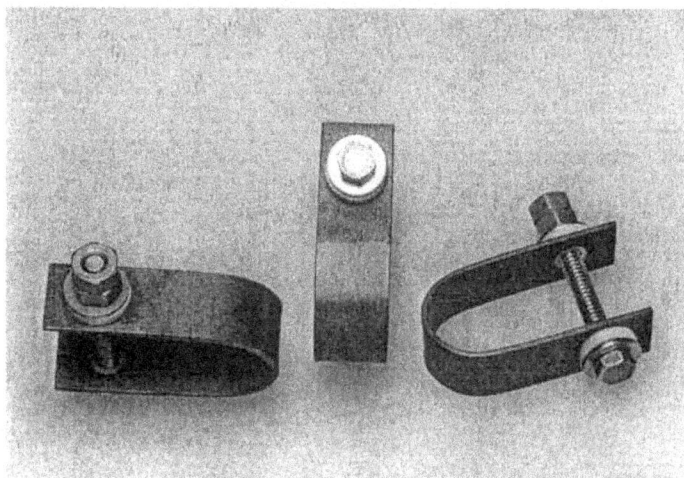

Fig. 102. Type 304LW tested in 40-wt% hydroxide solution with halides at 90°C (DOE 2003a).

Additional experiments were completed to evaluate the effects of halides in the test solution and to assess any increased susceptibility to cracking from the welding operation. Three 304LW U-bend coupons were exposed to 40-wt% hydroxide solution without halides at 90°C for 5 months. The results were very similar to the tests with halides in solution; small weight losses; and no visible indications of localized attack such as cracking. Figure 103 is a photograph of the coupons after exposure. Unwelded Type 304L coupons also showed no differences in degradation after 5.5 months in the 40-wt% hydroxide solution with halides at 90°C. Microscopic examination and weight-loss measurements were essentially identical between the welded and unwelded coupons. Final conditions are shown in Fig. 104.

Fig. 103. Type 304LW tested in 40-wt% hydroxide solution without halides at 90°C (DOE 2003a).

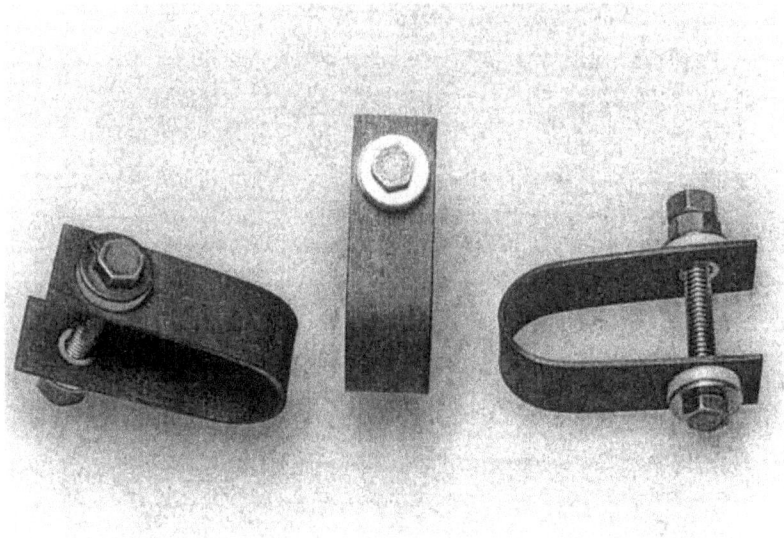

Fig. 104. Type 304L (unwelded) tested in 40-wt% hydroxide solution with halides at 90°C (DOE 2003a).

Type 304L slow strain rate tests

Tensile coupons with Type 304L were prepared with the same weld configurations as the Type 316L coupons—a transverse weld across the gage area and a weld longitudinally down the test coupon. Baseline tests with deionized water were performed first with both weld configurations. One longitudinal coupon and one transverse-welded coupon were tested with the 20-wt% and 40-wt% hydroxide solutions containing halides. The time to failure from each test is shown in Table 36.

Table 36. SSRT results of Type 304LW testing (DOE 2003a).

Slow Strain Rate Test Results of Type 304LW Transverse Welded Coupons	
Test Solution	Time to Failure (hours)
Baseline deionized water	203
20-wt% hydroxide, (w/ Bromine and Iodine)	194
40-wt% hydroxide, (w/ Bromine and Iodine)	187
Slow Strain Rate Test Results of Type 304LW Longitudinally Welded Coupons	
Test Solution	Time to Failure (hours)
Baseline deionized water	187
20-wt% hydroxide, (w/ Bromine and Iodine)	184
40-wt% hydroxide, (w/ Bromine and Iodine)	195

The results from the SSRTs indicate that Type 304LW stainless steel will probably not fail by stress corrosion cracking (SCC) in cesium/rubidium hydroxide solutions. All but one of the times to failure deviated less than 5% from the baseline tests, which is within the repeatability of the test equipment. Although the transversely-welded 304LW coupon in 40-wt% hydroxide solution deviated 7.9%, visual inspection of the sample found no indications of cracking. In addition, visual observations under a microscope of all coupons and scanning electron microscopy of selected samples revealed ductile fractures. Figure 105 shows the dimpled surface of the fracture face typical of a ductile failure.

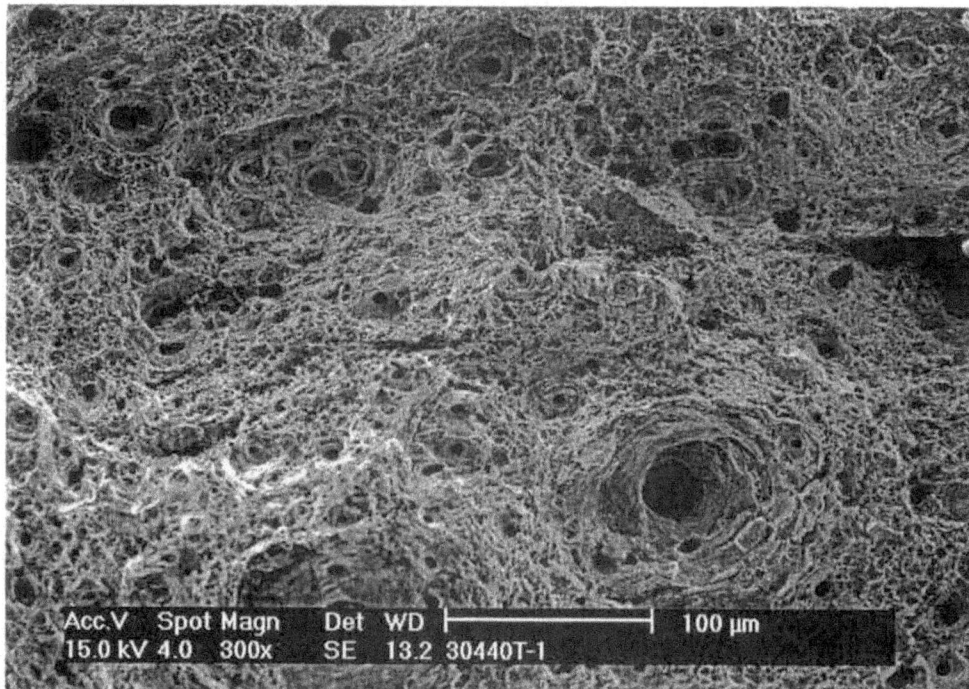

Fig. 105. SEM photo of type 304L, transverse weld, fracture surface after exposure to 40% caustic solution in a slow strain rate test (DOE 2003a).

Based on the results of the crevice, U-bend, and SSRT methods, the potential for Type 304L stainless steel to fail by SCC in cesium and rubidium hydroxide solutions is minimal. Corrosion rates were low and did not differ significantly from the control in deionized water, and the weight losses from the U-bend tests were small. Microscopic examinations found few indications of cracking or pitting, though some degradation effects were seen. Most of the slow strain rate data indicated no decrease in ductility or failure rates. One test (40-wt% hydroxide with the transverse weld) had a slightly low time to failure measurement. The presence of halides did not alter the degradation rates or increase the susceptibility of cracking. Nor did heat input from welding increase localized or preferential attack. Although a few results indicated that the risk of SCC with Type 304L stainless steel is slightly greater than with Type 316L, failure because of the presence of cesium and rubidium hydroxide is not expected.

CONCLUSIONS

No indications of accelerated general corrosion or localized corrosion (including cracking) were found in the crevice or U-bend tests with Type 316L stainless steel. Likewise, tensile failure times in the hydroxide solutions did not vary from the deionized water tests more than the equipment variability. Microscopic examination of the coupons indicated a ductile fracture. Some Type 304L coupons showed possible negative effects. Slight crevice attack was seen; one coupon had a corrosion rate above 2 mpy (0.51 μm/y) which was likely a measurement error. In addition, one tensile specimen had a failure time outside the equipment variability. Discoloration was seen to some degree on all coupons, but was very thin (on the order of less than nm). Unwelded coupons were tested under a few selected conditions, but no differences were found in the degradation as compared to welded samples. The effects of halides (iodine and bromine) were also assessed in the crevice and U-bend tests. Though they may provide some measure of inhibition to attack, the effect is not significant.

Neither the proposed SNF standardized canister constructed of Type 316L stainless steel nor the MCO made with Type 304L stainless steel is expected to be susceptible to the presence of cesium and rubidium from the fuel. Liquid Metal Embrittlement (LME) will not be a concern because austenitic stainless steels are not susceptible to those two elements. The metals should also be immune to SCC from cesium/rubidium hydroxide. However, Type 304L appears to be slightly more susceptible to localized corrosion in this environment than Type 316L stainless steel.

ESTIMATE OF PROBABILITY OF FABRICATION FLAW AND DROP FOR A STANDARDIZED DOE SPENT NUCLEAR FUEL CANISTER

INTRODUCTION

The spent nuclear fuel (SNF) owned by the U.S. Department of Energy (DOE) will be eventually transported to the designated national repository for disposal. The standardized DOE SNF canister is designed to accommodate this task.

The combined likelihood of dropping a standardized DOE SNF canister and the conditional probability that it breaches if dropped is required to be less than 10^{-4} (DOE 2002b). The likelihood of dropping a standardized DOE SNF canister will be addressed by the Yucca Mountain Site Characterization Office (YMSCO) (DOE 2002b) in its surface facility design analyses. The following analysis provides the conditional probability of a breach given a postulated drop. The Multi-Canister Overpacks (MCOs) used for the N-Reactor SNF are not addressed by this analysis. A separate analysis will be necessary to determine the conditional probability for the MCOs.

POTENTIAL DROPS

The Monitored Geologic Repository Canister Transfer System is being designed to minimize the likelihood that a standardized DOE SNF canister will be dropped, and if dropped, to ensure the drop height is as low as reasonably possible. The event trees developed by the YMSCO have identified several credible drops as follows:

1. For the brief period when the canister is removed from the transportation cask or loaded into the waste package, it is possible for the canister to drop back into the cask or waste package. This maximum vertical drop is bounded by the 23-ft (6.9 m) drop analyzed. A slap-down is not possible in this configuration, because the canister will be lifted straight up from a cask whose upper surface will be at floor level.
2. It is possible for a drop to occur when the canister is handled in the facility (when not over a transportation cask or a waste package). For scenarios other than the two-block, the drop height will be about 6 in (0.15 m). The maximum drop height due to a two-block scenario is 2 ft (0.6 m). While this scenario is also likely to be a vertical drop, a slap-down may be possible. It is difficult to conceive of a mechanism that could result in a slap-down given the vertical handling with a low-lift height, the low center of gravity, and the long aspect ratio of the canisters. Nonetheless, a slap-down is conservatively assumed.
3. Toppling of the canister is also possible. Analyses have shown that the stresses from a toppling event are less deleterious from a fracture mechanics viewpoint than the stresses resulting from either a 2-ft (0.6 m) slap-down or a 23-ft (6.9 m) drop.

For this analysis (DOE 2000b), a 23-ft (6.9 m) vertical drop and a 2-ft (0.6-m) slap-down are considered. These two scenarios bound all other drops, and most drops would be substantially less impacting. The 23-ft (6.9 m) vertical drop results in slightly higher peak stresses than the slap-down.

AS-DESIGNED SURVIVAL

A number of measures have been taken by the National Spent Nuclear Fuel Program (NSNFP) to ensure that the standardized canister design will provide containment during potential drop scenarios. The following list identifies tests and controls implemented by the NSNFP:

1. The standardized canister design includes an energy absorbing skirt.
2. The standardized canisters will be designed, fabricated, and examined per the requirements of the ASME (2001) B&PV Code, Section III, Division 3, Subsections WA and WB for the loads and environments identified in the design specification (DOE 1999).
3. Morton et al. (2000) performed dynamic analyses to determine the ability of the standardized canister to survive 30-ft (9-m) drops at various impact orientations. These analyses determined that the Standardized Canister is capable of surviving these drops without breach.
4. A series of drop-tests were performed at the Idaho National Engineering and Environmental Laboratory (INEEL) to assist in verification of the dynamic analysis.
5. A series of nine drop-tests from 30 ft (9-m) were performed by Morton et al. (2000) at Sandia National Laboratories under quality assurance controls for various drop orientations. These drop-tests were used to validate the dynamic analyses and to provide objective evidence that the standardized canisters will withstand such drops without breaching. Canisters successfully passed pressure tests following all nine of the drop tests including a canister with detected but uncorrected anomalies in the longitudinal seam weld. In addition, four of the most heavily-damaged canisters were helium leak-tested with leak rates less than $10^{-7} cm^3/s$.

The above activities support the conclusion that the standardized canisters, as designed, are capable of fulfilling their design requirements and surviving potential drops of up to 30 ft (9-m) in various orientations.

AGED SURVIVAL

Before the canister arrives at the repository surface facility, it will see product storage; loading operations; drying; interim storage with fuel; cask loading; and transport to the repository. Weld aging; corrosive action; internal pressure; and canister handling could reduce the canister's ability to withstand a drop event. DOE (2000b) addresses the as-fabricated canister condition. Studies of the potential severity and net effects of aging and corrosive processes are in progress to help quantify the extent of these issues and to enable effective prevention. The issues include the following:

1. The combined effects of welding heat; drying heat; fuel decay heat; and residual stress may affect the corrosion resistance and ductility of canister welds as a function of time.
2. Residual water will be present after drying of the SNF. Although the water may be physically or chemically bound to the fuel, the action of radiolysis may release the water within the canister over time resulting in corrosion and the potential buildup of hydrogen pressure. The hydrogen will also tend to collect at weld-flaws to accelerate crack propagation.
3. Some fission products are reactive liquid metals at SNF storage temperatures. If they are released from the SNF, they may promote stress corrosion cracking (SCC) of the canister welds.
4. External storage environments in some locations may be high in halide salts or acid forming microbes. Measures will be needed to ensure that the canisters are protected from these environmental corrosion sources during interim storage.

CRITICAL FLAW SIZE

The first step is determining the critical flaw size that has the potential to result in a breach if the canister is dropped. The American Petroleum Institute API-579 procedure was used to estimate the critical flaw size for the 8-in (203.1 mm) x 15-ft (4.5 m) standardized canister. The resulting flaw size was compared and found to be conservative relative to that calculated by extending the ASME (2001) Section IWB-3600 approach as well as to existing test data.

The peak stress for the 23-ft (6.9 m) vertical drop occurs in the base metal and exceeds the peak stresses for the 2-ft (0.6-m) slap-down. For the 23-ft (6.9 m) vertical drop, the critical flaw size for the base metal was estimated to be a 0.23-in (0.58-cm) deep flaw that penetrates the peak-stress surface and is approximately 2.3 in (5.8 cm) long. The peak-stress surface for the vertical drops is the outer surface.

Consistent with the reasoning given by Macheret (2000), the critical flaw was assumed to be an outer surface flaw with a 10:1 aspect ratio. For the purposes of this analysis, the critical flaw size will be conservatively rounded to an outer surface flaw that is 0.20 in (0.51 cm) deep and 2.0 in (5.1 cm) long for both the base metal and the weld metal. It should be noted that this critical flaw is greater than half of the wall thickness.

PROBABILITY OF FLAW

The following assumptions specifically apply to the standardized canister for this analysis:

1. The critical flaw is an outer surface penetrating flaw 0.20 in (0.51 cm) deep and 2.0 in (5.1 cm) long.
2. Weld procedures employed for the standardized canister will provide equivalent reliability and inspectability to the Tungsten Inert Gas (TIG) welding technology used as the basis for the probabilistic assessments developed by Macheret (2000). The weld technology has some affect on the weld-flaw density, but this assumption is a reasonable basis.
3. All welds except the final closure weld are performed in the shop and subjected to radiographic (RT) and dye-penetrant examinations by the fabricator.
4. The top lid closure weld will be performed in a hot cell and subjected to a multi-angle ultrasonic (UT) examination.

Macheret (2000) considered both base metal and weld metal flaws, and determined that flaw densities in the base metal region are about an order of magnitude lower than flaw densities in the weld region. The waste package analysis concluded that base-metal flaws are a secondary concern relative to weld-metal flaws and do not warrant further consideration. On this same basis, base-metal flaws for the standardized canister are considered enveloped by weld-metal flaws.

Macheret (2000) performed an analysis of the weld-flaw probability for the Monitored Geologic Repository (MGR) disposal waste package. There are many similarities between the waste package and the standardized canister including the shapes; material of construction; and likely fabrication and examination methods for the shell and closure welds. Because of the similarities, the waste package analysis was used as the model for this standardized canister analysis. Macheret (2000) provides a detailed discussion of the methodology used.

The spreadsheet used by Macheret (2000) was also used for the standardized canisters after modifying some inputs to reflect the differences in the designs of the waste packages and the standardized canister:

1. Dimensions
2. Flaw density relative to 1-in (2.5 cm) thick welds to reflect the thinner wall thickness
3. Median flaw size and shape parameter were adjusted to reflect the thinner wall thickness

Several steps were taken to confirm that the revised spreadsheet was correctly modeling the standardized canister. This confirmation included:

1. Performing manual calculations for the weld lengths; flaw density; median flaw size; and shape parameter to confirm the spreadsheet was calculating them correctly. The manual calculations were compared to the spreadsheet values and found to be in agreement.
2. The cell references in the spreadsheet were checked to confirm they were referring to the correct cells for the case being analyzed.

The spreadsheet calculates the probabilities of a weld flaw greater than a specified size for the full range of potential outer surface weld flaw depths. The spreadsheet is based on a flaw length with a fixed 10:1 ratio to the flaw depth. Typically, flaws were found to be uniformly distributed between ratios of 2:1 to 10:1. This 10:1 ratio results in the flaw lengths that are at the upper extreme of the distribution. The conditional probability of undetected critical flaws was calculated separately for the shell (the lower head and seam weld of the canister barrel) and the top lid closure welds. Figure 106 provides the conditional probability of undetected critical flaws in the shell, and Fig. 107 provides the results for the top lid.

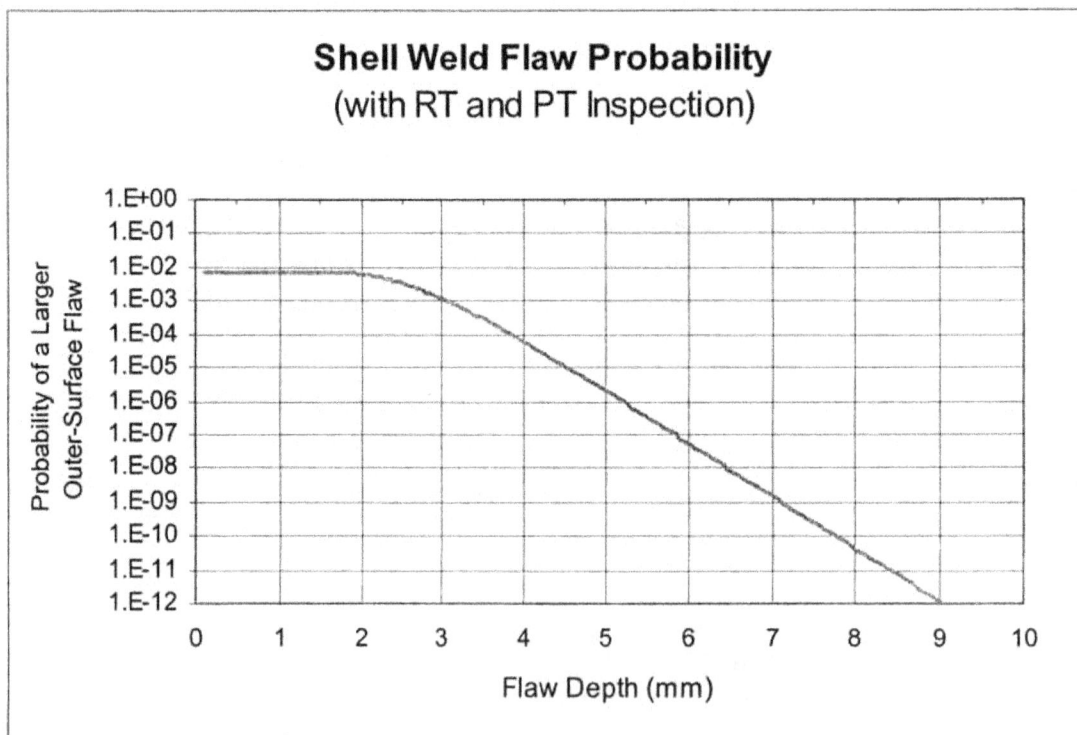

Fig. 106. Standardized canister weld flaw probability in the shell (DOE 2000b).

190

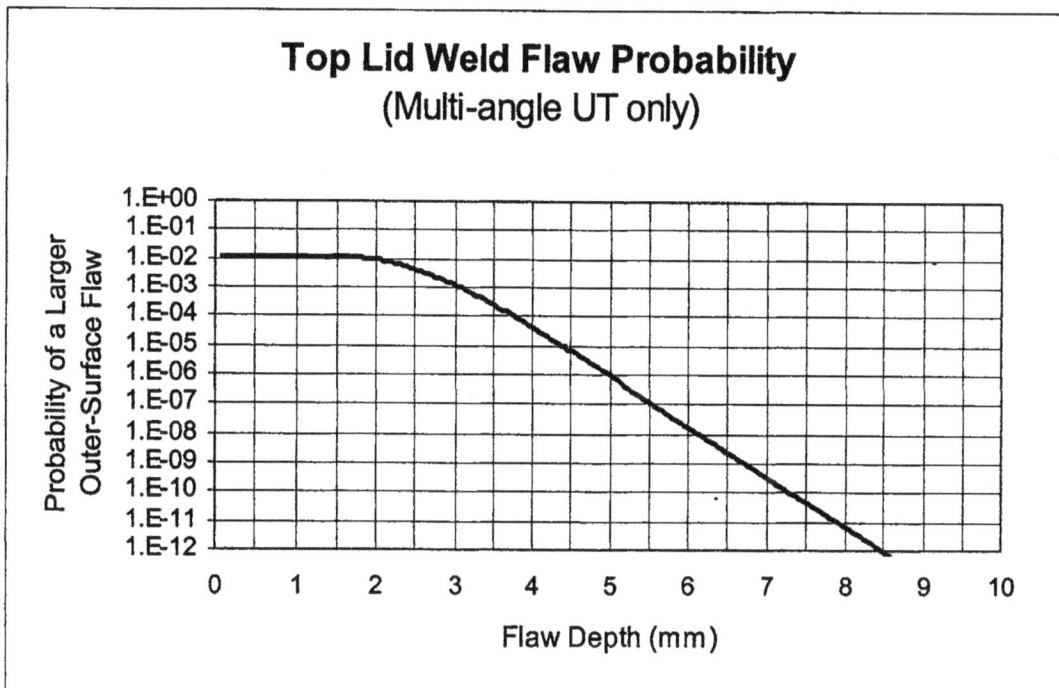

Fig. 107. Standardized canister weld flaw probability in the upper lid (DOE 2000b).

Using the spreadsheet results, the conditional probability for an outer surface weld flaw deeper than 0.20 in (5.1 mm) is 2×10^{-6} for the shell and 1×10^{-6} for the top lid closure weld. Therefore, the conditional probability of a weld flaw greater than 0.20 in (0.51 cm) deep anywhere in the standardized canister is 3×10^{-6}. This probability can be used as an estimate of the conditional probability that a standardized canister will breach if dropped.

CONSERVATISM

Sources of uncertainty in the methodology include:

1. Determination of peak stresses by using actual material properties from the test canisters in the structural analyses models rather than the minimums allowed by the design
2. A lack of test data specific to the weldments
3. The actual size, aspect ratio, and location of potential flaws

The methodology includes conservatism in several areas that are expected to be sufficient to offset potential uncertainties. Because of this conservatism, additional testing will not be pursued at this time.

This conditional probability calculation is based on a drop that will result in a through-wall crack, but the actual criterion is a drop that would result in a radiological release sufficient to exceed dose requirements. Consequently, the critical flaw size would actually be larger, and the associated conditional probability would be smaller.

191

No attempt was made to distinguish between the various drop heights and orientations, including 6-in (152 mm) vertical drop; 2-ft (0.6 m) slap-down; 23-ft (6.9 m) vertical drop; etc. The worst case was used for all drops. The peak stresses for other drops will be significantly lower than those of the worst-case drop.

In order for a flaw greater than the critical size to result in a breach, the flaw must be located at a point of high stress/strain in the standardized canister. No attempt was made to account for this orientation requirement because:

1. For a vertical drop onto the bottom skirt (the most likely orientation), flaws in the top lid closure weld and the seam weld might not result in a breach. The total weld length (seam and top circumferential) is about 5 times longer than the circumferential weld length of the bottom lid alone, so the probability of breach could be over-predicted by a factor of 5.
2. For a slap-down drop, flaws in the seam weld might only be at risk if the seam is properly aligned with the point of impact. If only a 30-degree rather than a 360-degree arc is at risk, then the probability could be over-predicted by an order of magnitude.

This analysis is based on a seam-welded pipe for the barrel of the canister. If a seamless pipe is used, then the weld length would be reduced by about 60%.

If the 18-in (457 mm) x 15-ft (4.5 m) standardized canister results are applied to the 18-in (457 mm) x 10-ft (3 m) standardized canister, this presents a conservatism. There is 34% more seam-weld length in the 15-ft (4.5 m) standardized canister than in the 10-ft (3 m) standardized canister. This would roughly result in a 20% over-prediction for the 10-ft (3 m) standardized canister.

The canister was assumed to contain the maximum allowable weight. Most canisters will be significantly lighter and have correspondingly-lower peak stresses.

The worst-case stresses in the base metal were also assumed to occur in the welds. Stresses are based on a drop onto an essentially unyielding surface. In reality, the facility floor will have some yield and reduce the peak stresses.

CONCLUSIONS

The conservative value of 3×10^{-6} should be used as the conditional probability of breach given the drop of a standardized canister for the preclosure event trees. This value is expected to substantially over-predict the likelihood of a canister breach given a drop.

Legal and Related Requirements

The development of standardized spent nuclear fuel canisters for the U.S. Department of Energy requires compliance with a number of laws, regulations and standards. This chapter contains information on these requirements.

THE NUCLEAR WASTE POLICY ACT AS AMENDED

This section contains excerpts from the Nuclear Waste Policy act (Public Law 97-425; 96 Stat. 2201), as amended by P.L. 100-203, Title V, Subtitle A (December 22, 1987), P.L. 100-507 (October 18, 1988), and P.L. 102-486 (The Energy Policy Act of 1992, October 24, 1992). The Act is codified at 42 U.S.C. 10101 et seq.

THE NUCLEAR WASTE POLICY ACT OF 1982

An Act to provide for the development of repositories for the disposal of high-level radioactive waste and spent nuclear fuel, to establish a program of research, development, and demonstration regarding the disposal of high-level radioactive waste and spent nuclear fuel, and for other purposes. [...]

DEFINITIONS

Sec. 2. For purposes of this Act

(1) The term "Administrator" means the Administrator of the Environmental Protection Agency. [...]
(3) The term "atomic energy defense activity" means any activity of the Secretary [of Department of Energy] performed in whole or in part in carrying out any of the following functions:
    (A) naval reactors development;
    (B) weapons activities including defense inertial confinement fusion;
    (C) verification and control technology;
    (D) defense nuclear materials production;
    (E) defense nuclear waste and materials by-products management;
    (F) defense nuclear materials security and safeguards and security investigations; and
    (G) defense research and development. [...]
(5) The term "civilian nuclear activity" means any atomic energy activity other than an atomic energy defense activity.
(6) The term "civilian nuclear power reactor" means a civilian nuclear powerplant required to be licensed under section 103 or 104 b. of the Atomic Energy Act of 1954 [42 U.S.C. 2133, 2134(b)].
(7) The term "Commission" means the Nuclear Regulatory Commission.
(8) The term "Department" means the Department of Energy.
(9) The term "disposal" means the emplacement in a repository of high-level radioactive waste, spent nuclear fuel, or other highly radioactive material with no foreseeable intent of recovery, whether or not such emplacement permits the recovery of such waste.
(10) The terms "disposal package" and "package" mean the primary container that holds, and is in contact with, solidified high-level radioactive waste, spent nuclear fuel, or other radioactive materials, and any overpacks that are emplaced at a repository.
(11) The term "engineered barriers" means manmade components of a disposal system designed to prevent the release of radionuclides into the geologic medium involved. Such term includes the high-level radioactive waste form, high-level radioactive waste canisters, and other materials placed over and around such canisters.

(12) The term "high-level radioactive waste" means—

(A) the highly radioactive material resulting from the reprocessing of spent nuclear fuel, including liquid waste produced directly in reprocessing and any solid material derived from such liquid waste that contains fission products in sufficient concentrations; and

(B) other highly radioactive material that the Commission, consistent with existing law, determines by rule requires permanent isolation. [...]

(23) The term "spent nuclear fuel" means fuel that has been withdrawn from a nuclear reactor following irradiation, the constituent elements of which have not been separated by reprocessing. [...]

(25) The term "storage" means retention of high-level radioactive waste, spent nuclear fuel, or transuranic waste with the intent to recover such waste or fuel for subsequent use, processing, or disposal. [...]

TRANSPORTATION

Sec. 137. (a)

(1) Transportation of spent nuclear fuel under section 136(a) [42 U.S.C. 10136(a)] shall be subject to licensing and regulation by the Commission and by the Secretary of Transportation as provided for transportation of commercial spent nuclear fuel under existing law.

(2) The Secretary, in providing for the transportation of spent nuclear fuel under this Act [42 U.S.C. 10101 et seq.], shall utilize by contract private industry to the fullest extent possible in each aspect of such transportation. The Secretary shall use direct Federal services for such transportation only upon a determination of the Secretary of Transportation, in consultation with the Secretary, that private industry is unable or unwilling to provide such transportation services at reasonable cost. [42 U.S.C. 10157]

SUBTITLE C—MONITORED RETRIEVABLE STORAGE MONITORED RETRIEVABLE STORAGE

Sec. 141. (a) Findings. The Congress finds that—

(1) long-term storage of high-level radioactive waste or spent nuclear fuel in monitored retrievable storage facilities is an option for providing safe and reliable management of such waste or spent fuel;

(2) the executive branch and the Congress should proceed as expeditiously as possible to consider fully a proposal for construction of one or more monitored retrievable storage facilities to provide such long-term storage;

(3) the Federal Government has the responsibility to ensure that site-specific designs for such facilities are available as provided in this section;

(4) the generators and owners of the high-level radioactive waste and spent nuclear fuel to be stored in such facilities have the responsibility to pay the costs of the long-term storage of such waste and spent fuel; and

(5) disposal of high-level radioactive waste and spent nuclear fuel in a repository developed under this Act [42 U.S.C. 10101 et seq.] should proceed regardless of any construction of a monitored retrievable storage facility pursuant to this section. [...]

REGULATION OF U.S. NUCLEAR REGULATORY COMMISSION ON DISPOSAL OF HIGH-LEVEL RADIOACTIVE WASTES IN GEOLOGIC REPOSITORIES

This section contains excerpts from regulations dealing with disposal of high-level radioactive waste (HLW) promulgated by the U.S. Nuclear Regulatory Commission and codified as 10 CFR PART 60.

§60.43 License specification.

(a) A license issued under this part shall include license conditions derived from the analyses and evaluations included in the application, including amendments made before a license is issued, together with such additional conditions as the Commission finds appropriate.

(b) License conditions shall include items in the following categories:

(1) Restrictions as to the physical and chemical form and radioisotopic content of radioactive waste.

(2) Restrictions as to size, shape, and materials and methods of construction of radioactive waste packaging.

(3) Restrictions as to the amount of waste permitted per unit volume of storage space considering the physical characteristics of both the waste and the host rock.

(4) Requirements relating to test, calibration, or inspection to assure that the foregoing restrictions are observed.

(5) Controls to be applied to restricted access and to avoid disturbance to the postclosure controlled area and to areas outside the controlled area where conditions may affect isolation within the controlled area.

(6) Administrative controls, which are the provisions relating to organization and management, procedures, recordkeeping, review and audit, and reporting necessary to assure that activities at the facility are conducted in a safe manner and in conformity with the other license specifications. [...]

§60.113 Performance of particular barriers after permanent closure.

(a) General provisions—(1) Engineered barrier system.

(i) The engineered barrier system shall be designed so that assuming anticipated processes and events:

(A) Containment of HLW will be substantially complete during the period when radiation and thermal conditions in the engineered barrier system are dominated by fission product decay; and

(B) any release of radionuclides from the engineered barrier system shall be a gradual process which results in small fractional releases to the geologic setting over long times. For disposal in the saturated zone, both the partial and complete filling with groundwater of available void spaces in the underground facility shall be appropriately considered and analysed among the anticipated processes and events in designing the engineered barrier system.

(ii) In satisfying the preceding requirement, the engineered barrier system shall be designed, assuming anticipated processes and events, so that:

(A) Containment of HLW within the waste packages will be substantially complete for a period to be determined by the Commission taking into account the factors specified in §60.113(b) provided, that such period shall be not less than 300 years nor more than 1,000 years after permanent closure of the geologic repository; and

(B) The release rate of any radionuclide from the engineered barrier system following the containment period shall not exceed one part in 100,000 per year of the inventory of that radionuclide calculated to be present at 1,000 years following permanent closure, or such other fraction of the inventory as may be approved or specified by the Commission; provided, that this requirement does not apply to any radionuclide which is released at a rate less than 0.1% of the calculated total release rate limit. The calculated total release rate limit shall be taken to be one part in 100,000 per year of the inventory of radioactive waste, originally emplaced in the underground facility, that remains after 1,000 years of radioactive decay.

§60.131 General design criteria for the geologic repository operations area.

(a) Radiological protection. The geologic repository operations area shall be designed to maintain radiation doses, levels, and concentrations of radioactive material in air in restricted areas within the limits specified in part 20 of this chapter. Design shall include:

(1) Means to limit concentrations of radioactive material in air;

(2) Means to limit the time required to perform work in the vicinity of radioactive materials, including, as appropriate, designing equipment for ease of repair and replacement and providing adequate space for ease of operation;

(3) Suitable shielding;

(4) Means to monitor and control the dispersal of radioactive contamination;

(5) Means to control access to high radiation areas or airborne radioactivity areas; and

(6) A radiation alarm system to warn of significant increases in radiation levels, concentrations of radioactive material in air, and of increased radioactivity released in effluents. The alarm system shall be designed with provisions for calibration and for testing its operability.

(b) Protection against design basis events. The structures, systems, and components important to safety shall be designed so that they will perform their necessary safety functions, assuming occurrence of design basis events.

(c) Protection against dynamic effects of equipment failure and similar events. The structures, systems, and components important to safety shall be designed to withstand dynamic effects such as missile impacts, that could result from equipment failure, and similar events and conditions that could lead to loss of their safety functions.

(d) Protection against fires and explosions.

(1) The structures, systems, and components important to safety shall be designed to perform their safety functions during and after credible fires or explosions in the geologic repository operations area.

(2) To the extent practicable, the geologic repository operations area shall be designed to incorporate the use of noncombustible and heat resistant materials.

(3) The geologic repository operations area shall be designed to include explosion and fire detection alarm systems and appropriate suppression systems with sufficient capacity and capability to reduce the adverse effects of fires and explosions on structures, systems, and components important to safety.

(4) The geologic repository operations area shall be designed to include means to protect systems, structures, and components important to safety against the adverse effects of either the operation or failure of the fire suppression systems.

(e) Emergency capability.

(1) The structures, systems, and components important to safety shall be designed to maintain control of radioactive waste and radioactive effluents, and permit prompt termination of operations and evacuation of personnel during an emergency.

(2) The geologic repository operations area shall be designed to include onsite facilities and services that ensure a safe and timely response to emergency conditions and that facilitate the use of available offsite services (such as fire, police, medical, and ambulance service) that may aid in recovery from emergencies.

(f) Utility services.

(1) Each utility service system that is important to safety shall be designed so that essential safety functions can be performed, assuming occurrence of the design basis events.

(2) The utility services important to safety shall include redundant systems to the extent necessary to maintain, with adequate capacity, the ability to perform their safety functions.

(3) Provisions shall be made so that, if there is a loss of the primary electric power source or circuit, reliable and timely emergency power can be provided to instruments, utility service systems, and operating systems, including alarm systems, important to safety.

(g) Inspection, testing, and maintenance. The structures, systems, and components important to safety shall be designed to permit periodic inspection, testing, and maintenance, as necessary, to ensure their continued functioning and readiness.

(h) Criticality control. All systems for processing, transporting, handling, storage, retrieval, emplacement, and isolation of radioactive waste shall be designed to ensure that nuclear criticality is not possible unless at least two unlikely, independent, and concurrent or sequential changes have occurred in the conditions essential to nuclear criticality safety. Each system must be designed for criticality safety assuming occurrence of design basis events. The calculated effective multiplication factor (keff) must be sufficiently below unity to show at least a 5 percent margin, after allowance for the bias in the method of calculation and the uncertainty in the experiments used to validate the method of calculation.

(i) Instrumentation and control systems. The design shall include provisions for instrumentation and control systems to monitor and control the behavior of systems important to safety, assuming occurrence of design basis events.

(j) Compliance with mining regulations. To the extent that DOE is not subject to the Federal Mine Safety and Health Act of 1977, as to the construction and operation of the geologic repository operations area, the design of the geologic repository operations area shall nevertheless include provisions for worker protection necessary to provide reasonable assurance that all structures, systems, and components important to safety can perform their intended functions. Any deviation from relevant design requirements in 30 CFR, chapter I, subchapters D, E, and N will give rise to a rebuttable presumption that this requirement has not been met.

(k) Shaft conveyances used in radioactive waste handling.

(1) Hoists important to safety shall be designed to preclude cage free fall.

(2) Hoists important to safety shall be designed with a reliable cage location system.

(3) Loading and unloading systems for hoists important to safety shall be designed with a reliable system of interlocks that will fail safely upon malfunction.

(4) Hoists important to safety shall be designed to include two independent indicators to indicate when waste packages are in place and ready for transfer. [...]

§60.130 General considerations.

Pursuant to the provisions of §60.21(c)(2)(i), an application to receive, possess, store, and dispose of high-level radioactive waste in the geologic repository operations area must include the principal design criteria for a proposed facility. The principal design criteria establish the necessary design, fabrication, construction, testing, maintenance, and performance requirements for structures, systems, and components important to safety and/or important to waste isolation. Sections 60.131 through 60.134 specify minimum requirements for the principal design criteria for the geologic repository operations area.

These design criteria are not intended to be exhaustive. However, omissions in §§60.131 through 60.134 do not relieve DOE from any obligation to provide such features in a specific facility needed to achieve the performance objectives. [...]

Design Criteria for the Geological Repository Operations Area

§60.133 Additional design criteria for the underground facility.

(a) General criteria for the underground facility.

(1) The orientation, geometry, layout, and depth of the underground facility, and the design of any engineered barriers that are part of the underground facility shall contribute to the containment and isolation of radionuclides.

(2) The underground facility shall be designed so that the effects of credible disruptive events during the period of operations, such as flooding, fires and explosions, will not spread through the facility.

(b) Flexibility of design. The underground facility shall be designed with sufficient flexibility to allow adjustments where necessary to accommodate specific site conditions identified through in situ monitoring, testing, or excavation.

(c) Retrieval of waste. The underground facility shall be designed to permit retrieval of waste in accordance with the performance objectives of §§60.111.

(d) Control of water and gas. The design of the underground facility shall provide for control of water or gas intrusion.

(e) Underground openings.

(1) Openings in the underground facility shall be designed so that operations can be carried out safely and the retrievability option maintained.

(2) Openings in the underground facility shall be designed to reduce the potential for deleterious rock movement or fracturing of overlying or surrounding rock.

(f) Rock excavation. The design of the underground facility shall incorporate excavation methods that will limit the potential for creating a preferential pathway for groundwater to contact the waste packages or radionuclide migration to the accessible environment.

(g) Underground facility ventilation. The ventilation system shall be designed to:

(1) Control the transport of radioactive particulates and gases within and releases from the underground facility in accordance with the performance objectives of §§60.111(a),

(2) Assure the ability to perform essential safety functions assuming occurrence of design basis events.

(3) Separate the ventilation of excavation and waste emplacement areas.

(h) Engineered barriers. Engineered barriers shall be designed to assist the geologic setting in meeting the performance objectives for the period following permanent closure.

(i) Thermal loads. The underground facility shall be designed so that the performance objectives will be met taking into account the predicted thermal and thermomechanical response of the host rock, and surrounding strata, groundwater system. [...]

§60.135 Criteria for the waste package and its components.

(a) High-level-waste package design in general.

(1) Packages for HLW shall be designed so that the in situ chemical, physical, and nuclear properties of the waste package and its interactions with the emplacement environment do not compromise the function of the waste packages or the performance of the underground facility or the geologic setting.

(2) The design shall include but not be limited to consideration of the following factors: solubility, oxidation/reduction reactions, corrosion, hydriding, gas generation, thermal effects, mechanical strength, mechanical stress, radiolysis, radiation damage, radionuclide retardation, leaching, fire and explosion hazards, thermal loads, and synergistic interactions.

(b) Specific criteria for HLW package design—

(1) Explosive, pyrophoric, and chemically reactive materials. The waste package shall not contain explosive or pyrophoric materials or chemically reactive materials in an amount that could compromise the ability of the underground facility to contribute to waste isolation or the ability of the geologic repository to satisfy the performance objectives.

(2) Free liquids. The waste package shall not contain free liquids in an amount that could compromise the ability of the waste packages to achieve the performance objectives relating to containment of HLW (because of chemical interactions or formation of pressurized vapor) or result in spillage and spread of contamination in the event of waste package perforation during the period through permanent closure.

(3) Handling. Waste packages shall be designed to maintain waste containment during transportation, emplacement, and retrieval.

(4) Unique identification. A label or other means of identification shall be provided for each waste package. The identification shall not impair the integrity of the waste package and shall be applied in such a way that the information shall be legible at least to the end of the period of retrievability. Each waste package identification shall be consistent with the waste package's permanent written records.

(c) Waste form criteria for HLW. High-level radioactive waste that is emplaced in the underground facility shall be designed to meet the following criteria:

(1) Solidification. All such radioactive wastes shall be in solid form and placed in sealed containers.

(2) Consolidation. Particulate waste forms shall be consolidated (for example, by incorporation into an encapsulating matrix) to limit the availability and generation of particulates.

(3) Combustibles. All combustible radioactive wastes shall be reduced to a noncombustible form unless it can be demonstrated that a fire involving the waste packages containing combustibles will not compromise the integrity of other waste packages, adversely affect any structures, systems, or components important to safety, or compromise the ability of the underground facility to contribute to waste isolation.

(d) Design criteria for other radioactive wastes. Design criteria for waste types other than HLW will be addressed on an individual basis if and when they are proposed for disposal in a geologic repository.

REGULATION OF U.S. NUCLEAR REGULATORY COMMISSION ON PACKAGING AND TRANSPORTATION OF RADIOACTIVE MATERIAL

This section contains excerpts from regulations on packaging and transportation of radioactive materials promulgated by the U.S. Nuclear Regulatory Commission and codified as 10 CFR PART 71.

§71.4 Definitions.

Package means the packaging together with its radioactive contents as presented for transport.

Subpart D—Application for Package Approval

§71.35 Package evaluation.

The application must include the following:

(a) A demonstration that the package satisfies the standards specified in subparts E and F of this part;
(b) For a fissile material package, the allowable number of packages that may be transported in the same vehicle in accordance with §71.59; and
(c) For a fissile material shipment, any proposed special controls and precautions for transport, loading, unloading, and handling and any proposed special controls in case of an accident or delay. [...]

§71.43 General standards for all packages.

(a) The smallest overall dimension of a package may not be less than 10 cm (4 in).
(b) The outside of a package must incorporate a feature, such as a seal, that is not readily breakable and that, while intact, would be evidence that the package has not been opened by unauthorized persons.
(c) Each package must include a containment system securely closed by a positive fastening device that cannot be opened unintentionally or by a pressure that may arise within the package.
(d) A package must be made of materials and construction that assure that there will be no significant chemical, galvanic, or other reaction among the packaging components, among package contents, or between the packaging components and the package contents, including possible reaction resulting from inleakage of water, to the maximum credible extent. Account must be taken of the behavior of materials under irradiation.
(e) A package valve or other device, the failure of which would allow radioactive contents to escape, must be protected against unauthorized operation and, except for a pressure relief device, must be provided with an enclosure to retain any leakage.
(f) A package must be designed, constructed, and prepared for shipment so that under the tests specified in §71.71 ("Normal conditions of transport") there would be no loss or dispersal of radioactive contents, no significant increase in external surface radiation levels, and no substantial reduction in the effectiveness of the packaging.
(g) A package must be designed, constructed, and prepared for transport so that in still air at 38°C (100°F) and in the shade, no accessible surface of a package would have a temperature exceeding 50°C (122°F) in a nonexclusive use shipment, or 85°C (185°F) in an exclusive use shipment.
(h) A package may not incorporate a feature intended to allow continuous venting during transport. [...]

§71.55 General requirements for fissile material packages.

(a) A package used for the shipment of fissile material must be designed and constructed in accordance with §§71.41 through 71.47. When required by the total amount of radioactive material, a package used for the shipment of fissile material must also be designed and constructed in accordance with §71.51.

(b) Except as provided in paragraph (c) of this section, a package used for the shipment of fissile material must be so designed and constructed and its contents so limited that it would be subcritical if water were to leak into the containment system, or liquid contents were to leak out of the containment system so that, under the following conditions, maximum reactivity of the fissile material would be attained:

(1) The most reactive credible configuration consistent with the chemical and physical form of the material;

(2) Moderation by water to the most reactive credible extent; and

(3) Close full reflection of the containment system by water on all sides, or such greater reflection of the containment system as may additionally be provided by the surrounding material of the packaging.

(c) The Commission may approve exceptions to the requirements of paragraph (b) of this section if the package incorporates special design features that ensure that no single packaging error would permit leakage, and if appropriate measures are taken before each shipment to ensure that the containment system does not leak.

(d) A package used for the shipment of fissile material must be so designed and constructed and its contents so limited that under the tests specified in §71.71 ("Normal conditions of transport")—

(1) The contents would be subcritical;

(2) The geometric form of the package contents would not be substantially altered;

(3) There would be no leakage of water into the containment system unless, in the evaluation of undamaged packages under §71.59(a)(1), it has been assumed that moderation is present to such an extent as to cause maximum reactivity consistent with the chemical and physical form of the material; and

(4) There will be no substantial reduction in the effectiveness of the packaging, including:

(i) No more than 5 percent reduction in the total effective volume of the packaging on which nuclear safety is assessed;

(ii) No more than 5 percent reduction in the effective spacing between the fissile contents and the outer surface of the packaging; and

(iii) No occurrence of an aperture in the outer surface of the packaging large enough to permit the entry of a 10 cm (4 in) cube.

(e) A package used for the shipment of fissile material must be so designed and constructed and its contents so limited that under the tests specified in §71.73 ("Hypothetical accident conditions"), the package would be subcritical. For this determination, it must be assumed that:

(1) The fissile material is in the most reactive credible configuration consistent with the damaged condition of the package and the chemical and physical form of the contents;

(2) Water moderation occurs to the most reactive credible extent consistent with the damaged condition of the package and the chemical and physical form of the contents; and

(3) There is full reflection by water on all sides, as close as is consistent with the damaged condition of the package.

REGULATION OF U.S. NUCLEAR REGULATORY COMMISSION ON LICENSING REQUIREMENTS FOR THE INDEPENDENT STORAGE OF CERTAIN WASTES

This section contains an excerpt from regulations on licensing requirements for the independent storage of spent nuclear fuel, high-level radioactive waste, and reactor-related radioactive class C waste on promulgated by the U.S. Nuclear Regulatory Commission and codified as 10 CFR PART 72.

72.3 Definitions.

High-level radioactive waste or HLW means (1) the highly radioactive material resulting from the reprocessing of spent nuclear fuel, including liquid waste produced directly in reprocessing and any solid material derived from such liquid waste that contains fission products in sufficient concentrations; and (2) other highly radioactive material that the Commission, consistent with existing law, determines by rule requires permanent isolation.

CODES AND STANDARDS OF THE
AMERICAN SOCIETY OF MECHANICAL ENGINEERS

The American Society of Mechanical Engineers (ASME), also known as ASME International is a major source for a number of industrial codes of standards. Numerous ASME coded and standards have been adopted by the U.S. government agencies by reference and are an integral part of enforceable regulations.

Codes and standards dealing with spent nuclear fuel canisters and related issues emerged from Pressure Vessel Codes, initially developed for fossil fuel power planets. Although many sections of ASME codes and standards are applicable to all relevant installation, nuclear-specific issues are included Section III of the Boiler and Pressure Vessel (B&PV) Code. In addition, the ASME Nuclear Accreditation Program assures strict conformance with the ASME Codes and Standards for manufacturing of a nuclear facility. It covers the quality assurance of construction materials, design, engineering, operation, inspection, and continuing maintenance of the site. The purpose of the Accreditation program is to ascertain that the applicant has, uses and abides by a quality assurance manual which is clear and understandable.

Nuclear Stamp

As in other areas of activities of the ASME Codes and Standards, Code Symbol Stamps are used to indicate that the stamped items conform to the ASME Code. The Stamp provides confidence that the stamped items conform to established safety standards.

An organization receives a Certificate of Accreditation when it is recognized that it has an acceptable quality assurance program. However, no stamp can be issued until ASME has conducted an implementation audit for conversion to a Certificate of Authorization which authorizes the use of one or more of the five Code Symbol Stamps.

N - Nuclear vessels, pumps, valves, piping systems, storage tanks, core support structures, concrete containments, and transport packaging.
NA - Field installation and shop assembly.
NPT - Fabrication, with or without design responsibility, for nuclear appurtenances and supports.
NV - Pressure relief valves.
N3 - Containment for spent fuel and high level radioactive waste

Content of Certain B&PV Code Sections

Section II. Materials

This Section consists of four parts. Each part is a service book to the other Code Sections.

Part A - provides material specifications for ferrous materials adequate for safety in the field of pressure equipment. These specifications contain requirements and mechanical properties, test specimens, and methods of testing. Covered are steel pipe; steel tubes: fittings, valves; steel plates; wrought iron as examples.

Part B - covers nonferrous materials adequate for safety in the field of pressure equipment. These specifications contain requirements for heat treatment, manufacture, chemical composition, heat and product analyses, test specimens and methods of testing. Covered as examples are; aluminum, copper, nickel, titanium, and zirconium, all in alloys and in bars plates, and rods.

Part C - covers specifications for welding rods, electrodes, and filler materials. Provided are specifications for manufacture, chemical composition, mechanical usability, testing requirements and procedures, and intended use. As examples the requirements are for steel and steel alloy rod and electrodes; aluminum and aluminum base alloy rods, copper, nickel titanium and zirconium and their alloys in rods and electrodes.

Part D - contains tables of design stress values, tensile and yield strength values, and tables and charts of material values. Part D facilitates ready identification of specific materials to specific Sections of the B&PV code.

Subsection SA-530 - Specifications for General Requirements for Specialized Carbon and Alloy Steel Pipe.

This specification covers a group of requirements which are mandatory requirements to the ASTM pipe product specifications. The specifications cover manufacturing process; chemical composition; mechanical requirements (such as method of mechanical testing); tensile requirements; permissible variations in weight, wall thickness, inside and outside diameters, length, and repairs by welding; test specimens, hydrostatic tests requirements; inspection, rejection and marking.

Section III. Rules for Construction of Nuclear Facility Components

The basic coverage of this Section is the providing of requirements for the materials, design, fabrication, testing, inspection, installation, certification, stamping, and overpressure protection of nuclear power plant components, piping supports, metal vessels and systems, pumps and valves. There are three divisions in Section III.

Division 1: This division has seven subsections. Each subsection contains the requirements for the material, design, fabrication, inspection, testing and overpressure protection of items in different categories.
Subsection NB - Class 1 components for Class 1 construction.
Subsection NC - Class 2 components for Class 2 construction
Subsection ND - Class 3 components for Class 3 construction
Subsection NE - Class MC components for metal containment vessels for Class MNC construction.
Subsection NF - Supports for Classes 1, 2, 3, and MC construction.
Subsection NG - Core Support structures are those structures which provide direct support or restraint of the core within the reactor pressure vessel.
Subsection NH - Class 1 components in elevated temperature service
Appendices - Appendices, both mandatory and nonmandatory for Section III, Division 1, including a listing of design and design analysis methods and information and data report forms.

Division 2: This division is the Code for concrete reactor vessels and containment.

Division 3: This division constitutes the requirements for the design and construction of the containment system of a nuclear spent fuel or high level radioactive waste transport packaging.

Section V. Nondestructive Examination

This section contains the requirements and methods for nondestructive examination which are referenced and required by other Code Sections. Included are manufacturers examination responsibilities, duties of authorized inspectors and requirements for qualification of personnel, inspection and examination. The requirements cover examination by radiographic, ultrasonic, liquid penetrant, and magnetic particle eddy current methods; visual examination; leak testing; and acoustic emission examination of fiber-reinforced plastic pressure vessels.

Peer Review Criteria, Findings, and Recommendations of the Review Panel

Criterion 1

Is the approach to the analysis and proof testing technically sound? In particular, are the underlying bases for the analysis and proof testing technically sound and defensible?

Findings of the RP

The approach to the analysis and proof testing by the Project Team (PT) is technically sound, and the underlying bases for the analysis and proof testing are technically sound and defensible.

The PT used a state-of-the-art finite element computer code—ABAQUS—to perform analyses of the canisters subjected to various postulated drop events. The analyses took into account elastic-plastic material behavior, using large deformation theory modeling. The testing was performed at Sandia National Laboratory—a nationally recognized facility for development and certification testing of radioactive material packaging. Such analysis and testing are the norm for qualification of transportation packaging for U.S. Nuclear Regulatory Commission (USNRC) certification, and for transportation within the U.S. Department of Energy (DOE) complex.

An objective of the test and analysis work was to demonstrate that the finite element techniques being used would generate results comparable to the test results. Similarity of the deformations was demonstrated; however, comparisons of plastic strains did not prove meaningful. The strain gages used were limited to less than 10% strain levels. The test program would have been more valuable if useful plastic strain measurements had been produced to fully validate the computational model.

The test canisters were loaded with continuous lengths of reinforcing bar (rebar) to simulate the weight of spent nuclear fuel (SNF). However, the rebar, which extended the full length of the canister, may have a stiffening effect which may not realistically represent the actual loading of the canister.

The conventional criterion used for certification of radioactive material containment systems is post test leak tightness (based on helium leak-testing performed according to ANSI N 14.5). Such a demonstration was provided for the four most deformed of the nine tested canisters. Results of the test program would have been more complete if all nine of the canisters were subjected to the leak-test in accordance with N 14.5. However, a pressure drop structural integrity test at 50 psi was performed for all nine canisters and the results were satisfactory.

Strain rate effects were introduced into the plasticity model by increasing stress-strain response by 20% in all finite elements. This simplification may not be appropriate for the canister finite element models.

Criterion 2

Are the underlying bases for the individual testing steps, as well as the integrated demonstration, technically sound and defensible?

Findings of the RP

The underlying bases for the individual testing steps, as well as the integrated demonstration, are technically sound and defensible. The basis for the computational modeling was the standard canister loaded with rebar simulating the maximum content weight. The integrated basis for canister analysis and testing is documented in the DOE SNF Canister Qualification Support Chart presented by the PT. This chart, which is updated as the program proceeds, identifies components of the qualification support program, including areas of technical concern regarding canister performance under postulated drop conditions.

The instrumentation procedures used during the drop tests did not permit measurements of large strains. The PT did not idenify a criterion that related breach of the cask to a maximal plastic strain. Also, the drop tests did not include simulated flaws in the cask structure or pressure boundary.

The drop tests have been conducted on new canisters, whereas the canisters in actual service will have aged to various degrees, some for many years. The aging effects could include hydrogen embrittlement; stress corrosion cracking; radiation degradation; helium deoxidation of metal surfaces; and other age-related phenomena, many of which are addressed in tasks yet to be completed in the DOE Spent Nuclear Fuel Canister Qualification program. The PT did not present sufficient data from a literature survey on the effects of aging on the materials of construction of the canisters.

The PT presented analysis results for various drop tests with a correlation of displacements to corresponding data from the test results. However, no conclusion could be reached about the adequacy of the component on the basis of the analysis alone, the stated reason being that the analytical results would not meet ASME (2001) Section III criteria. In ASME (2001), Section III, Appendix F (F-1340), limit on general primary membrane stress intensity for the 316L stainless steel tube material appears to be about 15%. The drop test analysis results satisfy this limit with a few isolated exceptions where the strains are about 20%. It should be possible to modify the design to achieve strains within the Appendix F limit.

Criterion 3

Is the execution of the program consistent with established scientific and engineering principles and standards?

Findings of the RP

The PT has, by and large, followed established scientific and engineering principles and standards. The PT has indeed thought in-depth about the steps required in the canister qualification support program. The overall plan for the canister qualification support includes four project tasks supporting the project goal, as follows:

1. Aging Studies
2. Accidental Drop Events
3. Flaw Evaluations and Testing
4. Material Considerations

Each of the four project tasks is divided into several sub-tasks. For example, under "Aging Studies" the following sub-tasks are listed:

1. Thermal Embrittlement
2. Radiation Degradation
3. SNF Interactions with DOE SNF Canister Material including liquid metal embrittlement; stress corrosion susceptibility to cesium/rubidium hydroxide; and hydrogen damage

Each of the other three project tasks are also divided into several sub-tasks. The status of the sub-tasks has been identified by color code. The PT did not have detailed plans for some of the incomplete sub-tasks. Consequently, the RP could not evaluate the degree to which success or failure of some of the incomplete sub-tasks would be achieved as a result of their effort.

The RP did find that some of the sub-tasks designated as "complete" lacked "completeness" in the sense that the experiments performed raised additional questions about the issues involved. For example: the friction factor dependence on the angle of drop test and the lack of strain measurement associated with the drop test.

Criterion 4

Has the Project Team collected sufficient data to meet the licensing requirements of the U.S. Nuclear Regulatory Commission and respond to stakeholder concerns?

Findings of the RP

The PT has developed a DOE Spent Nuclear Fuel Canister Qualification Support program to demonstrate that the canisters will meet the licensing requirements of the U.S. Nuclear Regulatory Commission (USNRC). This ongoing program has already demonstrated that new canisters will withstand accidental drops without rupture, and is being extended to demonstrate the effects of aging, flaws, and material considerations. When complete, the program will have collected sufficient data to meet the licensing requirements of the USNRC.

RECOMMENDATIONS

Based on an assessment of the information presented to the Review Panel (RP) and the Findings developed in response to the Review Criteria, the RP makes the following recommendations:

1. The Project Team (PT) should consider analyzing the response of a canister filled with rebar cut into segments subjected to an off-vertical drop test. This analysis would better simulate typical loadings of spent nuclear fuel (SNF) in the canister. The interaction of several rebar segments should be modeled using contact algorithm available in ABAQUS (1998). Very high-friction parameters should be considered due to possible interlocking of rebar segments. If the analyses show a potential problem, the PT should consider one or more off-vertical drop tests of canisters filled with segmented rebar. If successful, the analyses or tests with segmented rebar would add confidence that the stiffness of the rebar used in the tests conducted to-date, realistically simulated the actual loadings of the canisters.
2. The PT should determine whether a drop test directly onto the canister closure weld area is enveloped by previous testing and/or calculations. Otherwise, the PT should consider a drop test directly onto the

canister closure weld area or justify not testing this orientation. The test could be accomplished as either a tip-over with the top-end closure weld region impacting a rail-like target, or a side drop to a rail-like target from a height giving the same impact velocity as a tip-over drop.

3. Any future tests should use appropriate methods to measure large strains. For example, strain gauges could be placed both in the canister interior and over a larger region than the finite element size. The experimental data would have to be compared with strain averages over a similar region in the finite element model.

4. The PT should develop a criterion that will relate canister integrity to both the maximal membrane and the surface plastic strain. Either the full-scale canister tests or tests on smaller specimens should be considered.

5. Strain-rate effects should not be introduced into the plasticity model in an artificial manner such as correcting stress-strain response by 20% in all elements. Instead, the rate-dependent plasticity model should be used. This rate-dependent model is available in ABAQUS (1998).

6. The PT should attempt to qualify the canister's containment boundary to ASME (2001) Section III, Appendix F rules under drop loadings.

7. A literature survey on the effects of aging (such as thermal embrittlement, radiation degradation, and hydrogen damage) on the materials of construction of the casks should be undertaken. The PT should consider committing to additional experimentation on the effects of aging only if the results of such a literature survey are inadequate to support the design analysis of the canister.

8. Determine a location and size of a critical flaw from the distribution of plastic strains. Place the critical flaw in the finite element model of the full-scale canister and carry out finite element analysis of the drop test. Compute stress intensity factors and compare them to fracture toughness of steel. If the results indicate instability of the crack, then full-scale testing is recommended.

9. Implement a hydrogen embrittlement model into the ABAQUS (1998) UMAT subroutine. There are models that provide a two-way coupling between diffusion and deformation processes known as stress-assisted diffusion models. Once those models are implemented, the PT can estimate the amount of damage induced by hydrogen diffusion. This damage can then be introduced into a deformation process using a Gurson-type model, which is also available in ABAQUS (1998).

10. Determine the influence of hydrogen embrittlement on fracture toughness and account for it in calculation of crack stability. This may involve a literature search and/or finite element calculation of hydrogen diffusion models.

11. The PT should be commended on the plan to investigate potential synergistic effects of aging, drop dynamics, flaw, and material consideration as a final qualification test of the canister.

References

ABAQUS. ABAQUS/Explicit user's manual. Volumes I and II. Version 5.8. Pawtucket, RI: Hibbitt, Karlson, and Sorensen; 1998.

ABAQUS. ABAQUS/Explicit user's manual. Volumes I and II. Version 6.3-3. Pawtucket, RI: Hibbitt, Karlson, and Sorensen; 2002.

Aderl, R.A.; Nagata, P.K. Helium permeability through austenitic stainless steel, technical report EGG-MS-8714. (September) 1989.

Anderson, P.A. Interaction of DOE SNF and packaging materials, DOE/SNF/REP-020. Rev. 0. Idaho Falls, ID: Idaho National Engineering and Environmental Laboratory; (September) 1998.

ANSI (American National Standards Institute). American national standard for radioactive materials—leakage tests on packages for shipment, ANSI N 14.5. New York, NY: American National Standards Institute; 1987.

ANSI (American National Standards Institute). Design criteria for an independent spent fuel storage installation (dry type), ANSI/ANS 57.9. New York, NY: American National Standards Institute; 1992.

ASNT (American Society of Nondestructive Testing). Recommended practice no. SNT-TC-1A. Columbus, OH: American Society of Nondestructive Testing; 1992.

ASME (American Society of Mechanical Engineers). Quality assurance program requirements for nuclear facilities, ANSI/ASME NQA-1. New York, NY: American Society of Mechanical Engineers; 1986.

ASME (American Society of Mechanical Engineers). Boiler and pressure vessel code. New York, NY: American Society of Mechanical Engineers; 2001.

ASME (The American Society of Mechanical Engineers). Manual for peer review. Washington, DC: ASME; 2003.

ASME/RSI (The American Society of Mechanical Engineers/Institute for Regulatory Science). Assessment of technologies supported by the U.S. Department of Energy; results of the peer review for fiscal year 1997, CRTD Vol. 47. New York: ASME; 1997.

ASME/RSI (The American Society of Mechanical Engineers/Institute for Regulatory Science). Assessment of technologies supported by the U.S. Department of Energy; results of the peer review for fiscal year 1998, CRTD Vol. 50. New York: ASME; 1998.

ASME/RSI (The American Society of Mechanical Engineers/Institute for Regulatory Science). Assessment of technologies supported by the U.S. Department of Energy; results of the peer review for fiscal year 1999, CRTD Vol. 56. New York: ASME; 1999.

ASME/RSI (The American Society of Mechanical Engineers/Institute for Regulatory Science). Assessment of technologies supported by the U.S. Department of Energy; results of the peer review for fiscal year 2000, CRTD Vol. 61. New York: ASME; 2000.

ASME/RSI (The American Society of Mechanical Engineers/Institute for Regulatory Science). Strategy for remediation of groundwater contamination at the Nevada Test Site; technical peer review report; report of the review panel, CRTD Vol. 62. New York: ASME; 2001a.

ASME/RSI (The American Society of Mechanical Engineers/Institute for Regulatory Science). Requirements for disposal of remote-handled transuranic wastes at the waste isolation pilot plant; technical peer review report; report of the review panel, CRTD Vol. 63. New York: ASME; 2001b.

ASME/RSI (The American Society of Mechanical Engineers/Institute for Regulatory Science). Assessment of technologies supported by the U.S. Department of Energy; results of the peer review for fiscal year 2001, CRTD Vol. 64. New York: ASME; 2001c.

ASME/RSI (The American Society of Mechanical Engineers/Institute for Regulatory Science). Waste isolation pilot plant initial report for polychlorinated biphenyl disposal authorization; technical peer review report; report of the review panel, CRTD Vol. 65. New York: ASME; 2002a.

ASME/RSI (The American Society of Mechanical Engineers/Institute for Regulatory Science). Airborne release fractions; technical peer review report; report of the review panel, CRTD Vol. 68. New York: ASME; 2002b.

ASME/RSI (The American Society of Mechanical Engineers/Institute for Regulatory Science). The beryllium oxide manufacturing process; technical peer review report; report of the review panel, CRTD Vol. 69. New York: ASME; 2002c.

ASME/RSI (The American Society of Mechanical Engineers/Institute for Regulatory Science). Assessment of technologies supported by the U.S. Department of Energy; results of the peer review for fiscal year 2002; CRTD Vol. 70. New York: ASME; 2002d.

ASME/RSI (The American Society of Mechanical Engineers/Institute for Regulatory Science). Radionuclide transport in the environment; technical peer review report; report of the review panel; CRTD Vol. 71. New York: ASME; 2002e.

ASME/RSI (The American Society of Mechanical Engineers/Institute for Regulatory Science). Review of selected nuclear safety programs at Savannah River Site; technical peer review report; report of the review panel, CRTD Vol. 72. New York: ASME; 2003a.

ASME/RSI (The American Society of Mechanical Engineers/Institute for Regulatory Science). Hanford Site 100 B/C Risk Assessment Pilot Project; technical peer review report; report of the review panel, CRTD Vol. 73. New York: ASME; 2003b.

ASME/RSI (The American Society of Mechanical Engineers/Institute for Regulatory Science). Salt waste processing facility technology readiness, CRTD Vol. 75. New York: ASME; 2003c.

ASME/RSI (The American Society of Mechanical Engineers/Institute for Regulatory Science). Spent Nuclear Fuel Canister Welding Concept, CRTD Vol. 76. New York: ASME; 2004.

ASTM (American Society of Testing and Materials). Standard practice for preparing, cleaning, and evaluating corrosion test specimens, G1. West Conshohocken, PA: American Society of Testing and Materials International; 1990.

ASTM (American Society of Testing and Materials). Standard practice for evaluating stress-corrosion-cracking resistance of metals and alloys in a boiling magnesium chloride solution, G36. West Conshohocken, PA: American Society of Testing and Materials International; 1994.

ASTM (American Society of Testing and Materials). Standard guide for crevice corrosion testing of iron-base and nickel-base stainless alloys in seawater and other chloride-containing aqueous environments, G78. West Conshohocken, PA: American Society of Testing and Materials International; 1995a.

ASTM (American Society of Testing and Materials). Standard practice for slow strain rate testing to evaluate the susceptibility of metallic materials to environmentally assisted cracking, G129. West Conshohocken, PA: American Society of Testing and Materials International; 1995b.

Bibler, N.E. Radiolytic gas production from concrete containing Savannah River Plant waste, DO-1464. Aiken, SC: Savannah River Plant; (January) 1978.

Blandford, R.K. Structural response evaluation of the 24-in diameter DOE standardized spent nuclear fuel canister, EDF-NSNF-026. Idaho Falls, ID: Idaho National Engineering and Environmental Laboratory; (August 21) 2003.

Blumenthal, W.B. The chemical behavior of zirconium. Princeton, NJ: Van Nostrand; 1958.

Caskey, Jr., G.R. Surface effects of tritium diffusion in materials in a radiation environment. Advances in chemistry series No. 158. Radiation effects on solid surfaces. Columbus, OH: American Chemical Society; 366-375; 1976.

Caskey, Jr., G.R. Fractography of hydrogen-embrittled stainless steel. Scripta Metallurgica 11:1077-1083; 1977.

Cotton, F.A.; Wilkinson, G. Advanced inorganic chemistry. Cambridge, MA: John Wiley & Sons; 1962.

Coyle, Jr., R.J.; Cargol, J.A.; Fiore, N.F. The effect of aging on hydrogen embrittlement of a nickel alloy. Metallurgical Transactions A 12A:653; (April) 1981.

Daniels, F.; Alberty, R.A. Physical chemistry. New York, NY: John Wiley & Sons; 1962.

DOE (U.S. Department of Energy). Technical strategy for the treatment, packaging, and disposal of aluminum-based spent nuclear fuel. Washington, DC: DOE, Office of Spent Fuel Management; (June) 1996.

DOE (U.S. Department of Energy). Preliminary design specification for department of energy spent nuclear fuel canisters, DOE/SNF/REP-011. Rev. 3. Volume I and II. Idaho Falls, ID: Idaho National Engineering and Environmental Laboratory; (August 17) 1999.

DOE (U.S. Department of Energy). Hydrogen damage in DOE spent nuclear fuel packages, DOE-SNF/REP-019. Rev. 0. Idaho Falls, ID: Idaho National Engineering and Environmental Laboratory; (August) 2000a.

DOE (U.S. Department of Energy). Conditional probability estimate for a standardized DOE SNF canister breach given a drop, DOE/SNF/REP-065. Rev. 0. Idaho Falls, ID: Idaho National Engineering and Environmental Laboratory; (December) 2000b.

DOE (U.S. Department of Energy). Waste acceptance system requirements document, DOE/RW-0351. Washington, DC: DOE; (January) 2002.

DOE (U.S. Department of Energy). Fuel canister stress corrosion cracking susceptibility experimental results, DOE/SNF/REP-082. Rev. 0. Idaho Falls, ID: Idaho National Engineering and Environmental Laboratory; (March) 2003a.

DOE (U.S. Department of Energy). Software report for ABAQUS/Explicit version 6.3-3., DOE/SNF/REP-085. Rev. 2. Idaho Falls, ID; Idaho National Engineering and Environmental Laboratory; (June) 2003b.

DOE (U.S. Department of Energy). Test plans for the Department of Energy spent nuclear fuel canister and basket development project, DOE/SNF/PP-039. Rev. 0. Idaho Falls, ID; Idaho National Engineering and Environmental Laboratory; (June) 2003c.

DOE (U.S. Department of Energy). Standard contract for disposal of spent nuclear fuel and/or high-level radioactive waste, 10 CFR 961, 2003d. Available at http://www.access.gpo.gov/nara/cfr/

Duncan, D.R.; Ball, D.E. K-Basins particulate water content, behavior, and impact, HNF-1523. Rev. 0. Charlotte, NC: Duke Engineering and Services; (November) 1997.

Garibov, A.A.; Melikzade, M.M.; Bakirov, M.Ya.; Ramazanova, M.Kh. Radiolysis of adsorbed water molecules on the oxides Al_2O_3, La_2O_3, and BeO. High Energy Chem. 16(3):117-179; 1982.

Garibov, A.A. et al. Heterogenous radiolysis of water: Effect of the concentration of water in the adsorbed phase on the hydrogen yield. High Energy Chem. 21(6):416-420; 1987.

Goldman, L.H. et al. Multi-Canister overpack design report, HNF-SD-SNF-DR-003. Rev. 3. Richland, WA: Fluor Hanford; (February) 2000a.

Goldman, L.H. et al. Multi-Canister overpack design report, HNF-SD-SNF-DR-003. Rev. 3A. Richland, WA: Fluor Hanford; (July) 2000b.

Goldman, L.H. et al. Multi-Canister overpack design report, HNF-SD-SNF-DR-003. Rev. 3B. Richland, WA: Fluor Hanford; (April) 2002.

Green, J.R. Progress report for the enhancement of Radcalc: Isotope database, gamma absorption fractions, and G(H2) values, WHC-SD-TP-RPT-014. Rev. 0. Richland, WA: Westinghouse Hanford Company; 1994.

Hazarabedian, A.; Overero-Garcia, J. Effects of strain rate and prior heat treatments on hydrogen embrittlement of 316L SS and 304 SS in aqueous sulfide environment. In: Moody, N.R.; Thompson, A.W., eds. Hydrogen effects on material behavior. Warrendale, PA: The Mineral, Metals, and Materials Society; 1990.

Hecker, S.S.; Stout, M.G.; Staudhammer, K.P.; Smith, J.L. Effects of strain state and strain rate on deformation-induced transformation in 304 stainless steel: Part 1. Magnetic measurements and mechanical behavior. Metallurgical Transactions A 13A:619-626; (April) 1982.

Hillner, E. Zircaloy corrosion technology, WAPD-MT(FZCT)-750. West Mifflin, PA: Bettis Atomic Power Laboratory; (October 30) 1985.

Holt, J.M., ed. Structural alloys handbook. West Lafayette, IN: CINDAS/Purdue University; 1995.

HSC Chemistry® for Windows. Version 3.0. Pori: Outokumpu Research Oy; (April 30) 1997.

Huang, G.L.; Matlock, D.K.; Krauss, G. Martensite formation, strain rate sensitivity and deformation behavior of type 304 stainless steel sheet. Metallurgical Transactions A 20A:1239-1246; (July) 1989.

Hyzak, J.M.; Rawl, D.E.; Louthan, M.R. Hydrogen affected impact of type 304L stainless steel. Scripta Metallurgica 15:930-939;1981.

Jones, R.H. Stress-Corrosion cracking materials performance and evaluation. Metals Park, OH: American Society for Metals; 1992.

Kesavan, S. The kinetics of hydrogen evolution and absorption on high nickel alloys at elevated temperatures. Ph.D. Dissertation. Columbus, OH: Ohio State University; 1991.

Lam, P.S.; Sindelar, R.L.; Peacock, Jr., H.B. Vapor corrosion of aluminum cladding alloys and aluminum-uranium fuel materials in storage environments (U), WSRC-TR-0120. Aiken, SC: Westinghouse Savannah River Company; (April) 1997.

Lange, N.A. Handbook of chemistry. 10th Edition. New York, NY: McGraw-Hill Book; 1961.

Lessing, P.A. Effects of water in canisters containing DOE spent nuclear fuel, DOE/SNF/REP-017. Idaho Falls, ID; Idaho National Engineering and Environmental Laboratory; (August) 1998.

LMITCO (Lockheed Martin Idaho Technologies Company). LMITCO technical procedure. Leak test procedure, TPR-4976. Rev. 0. Idaho Falls, ID: LMITCO; (May 1) 1996a.

LMITCO (Lockheed Martin Idaho Technologies Company). LMITCO technical procedure. Radiographic examination, TPR-4970. Rev. 0. Idaho Falls, ID: LMITCO; (August 19) 1996b.

LMITCO (Lockheed Martin Idaho Technologies Company). LMITCO technical procedure. Liquid penetrant examination, TPR-4975. Rev. 1. Idaho Falls, ID: LMITCO; (June 15) 1998.

LMITCO (Lockheed Martin Idaho Technologies Company). LMITCO technical procedure. Manual ultrasonic examination, TPR-4974. Rev. 2. Idaho Falls, ID: LMITCO; (January 15) 1999.

Logan, H.L. The stress corrosion of metals. New York, NY: John Wiley; 1966.

Lorenz, B.D. et al. Multi-Canister overpack topical report, HNF-SD-SARR-005. Rev. 1. Richland, WA: Fluor Hanford; (November) 1999.

Macheret, P. Analysis of mechanisms for early waste package failure. Office of civilian radioactive waste management, ANL-EBS-MD-000023. Rev. 01. Chicago, IL: Argonne National Laboratory; (January 01) 2000.

Massalski, T.B. Binary alloy phase diagrams. Metals Park, OH: American Society for Metals; 1986.

Mills, W.J. Fracture toughness of type 304 and 316 stainless steels and their welds. International Materials Reviews 42(2):45; 1997.

Morton, D.K.; Snow, S.D.; Rahl, T.E. FY 1999 drop testing report for the 18-in standardized DOE SNF canisters, EDF-NSNF-007. Rev. 2. Idaho Falls, ID: Idaho National Engineering and Environmental Laboratory; (March 16) 2000.

Nebergall, W.H.; Schmidt, F.C. General chemistry. Boston, MA: D.C. Heath; 1959.

NWPA (Nuclear Waste Policy Act). Public law 97-425; 1982.

NWPAA (Nuclear Waste Policy Amendments Act). Public law 100-203; 1987.

Pajunen, A.L. Evaluation of radiolytic gas generation from water dissociation in a multicanister overpack, HDF-SD-SNF-CN-006. Rev. 0. Charlotte, NC: Duke Engineering and Services. (April) 1997.

Peckner, D.; Bernstein, I.M. Handbook of stainless steels. New York, NY: McGraw-Hill Book; 1977.

Perng, T.; Alstetter, C. Effects of deformation on hydrogen permeation in austenitic stainless steels. Acta Metallurgica 34(9):1771-1781; 1986.

Presidential Memorandum. Disposal of defense waste in a commercially repository. (April 30) 1985.

Reed-Hill, R.E. Physical metallurgy principles. 2nd Edition. New York, NY: Van Nostrand; 1973.

Rozenak, P.; Robertson, I.M.; Birnbaum, H.K. HVEM studies of the effects of hydrogen on the deformation and fracture of AISI 316 austenitic stainless steel. In: Moody, N.R., Thompson, A.W., eds. Hydrogen effects on material behavior. Warrendale, PA: The Mineral, Metals, and Materials Society; 1990.

RSI (Institute for Regulatory Science). Handbook of peer review. Columbia, MD: RSI; 2003.

Sargent-Welch. Table of periodic properties of the elements (comprehensive). Catalog Number S-18806. Skokie, IL: Sargent-Welch Scientific Company; 1979.

Sindelar, R.L.; Basden, J.N. SRS transfer and storage services technical functional performance requirements for dry storage of Al-based SNF (U), WSRC-TR-97-00262. Aiken, SC: Westinghouse Savannah River Company; (September) 1997.

Singh, S.; Altstetter, C. Effects of hydrogen concentration on slow crack growth in stainless steels. Metallurgical Transactions A 13:1799-1808; (October) 1982.

Snow, S.D. et al. Preliminary drop testing results to validate an analysis methodology for accidental drop events of containers for radioactive materials, ASME PVP-Vol. 425. New York, NY: American Society of Mechanical Engineers; 63-68; (July 22-26) 2001.

Snow, S.D. Analytical evaluation of the MCO for repository-defined and other related drop events, EDF-NSNF-029. Rev. 0. Idaho Falls, ID: Idaho National Engineering and Environmental Laboratory; (September 30) 2003.

Snow, S.D.; Morton, D.K. Analytical evaluation of the Idaho spent fuel project canister for accidental drop event, EDF-NSNF-027. Rev. 0. Idaho Falls, ID: Idaho National Engineering and Environmental Laboratory; (September 30) 2003.

Snow, S.D.; Rahl, T.E. Analytical evaluation of preliminary drop tests performed to develop a robust and drop resistant design concept for the standardized DOE spent nuclear fuel canister, EDF AMG-06-98. Idaho Falls, ID: Idaho National Engineering and Environmental Laboratory; (September 30) 1998.

Snow, S.D.; Rahl, T.E. Analytical evaluation of preliminary puncture-drop tests performed to develop a robust and drop resistant design concept for the Standardized DOE spent nuclear fuel canister, EDF AMG-01-99. Idaho Falls, ID: Idaho National Engineering and Environmental Laboratory; (June 17) 1999.

Sridhar, N.; Wilde, B.E.; Manfredi, C.; Kesevan, S.; Miller, C. Hydrogen absorption and embrittlement of candidate container materials, CNWRA 91-008. San Antonio, TX: Center for Nuclear Waste Regulatory Analysis; (June) 1991.

Sridhar, N.; Cragnolino, G.A.; Dunn, D.D. Experimental investigations of failure processes of high-level radioactive waste container materials, CNWRA 95-010. San Antonio, TX: Center for Nuclear Waste Container Analysis; (May) 1995.

Toh, T.; Baldwin, W.M. Stress Corrosion Cracking and Embrittlement. New York, NY: Wiley and Sons; 1956.

Tromp, R.L. Generic fission product and actinide content estimates for ATR fuel, RLT-1-91. Westinghouse Idaho Nuclear Company letter to J.D. Christian. Idaho Falls, ID: Westinghouse Idaho Nuclear Company; (February 4) 1991.

USNRC (U.S. Nuclear Regulatory Commission). Standard review plan for transformation packages for spent nuclear fuel, NUREG-1617. Washington, DC: USNRC; (March) 2000.

225

USNRC (U.S. Nuclear Regulatory Commission). Disposal of high-level radioactive wastes in geologic repositories, 10 CFR 60, 2003a. Available at http://www.access.gpo.gov/nara/cfr/

USNRC (U.S. Nuclear Regulatory Commission). Packaging and transportation of radioactive material, 10 CFR 71, 2003b. Available at http://www.access.gpo.gov/nara/cfr/

USNRC (U.S. Nuclear Regulatory Commission). Licensing requirements for the independent storage of spent nuclear fuel, high-level radioactive waste, and reactor-related greater than class C waste, 10 CFR 72, 2003c. Available at http://www.access.gpo.gov/nara/cfr/

Walker, F.W. et al. GE nuclear energy. 14th edition. Menlo Park, NJ: General Electric Company; 1989.

Wefers, K.; Misra, C. Oxides and Hydroxides of Aluminum. Alcoa Laboratories. Pittsburgh, PA: Alcoa; 1987.

Biographical
Summaries

Gary A. Benda is President of U.S. Energy Corp.—an environmental management firm specializing in radioactive mixed waste management, health physics, decontamination and decommissioning, and technology development. Previously, he was Vice-President, General Manager of the Programs Division for NUKEM Nuclear Technologies, Inc. His responsibilities included developing and maintaining federal programs in North America that specialized in engineering and waste-processing services. Prior to NUKEM, he spent over 17 years with Chem-Nuclear Systems/WMX Technologies in various management roles. He also directed the site investigation, geophysical analysis, site screening, and license application, as well as managed the public hearings and licensing operations associated with local and national regulatory agencies for new low-level waste sites. He has over 20 years of experience in environmental restoration, technology development, and waste management, and has instructed over 20 national and international professional courses on radioactive waste management, mixed waste, and technology development. He is a member of the American Society of Mechanical Engineers (ASME), American Nuclear Society, and Health Physics Society. He has served as Chair of the ASME National Mixed Waste Committee, Environmental Remediation Committee, and Environmental Engineering Division. He has also chaired over 100 technical sessions at numerous national and international conferences on environmental management. He has authored and coauthored various scientific papers, reports, book chapters, and articles on the nuclear environment. Gary Benda is a Certified Health Physicist. He received a B.S. in Health Physics from Oklahoma State University, an M.S. in Applied Nuclear Science from Georgia Institute of Technology, and an M.B.A. from Seattle City University.

Erich W. Bretthauer is a consultant. Previously, he held the position of research professor at the University of Nevada-Las Vegas from January 1993 to March 1995. In that capacity, he served as Executive Director of Nevada Industry, Science Engineering & Technology, a public-private partnership which developed programs to enhance the scientific infrastructure of the state of Nevada. He was also the Assistant Administrator for Research and Development at the U.S. Environmental Protection Agency (EPA) from March 1990 until January 1993. In that capacity, he managed the Research and Development activities of a large and multi-disciplinary agency. Erich Bretthauer rose through the ranks of the EPA and served in a number of capacities ranging from a bench scientist to policy manager at national and international levels. He directed the EPA's emergency and long-term monitoring program after the accident at Three Mile Island, as well as its bioremediation program in Prince William Sound after the Valdez oil spill. Erich Bretthauer was the leader of the U.S. delegation and co-leader of a five-year North Atlantic Treaty Organization project which focused on exposures, risks, and measures to control dioxins. He also directed the EPA's ecological research program, and was Director of EPA's Environmental Monitoring Systems Laboratory in Las Vegas. He is a member of Sigma Xi; the American Chemical Society; and the American Water Works Association; and has served on the Federal Advisory Committee to the Civil Engineering Research Foundation. Erich Bretthauer is the author and coauthor of numerous papers, reports, and other publications. He received his B.S. and M.S. in chemistry from the University of Nevada, Reno, NV.

Ernest L. Daman is Chairman Emeritus of Foster Wheeler Development Corporation where he previously served as Director of Research and Chairman of the Board. He also held the position of Senior Vice President at the parent company, FWC. He is a Past President of American Society of Mechanical Engineers and was elected to the National Academy of Engineering. Ernest Daman is a Fellow of the Institute of Energy (England) and the American Association for the Advancement of Science, and Past Chairman of the American Association of Engineering Societies. He served on several American Society of Mechanical Engineers committees as member or chairman. Ernest Daman is the author of numerous papers and holds 18 patents. He was responsible for the design and development of a combined steam gas turbine plant, fluidized bed combustion, fast breeder reactor components, supercritical steam generators, environmental control processes, and advanced high-efficiency power generation systems. Ernest Daman received his B.M.E. degree from the Polytechnic Institute of Brooklyn.

Irwin Feller is Director of the Institute for Policy Research and Evaluation and Professor of Economics at The Pennsylvania State University, where he has been on the faculty since 1963. His current research interests include the economics of academic research, the University's role in technology-based economic development, and the evaluation of federal and state technology programs. He is the author of *Universities and State Governments*: A Study in Policy Analysis, and over 100 refereed journal articles, final research reports, book chapters, reviews, and numerous papers presented to academic, professional, and private organizations. He is former Chair of the Committee on Science, Engineering, and Public Policy, American Association for the Advancement of Science. Irwin Feller was the American Society of Mechanical Engineers Pennsylvania State Fellow in 1996-1997. He has been appointed to the National Research Council's Committee on Science, Engineering, and Public Policy; International Benchmarking of U.S. International Competitiveness-Immunology; Transportation Research Board, Research and Technology Coordinating Committee, National Research Council; and National Institute of Standards and Technology-Manufacturing Extension Partnership National Advisory Board. Irwin Feller is Chair of the National Science Foundation's Advisory Committee on Social, Behavioral, and Economic Sciences. He received a B.B.A. in Economics from the City University of New York and a Ph.D. in Economics from the University of Minnesota.

Jacob Fish is currently Professor of Civil Engineering; Mechanical and Aerospace Engineering; and Information Technology at the Rensselaer Polytechnic Institute. He is also directing the National Science Foundation Nanoscale Interdisciplinary Research Teams (NIRT) program at Rensselaer. He has 20 years of experience (both in industry and academia) in the field of multiscale computational engineering. He has served as a consultant to the NY Department of Law; GE CRD; Lockheed Missiles & Space Company; ANSYS; SDRC; and EMRC software houses. He is President of the United States Association for Computational Mechanics; a member of the National Research Council for the Air and Ground Vehicle Technology; and is a Fellow of both the United States Association for Computational Mechanics and the International Association for Computational Mechanics. He is Editor-in-Chief of the *International Journal for Multiscale Computational Engineering*; Editorial Board Member of the *International Journal for Numerical Methods in Engineering*; and Editorial Board Member of the *International Journal of Computational Engineering Science*. He has also served as the Associate Editor of the *Journal of Engineering Mechanics*. He has been the recipient of several awards, including the National Science Foundation (NSF) Presidential Young Investigator Award and the Best Paper Award. He has published over 100 journal articles and book chapters. Jacob Fish has a B.S. degree in Structural Engineering, and an M.S. degree in Structural Mechanics from Technion - Israel Institute of Technology; and a Ph.D. in Theoretical and Applied Mechanics from Northwestern University.

Robert A. Fjeld is Dempsey Professor of Environmental Engineering and Science at Clemson University. He coordinates the Department's nuclear environmental focus area, which is concerned with the environmental aspects of nuclear technologies including health physics, radioactive waste management, and risk assessment. Previously, he served as a faculty member in the Nuclear Engineering Department at Texas A&M University. He has active research on actinide transport in soils, instrumentation for measuring radioactivity in environmental samples, and environmental risk assessment. Robert Fjeld is a member of the Health Physics Society, American Nuclear Society, Society for Risk Analysis, and the American Society of Mechanical Engineers, where he serves as newsletter editor for the Mixed Waste Committee. He has served on two NRC Committees studying decontamination and decommissioning issues. Robert Fjeld is the author or coauthor of over 80 technical publications and presentations on topics such as radiation measurements, environmental transport of radionuclides, risk assessment, and aerosol physics. Robert Fjeld received a B.S. degree from North Carolina State University; and an M.S. degree and a Ph.D. from The Pennsylvania State University, all in Nuclear Engineering. He is a registered Professional Engineer.

Norman J. Gerstein is President and Technical Director of NJG, Inc., located in Potomac, MD. NJG provides engineering and management support to DOE and aerospace programs. NJG also provides nuclear and fossil fuel technologies primarily related to National Defense Programs through the U.S. Department of Energy (DOE) and the U.S. Army Corps of Engineers. He has over 40 years of experience in safety oversight, accident investigation, and facility industry processes. He has a broad knowledge and understanding of energy technology systems; components; and facilities with emphasis on design, development, fabrication, and operation of government-owned plants. Norman Gerstein was involved in Strategic Defense Initiative (SDI) activities related to energy conversion systems and power supplies for directed and kinetic energy weapons. He was instrumental in the evolving technologies for application and support of the DOE activities that related to spill test technology and environmentally germain technology requirements that utilized hazardous fluids. Specific systems include Magneto Hydro Dynamics; open and closed cycle gas turbines; and fuel cells for energy conversion power sources. Previously, Norman Gerstein worked as Principal Officer at Gencon. His specific assignments included engineering services associated with evaluating DOE facilities for ground testing of space reactor systems; cross-comparison studies of space reactor test facilities; and preparing management and environmental health and safety plans for the Idaho Chemical Processing Plant. Upon conversion of the U.S. Energy Research and Development Administration to DOE, Norman Gerstein served as Program Manager of the Residual Energy Applications Program tasked with taking the waste heat from the gaseous diffusion plants and converting it into useful energy applications. Norman Gerstein received the U.S. Atomic Energy Commission's Special Act and Service Award. He was a panel member for the External Review Committee which did a report "*Unusual Occurrence*" on the conduct of operations at Rockwell Hanford Operations Facilities. He has co-authored technical papers associated with his expertise. Norman J. Gerstein attended the U.S. Military Academy. He received a B.S. degree in General Engineering/Chemistry at Grinnell College; and an M.S. degree in Chemistry from the University of Basel in Switzerland. He is a Registered Professional Engineer in California.

William T. Gregory, III is currently Principal of Vinculum Marketing Solutions. Prior to forming Vinculum, he was Director of Government Programs for Foster Wheeler Environmental Corporation, an engineering and construction firm providing environmental and waste management services to government and private sector clients worldwide. Previously, he held a number of operational and business development positions at equipment manufacturing and service provision firms supporting nuclear utilities, industrial and process industries, and government agencies. His work has involved the management, processing, and disposition of hazardous, radioactive, and mixed wastes. He has also worked on the decontamination and decommissioning of nuclear facilities and on providing a wide range of environmental services in response to regulatory drivers. Prior to entering the private sector, he served with the U.S. Navy on nuclear submarines and at the operational command center for submarine operations in the Atlantic Fleet. William Gregory is actively involved with a number of international, national, and local organizations including the American Society of Mechanical Engineers and the American Nuclear Society. He is a founding member of the Board of Directors for the annual international Waste Management Symposium. William Gregory has served as an elected officer of several American Society of Mechanical Engineers divisions. He received a B.S. degree in Geology from the University of New Mexico, and an M.B.A. degree from Lamar University. He also attended naval nuclear power, nuclear weapons, and engineering schools as a U.S. Navy officer.

Tom A. Hendrickson is currently an independent consultant in the fields of energy, engineering, and technology. His career encompassed service to both government and industry. He was a Senior Executive of Raytheon Federal Engineers & Constructors, developing high technology projects. He was Principal Deputy Assistant Secretary of the Office of Nuclear Energy at the U.S. Department of Energy, where he oversaw programs including: Civilian Reactor Development; the Naval Nuclear Propulsion Program; Uranium Enrichment; Space and Defense Power Systems; Isotope Production; and Nuclear Safety Policy. He later became the Director of the New Production Reactors for the U.S. Department of Energy, responsible for designing and building new tritium production capacity for nuclear weapons; research and development; safety and environmental compliance; and construction. Concurrently, he served as Acting Under Secretary of Energy responsible for all defense and nuclear energy activities of the department. Early in his career, he served on the staff of the Atomic Energy Commission in Washington, DC. He directed the headquarters staff and contractors involved in submarine nuclear propulsion engineering, including research, development, design, and construction of all new design nuclear powered submarines and land-based prototypes. During this period, he also served as Project Officer for all new submarine developments including the NR-1; the USS Los Angeles SSN-688 class of 62 attack submarines, and the electric drive submarine. He helped with the development of port-entry safety procedures and sea trials of the United States' first nuclear-powered surface ships, the USS Long Beach and the USS Enterprise; as well as the first refueling of the Shippingport Atomic Power Station. He is a member of the American Nuclear Society, the American Society of Mechanical Engineers, and the American Physical Society. Tom Hendrickson received a B.A. degree in Physics from Harvard College and an M.S. degree in Physics from Georgetown University. He is a licensed Professional Engineer.

Nathan H. Hurt is a consultant in management and engineering with Technical and Management Consulting. He provides services to industrial firms and government agencies involved in environmental clean-up and waste management—both chemical and radioactive. He has extensive experience in the areas of executive management; plant management; engineering management; project management; marketing; and sales. He specializes in the areas of: uranium enrichment/production; engineering; development and marketing; plant management of rubber chemicals; petrochemicals; and thermoplastics. He also specializes in the engineering management of synthetic rubber and lattices; vinyl monomers and copolymers; polyesters; U.S. Department of Energy (DOE) weapons plants; quality assurance management; and operational readiness review. Nathan Hurt has been involved with the decommissioning of nuclear facilities. He was the Corporate Sponsor or Program Manager for seven decommissioning contracts at the DOE Complexes in Oak Ridge, TN; and Pinellas, FL. Previously, Nathan Hurt worked for Sharp and Associates, Inc. as the Director and Project Manager at the Oak Ridge Office. He was Vice President and Director of Oak Ridge Operations for IDM Environmental Corp., where he was responsible for the marketing and sales of decontamination, decommissioning, and waste management. He served as Project Manager for the laboratory quality assurance program at Westinghouse Hanford; DOE's Rocky Flats Plant—plant-wide identification of electrical equipment. He managed a study for a waste treatment and storage facility at the Portsmouth Area Uranium Enrichment Facility which included incineration and compaction of low-level radioactive wastes. He also worked for The Goodyear Tire and Rubber Company, including Goodyear Atomic, as Director of Research and Development, and President, where he was responsible for the operation of the Portsmouth Area Uranium Enrichment Facility. Nathan Hurt is a Past President of the American Society of Mechanical Engineers. He has been a member of: the American Association of Engineering Societies' Board of Governors; the American Institute of Chemical Engineers; and the Institute of Nuclear Materials Management. He is also a member of Tau Beta Pi Honorary Engineering Society; Pi Tau Sigma Honorary Mechanical Engineering Society; and was a member of The Nuclear Engineering Advisory Board of Worcester Polytechnic Institute. Nathan Hurt received a B.S. degree in Mechanical Engineering from the University of Colorado and has done Graduate, Technical, and Management course work at Pennsylvania State University. He is a registered Professional Engineer in Ohio.

Ratib Karam is currently Executive Director of the Education, Research, and Development Association of Georgia Universities. He is also Professor Emeritus of Nuclear Engineering and Health Physics (NE/HP) at the Georgia Institute of Technology (Georgia Tech). He chaired the NE/HP program for five years. He was also Director of the Neely Nuclear Research Center at Georgia Tech and administered the 5 MW Georgia Tech Research Reactor. During that time, he developed a plan and obtained U.S. Nuclear Regulatory Commission approval for the dismantlement and decommissioning of the Georgia Tech AGN-201 reactor. The conversion of the Georgia Tech Research Reactor from high to low enrichment necessitated a new safety analysis report which he wrote and for which he obtained U.S. Nuclear Regulatory Commission approval. He was also a Scientist and Engineer at Argonne National Laboratory. During his career he consulted on numerous nuclear engineering projects. His current research interests include nuclear reactor safety; operational safety; emergency planning; procedure development; environmental health; measuring techniques and methods; radiation transport; and shielding. He is a Fellow of the American Nuclear Society. He served on the National Technical Program Committee, and as Secretary and Executive Committee member of the Reactor Physics Division of the American Nuclear Society. He is a member of the editorial board of *Annals of Nuclear Energy*. He is author, coauthor, or editor of over 80 publications. Ratib Karam received a B.S. degree in Chemical Engineering, and an M.S. degree and a Ph.D. in Nuclear Engineering from the University of Florida.

Michael C. (M.C.) Kirkland is Vice President for the Southeastern Region of the Institute for Regulatory Science (RSI). In that capacity he leads various RSI projects related to the RSI mission in the southeastern U.S. Previously he was an independent consultant involved in peer review and various independent studies. For example, he led a team that performed an External Independent Review of the $1.3 billion Spallation Neutron Source Project at Oak Ridge, TN. He assisted in the planning and review of a management assessment at a U.S. Department of Energy (DOE) Site that involved the restart of a plutonium facility. He participated in planning, procurement, and review activities in the environmental remediation area that included decommissioning activities at a shut down nuclear test reactor; and designed and installed a ground water cleanup technology. M.C. Kirkland managed several environmental and construction projects that employed many soil investigative techniques including significant work with cone penetrometers. Additionally, he provided consulting services to a large environmental remediation services company regarding Dense Non-Aqueous Phase Liquid locating and removal techniques. During his tenure at the Savannah River Site (SRS) of DOE, M.C. Kirkland was a Technical Advisor, Project Manager, and Director of the Project Engineering Division. He evaluated nuclear and mixed waste conditions and aspects of high level wastes and spent nuclear fuel; determined material inventories; performed pollution prevention and environmental health and safety evaluations for a proposed waste treatment facility; served as technical advisor to a study administered by the Savannah River Operations Office; and developed integrated schedules defined for this project. M.C. Kirkland was director of the Project Engineering Division and managed the SRS design and construction program. He has been involved with waste management and environmental projects; cutting-edge technology programs; and worked with lasers and magnetic containment. He served as Director of the Waste and Fuel Cycle Technology Office, and planned and coordinated the programs of the DOE National High Level Waste Technology Office; the SR Fuel Cycle Technology Program; and the Commercial Interim Spent Fuel Management Program. M.C. Kirkland holds a B.S. in Mechanical Engineering from the University of South Carolina. He is registered as a Professional Engineer in South Carolina.

Peter B. Lederman is a consultant with over 48 years of experience in all facets of process engineering, environmental management, control, and policy development. This includes hazardous substance management; environmental remediation; environmental audit; pollution prevention; development of air pollution control devices; and reuse of waste products. He recently retired as Executive Director of the Center for Environmental Engineering & Science, Executive Director for Patents and Licensing, and Research Professor of Chemical Engineering and Environmental Policy at the New Jersey Institute of Technology. Peter Lederman managed major programs in industrial waste treatment research and development, and in oil and hazardous material spill control and remediation. Most recently, he was responsible for a study of the Economic Impact of Environmental Regulations. He has been responsible for technology transfer efforts including the maturing and licensing of innovative environmental technologies. He is a Fellow of the American Institute of Chemical Engineers; a Diplomat of the American Academy of Environmental Engineers; and a member of the ASME. He has served on several committees of the NRC and is the chair of the NRC Committee on Review and Evaluation of the Army Chemical Stockpile Disposal Program. He chaired the American Institute of Chemical Engineer's Environmental Division and is currently chair of its Societal Impacts Operating Council. Peter Lederman received a B.S.E., M.S.E., and Ph.D. (all in Chemical Engineering) from the University of Michigan in Ann Arbor, MI and is a registered Professional Engineer.

Betty R. Love is currently Executive Vice President of the Institute for Regulatory Science. In that capacity, she is responsible for the management of day-to-day operations of the Institute, and for administration of several projects. She is the Administrative Manager of a large-scale peer review program in collaboration with the American Society of Mechanical Engineers for a number of organizations including the U.S. Department of Energy. Her current research activities center around the development and implementation of a systematic approach to stakeholder participation, notably in scientific meetings. Previously, Betty Love was Director, Department of Training and Information within the Office of Environmental Health and Safety of Temple University in Philadelphia, PA. During that period she was instrumental in the development of a "Handbook of Environmental Health and Safety". She also developed and implemented a large-scale training program not only for the faculty and staff of the University but also for others. Betty Love is currently Managing Editor of *Technology*. She has published several papers in peer-reviewed journals; has edited a number of compendia; and is the primary author of *Manual for Public and Stakeholder Participation*. Betty Love received a B.S. degree in Business Administration from Virginia State University in Petersburg, VA, and an M.S. degree in Developmental Clinical Psychology from Antioch College in Yellow Springs, OH.

Jeffrey A. Marqusee is currently the Technical Director of the Strategic Environmental Research and Development Program (SERDP), and the Director of the Environmental Security Technology Certification Program (ESTCP). SERDP is a tri-agency (U.S. Department of Defense [DOD], U.S. Department of Energy, and U.S. Environmental Protection Agency) environmental research and development program managed by the DOD. SERDP supports research and development to solve environmental issues of relevance to DoD in the areas of cleanup, compliance, conservation and pollution prevention. ESTCP is a DOD-wide program designed to demonstrate innovative environmental technologies at DoD facilities. ESTCP provides for rigorous validation of the cost and performance of new environmental technologies in cooperation with the regulatory and end-user communities. Prior to his current position, Jeffrey Marqusee served as a program manager for environmental technology in the Office of the Deputy Under Secretary of Defense for Environmental Security. He was the principal advisor to the Deputy Under Secretary on environmental technology issues. Before joining DoD, he worked at the Institute for Defense Analyses, where he advised both DOD and National Aeronautics and Space Administration in the areas of remote sensing, environmental matters and military surveillance. Jeffrey Marqusee has worked at Stanford University, the University of

California and the National Institute of Standards and Technology. He has a Ph.D. in Physical Chemistry from the Massachusetts Institute of Technology.

A. Alan Moghissi is currently President of the Institute for Regulatory Science (RSI), a non-profit organization dedicated to the idea that societal decisions must be based on best available scientific information. The activities of the Institute include research, scientific assessment, and science education at all levels—particularly the education of minorities. Previously, Alan Moghissi was Associate Vice President for Environmental Health and Safety at Temple University in Philadelphia, PA and Assistant Vice President for Environmental Health and Safety at the University of Maryland at Baltimore. In both positions, he established an environmental health and safety program and resolved a number of relevant existing problems in those institutions. As a charter member of the U.S. Environmental Protection Agency (EPA), he served in a number of capacities, including Director of the Bioenvironmental/Radiological Research Division; Principal Science Advisor for Radiation and Hazardous Materials; and Manager of the Health and Environmental Risk Analysis Program. Alan Moghissi has been affiliated with a number of universities. He was a visiting professor at Georgia Tech and the University of Virginia, and was also affiliated with the University of Nevada and the Catholic University of America. Alan Moghissi's research has dealt with diverse subjects ranging from measurement of pollutants to biological effects of environmental agents. A major segment of his research has been on scientific information upon which laws, regulations, and judicial decisions are based—notably risk assessment. He has published nearly 400 papers, including several books. He is the Editor-in-Chief of *Technology: A Journal of Science Serving Legislative, Regulatory, and Judicial Systems*, which traces its roots to the *Journal of the Franklin Institute*—one of America's oldest continuously published journals of science and technology. Alan Moghissi is a member of the editorial board of several other scientific journals and is active in a number of civic, academic, and scientific organizations. He has served on a number of national and international committees and panels. He is a member of a number of professional societies including the American Society of Mechanical Engineers and is past chair of its Environmental Engineering Division. He is also an academic councilor of the Russian Academy of Engineering. Alan Moghissi received his education at the University of Zurich, Switzerland, and Technical University of Karlsruhe in Germany, where he received a doctorate degree in physical chemistry.

Lawrence C. Mohr, Jr., is currently Professor of Medicine, Biometry, and Epidemiology; and Director of the Environmental Biosciences Program at the Medical University of South Carolina. His areas of research and special interest include internal medicine and pulmonary disease—specifically diseases of the chest and respiratory system. An area of particular interest to Lawrence Mohr is environmental medicine, including molecular epidemiology and biomarker applications. He has been involved in studies related to environmental lung disease; pathophysiology; prevention and treatment of high altitude illness; high altitude physiology; risk assessment of environmental hazards and clinical epidemiology. Other areas of considerate interest to Lawrence Mohr are assessment of clinical outcomes; health policy analysis; and international health. This latter area includes: global epidemiology; medical relief operations; and health care in Central and Eastern Europe, as well as medical history—the impact of illness on world leaders. Previously, he held academic appointments as a Teaching Fellow in Medicine at the Uniformed Services University of the Health Sciences in Bethesda, MD. He was Associate Clinical Professor of Medicine and Emergency Medicine at George Washington University, Washington, DC. While in these institutions, he was a staff member of the Medical Support Group for the President of the United States. Lawrence Mohr was on the Medical Staff of Walter Reed Army Medical Center—where he completed his Internship and Residency in Internal Medicine—as well as George Washington University Hospital, both in Washington, DC. He has held Visiting Professorships at various universities. He served as Visiting Chief Resident at Presbyterian Hospital and Visiting Professor at the School of Nursing, both at Columbia University. Additionally, Lawrence Mohr was Visiting Professor of: William Beaumont Army Medical Center, Tulane University, University of Cincinnati, New York University, Brown University, East Carolina University, and the Mayo Clinic. Lawrence Mohr is a Fellow of the American College of Physicians and the American College of Chest Physicians. He is a member of several professional societies including: the American Federation for Medical Research; the Society for Risk Analysis; and the Wilderness Medical Society. Previously, he was on the Scientific Advisory Board for the Consortium in Environmental Risk Evaluation and the Savannah River Health Information System. He has authored or coauthored more than 60 articles, books, or technical publications. He received an A.B. degree in Chemistry as well as an M.D. degree, both from the University of North Carolina, Chapel Hill. Lawrence Mohr, Jr., is certified by the American Board of Internal Medicine.

Goetz K. Oertel's career in engineering, physics, chemistry, astronomy, and technical program management spans more than 40 years. He consults for industrial, academic, and governmental organizations in North and South America. As President and CEO of the Association of Universities for Research Astronomy, a nonprofit corporation, he engineered the initiation and completion of two 8-m aperture optical telescopes, and oversaw the Space Telescope Science Institute from before launch, through repair of the "Hubble flaw", to its successful operation. He initiated the conceptional phase of the Next Generation Space Telescope that will succeed Hubble as well as the Advanced Solar Telescope, and he oversaw the completion of ambitious ground-based astronomy facilities. He held technical and management positions in the U.S. Department of Energy, including Director of Defense Waste Management; Acting Manager of the Savannah River Operations Office; Deputy Manager of Albuquerque Operations Office; and Deputy Assistant Secretary for Safety, Health, and Quality Assurance. He had primary responsibility for the congressionally-mandated Defense Waste Management Plan, and for managing the related technology development, operations, and projects. He led the initiation of the Defense Waste Processing Facility, and saw it and the Waste Isolation Pilot Plant through technical, managerial, stakeholder, and political challenges. He was National Aeronautics and Space Administration Space Science Chief and Program Manager, and Aerospace Engineer at Langley. He was a Fellow in the White House with the President's Science Advisor and the Office of Management and Budget's Space and Energy branch. He chaired the Westinghouse West Valley Corporation Technical Advisory Group for high-level nuclear waste vitrification and management before, during, and after that project's successful vitrification campaign. He was appointed as Associate Member for life of the National Academies. He is a member of the American Physical Society, Sigma Xi, and other professional organizations. He is a Fellow of the American Association for the Advancement of Science. He is Chair or member of boards and committees of the National Research Council; George Mason University; the American Society of Mechanical Engineers; International University Exchange; and Westinghouse West Valley Corporation. He is a founding member of the Editorial Board for "Data Science", the new international on-line journal of Codata. He published numerous peer-reviewed papers and was awarded two patents. Trained as electrical engineer and physicist, he received a Vordiplom in Physics and Chemistry from the University of Kiel while on German industrial and governmental scholarships, and a Ph.D. in Physics from University of Maryland at College Park under a Fulbright scholarship.

Sorin R. Straja is currently Vice President for Science and Technology of the Institute for Regulatory Science. He has over 20 years of expertise in mathematical modeling and software development as applied in chemical engineering and risk assessment. Previously he served as Assistant Professor of Biostatistics with Temple University, Philadelphia; as Director of the Department of Occupational Health and Safety of Temple University, Philadelphia; and as a chemist with University of Maryland at Baltimore. Sorin Straja has extensive experience in the chemical industry where he worked as a senior R&D consultant with the Chemical and Biochemical Energetics Institute, and as a plant manager with Chemicals Enterprise Dudesti and Plastics Processing Bucharest from Romania. He was an Assistant/Adjunct Professor of Chemical

Engineering with the Polytechnic Institute Bucharest. Sorin Straja is the author of two books and 44 scientific papers published in internationally recognized and peer-reviewed journals. He was an editor of *Environment International,* and currently is a contributing editor of *Technology.* Sorin Straja received a Certificate of Appreciation for Teaching from Temple University, the "Nicolae Teclu" Prize of the Romanian Academy, and a Certificate of Appreciation from U.S. Department of Agriculture for significant volunteer contributions. He is a Fellow of the Global Association of Risk Professionals, and a member of the American Chemical Society, American Institute of Chemical Engineers, Society for Risk Analysis, and New York Academy of Sciences. Sorin Straja holds a M.S. in Industrial Chemistry and a Ph.D. in Chemical Engineering both from Polytechnic Institute Bucharest.

Glenn W. Suter, II is currently Science Advisor at the National Center for Environmental Assessment of the U.S. Environmental Protection Agency (EPA) in Cincinnati, OH. Previous to his current position, he was at Oak Ridge National Laboratory, initially as Research Associate and gradually rising to Science Leader at the Environment Science Division of the Laboratory. His interest has focused on Ecotoxicology in general and Ecological Risk Assessment in particular. He is one of the developers of the most widely-used methodology for Ecological Risk Assessment. This method has been applied to the impact of pollutants on fish, contaminated soils, production of synthetic fuels, and various other ecosystems. Glenn Suter has lectured widely, both nationally and internationally on Ecological Risk Assessment. He is currently a member of the U.S. EPA's Risk Assessment Forum. He has been a member of numerous panels and has consulted with various governmental agencies and private organizations, including the Council of Environmental Quality. He was a member of the Scientific Review Panel for Savannah River Ecology Laboratory; the National Science Foundation Panel on Decision Making and Valuation for Environmental Policy; and the U.S. EPA Science Advisory Board and Conservation Foundation, Ecosystem Valuation Forum. In addition, he was a member of the International Institute of Applied Systems Analysis Task Force on Risk and Policy Analysis and the Council on Environmental Quality. He was a member of the Board of Directors, for the Society for Environmental Toxicology and Chemistry. Glenn Suter is presently on the Editorial Board of *Environmental Health Perspectives* and *Human and Ecological Risk Assessment.* Previously, he was on the Editorial Board of *Handbook of Environmental Risk Assessment and Management* and *Environmental Toxicology and Chemistry.* Glenn Suter is the author of three books and is author and coauthor of over 200 publications. He received a B.S. degree in Biology from Virginia Polytechnic Institute and a Ph.D. in Ecology from the University of California, Davis.

Cheryl A. Trottier is currently Chief of the Radiation Protection, Environmental Risk, and Waste Management Branch of the Office of Nuclear Regulatory Research at the U.S. Nuclear Regulatory Commission. In that capacity, she is responsible for the management of research programs and the development of technical bases to support rulemaking. This includes the development of models for assessing the maximum doses likely to be received from lands and materials cleared from regulatory control; evaluating hydrologic model and parameter uncertainty; the development of realistic parameters for assessing sorption processes in geochemical models; and refining evaluations of radionuclide transport mechanism in the environment. In her 25 years of experience in the field of radiation protection, she has been involved in the management of environmental radiation protection monitoring programs and laboratory measurements, and the emergency preparedness coordination for an electric utility. She was also involved in the areas of materials use regulation oversight; development of regulations; and the development of guidance for use of byproduct and special nuclear materials. This included finalization of regulations and the development of regulations to certify the gaseous diffusion plants. Cheryl Trottier serves as one of the U.S. representatives to the Nuclear Energy Agency Committee on Radiation Protection and Public Health. She is a member of the American

Nuclear Society, and serves as a member of the Committee on Site Clean-Up Restoration Standards. She received her B.A. degree in Biology from Rutgers University.

Peter Turula is currently a private consultant with experience in structural engineering and radioactive material transportation packagings. His special expertise is in structural systems and stress analysis, and the dynamic response of packaging structures. Since 1985, he has been advisor to the U.S. Department of Energy, where he evaluated radioactive transportation packagings to assess their safety and compliance with national and international regulations. He also worked for Portland Cement Association; Pioneer Service and Engineering; Argonne National Laboratory; and has provided services to the USNRC and Los Alamos National Laboratory. Before retiring from Argonne National Laboratory, he gained 40 years of national, industry, and university experience in areas such as: design and evaluation of radioactive material transportation packagings; nuclear power plant facility structures and reactor components; development and application of computer codes for structural analysis and design; management of the development and design of a prototype sodium-based nuclear steam supply system pumps and steam generators; and the formulation and execution of internationally-collaborative projects involving radioactive material transportation with engineers from the former Soviet Union. He has also been instrumental in the development of the American Society of Mechanical Engineers (ASME) rules for nuclear packaging containment vessels, and ANSI N14 transportation standards. He has authored 60 publications on structural analysis and design; radioactive materials packaging; and computer applications. Peter Turula received a B.S. degree, an M.S. degree, and a Ph.D. in Civil Engineering from the Illinois Institute of Technology; and an MBA in the Graduate School of Business from the University of Chicago. He is a licensed Professional Engineer.

Charles O. Velzy is a consultant in the field of waste treatment and disposal. Previously, he held increasingly responsible positions with the environmental consulting engineering firm, Charles R. Velzy Associates, Inc., becoming President in 1976. In 1987, when Velzy Associates merged with Roy F. Weston, Inc., Charles Velzy became Vice President of Weston, a position which he held until retiring in 1992. He has over 35 years of experience as an environmental engineering consultant specializing in: the analysis of waste management problems; design of wastewater treatment and waste disposal systems; and design of new, retrofit of existing, testing, and permitting of waste combustion facilities. He has authored or co-authored over 80 publications—primarily in the field of solid waste management. He has served on the Science Advisory Board of the U.S. Environmental Protection Agency; as President of the American Society of Mechanical Engineers (ASME); Chair of the ASME Peer Review Committee; and as Treasurer of the American Academy of Environmental Engineers (AAEE). He has served on numerous committees of the ASME, the AAEE, the American National Standards Institute, and the American Society for Testing and Materials. He is a registered professional engineer in New York and eleven other states. Charles Velzy received B.S. degrees in Mechanical and Civil Engineering, and an M.S. in Sanitary Engineering from the University of Illinois at Urbana-Champaign.

Acronyms

ANS	American Nuclear Society
ASME	American Society of Mechanical Engineers
ANSI	American National Standards Institute
ASM	American Society for Metals
ASME	American Society of Mechanical Engineers
ASNT	American Society of Nondestructive Testing
ASTM	American Society of Testing and Materials
ATR	Advanced Test Reactor at INEEL
bcc	body centered cubic
BOL	Beginning of Life
B&PV	Boiler and Pressure Vessel
CAM	corrosion allowance material
CFR	Code of Federal Regulations
CGOC	canister center-of-gravity-over-the-corner drop orientation
CNWRA	Center for Nuclear Waste Regulatory Analysis
COF	Coefficient of friction
CRM	corrosion resistant material
CRWMS	Civilian Radioactive Waste Management System.
DBA	design basis accident
DI	deionized
DIS	Disposability Interface Specification
DOE	U.S. Department of Energy
DOT	U.S. Department of Transportation
EM	Office of Environmental Management (DOE)
EOL	End of Life
EP	Executive Panel
FE	finite element
FRR	foreign research reactors
FSR	fast strain rate
FWENC	Foster Wheeler Environmental Corporation
GTAW	gas tungsten arc weld
HAC	hydrogen-assisted cracking
HAZ	heat-affected zone
HEPA	high efficiency particulate air
HERF	high-energy rate forged
HEU	highly enriched uranium
HIC	high-integrity cans or container
HLW	high-level radioactive waste
HSC	hydrogen stress cracking
HSCC	hydrogen-assisted stress corrosion cracking
ICD	Interface Control Document
INEEL	Idaho National Engineering and Environmental Laboratory
ISFP	Idaho Spent Fuel Project
ISFSI	independent spent fuel storage installation
LLNL	Lawrence Livermore National Laboratory
LME	Liquid Metal Embrittlement
LMITCO	Lockheed Martin Idaho Technologies Company

MCO	Multi-Canister Overpack
MGDS	Mined Geologic Disposal System
MIC	microbially influenced corrosion
MNIP	Maximum Normal In-plant Handling Pressure
MNOP	Maximum Normal Operating Pressure
MRS	Monitored Retrieval Storage
MTR	Materials Test Reactor
NSNFP	National Spent Nuclear Fuel Program
NWPA	Nuclear Waste Policy Act of 1982 and its Subsequent Amendments
OCRWM	Office of Civilian Radioactive Waste Management of DOE
OD	outer diameter
PRCEE	Peer Review Committee for Energy and Environment
PT	Project Team
QA	quality assurance
QAPM	quality assurance program manager
RT	radiographic test
RP	Review Panel
RSI	Institute for Regulatory Science
SCC	stress corrosion cracking
SCG	slow crack growth
SHIC	standard high integrity can
SNF	spent nuclear fuel
SSR	slower strain rate
SRS	Savannah River Site
SSRT	slow strain rate test
SST	stainless steel
STP	standard temperature and pressure
TIG	Tungsten Inert Gas
USNRC	United States Nuclear Regulatory Commission
USEPA	United States Environmental Protection Agency
UT	ultrasonic testing
VLEU	Very Low Enriched Uranium
WASRD	Waste Acceptance System Requirements Document

Symbols

ksi	thousand pounds per square in
lbs	pounds force
mpy	mils per year
psi	pounds per square in
psig	pounds per square in gage pressure

Definitions

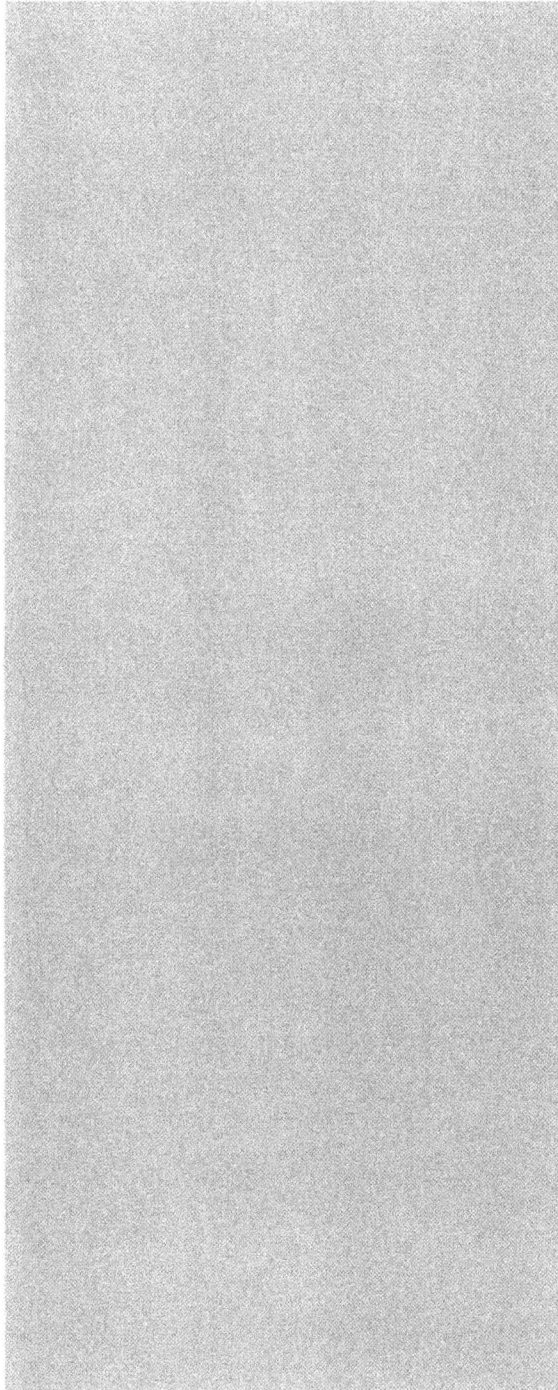

Can	a smaller structure surrounding SNF or other highly radioactive components that can be placed into a canister
Canister	a large structure (approximately 18-in [457 mm] to 24-in [609 mm] diameter) surrounding SNF or other highly radioactive components (bare or in cans) that facilitates handling, storage, transportation, and/or disposal and can be placed into a disposal container or cask
Cask	a structure used for the transportation or storage of SNF and/or HLW comprised of components intended to provide radiation shielding and retention of spent nuclear fuel and radioactive material contents during storage or transportation that meets all applicable regulatory requirements.
Confinement	retention of a material within an area from which releases or leakage are permitted, but controlled, and leakage of other substances into the area may also occur.
Containment	the complete and absolute retention of any substance within a closed area and no other substance may gain access inside the closed area.
Damaged SNF	SNF with cladding defects greater than hairline cracks or pinhole leaks
Disposal Container	the container in which the DOE SNF canisters are to be placed at the geo-logic repository for disposal, prior to acceptance of the final barrier weld at which time the disposal container becomes a waste package
Failed SNF	SNF with hairline cracks or pinhole leaks in the cladding (as defined for this document only)
Intact SNF	SNF with no hairline cracks or pinhole leaks in the cladding
Internals	items (baskets, spacers, sleeves, dividers, cans, etc.) placed inside the DOE SNF canister along with the SNF for supporting and positioning the SNF and to also prevent criticality if necessary
Package	the packaging together with its radioactive contents as presented for transport
Packaging	assembly of components necessary to ensure compliance with the requirements of 10 CFR Part 71
Repository	synonymous with geologic repository, a system that is intended to be used for the disposal of radioactive wastes in excavated geologic media
Storage Industry Canister	large outer container (approximately 5 or 6 ft in diameter) used in storage of SNF
Waste Package	the name of the disposal container after the final barrier weld is accepted